Integral Evaluations Using the Gamma and Beta Functions and Elliptic Integrals in Engineering

Integral Evaluations Using the Gamma and Beta Functions and Elliptic Integrals in Engineering

A Self-Study Approach

C. C. Maican, P.Eng.

International Press
www.intlpress.com

Integral Evaluations Using the Gamma and Beta Functions and Elliptic Integrals in Engineering: A Self-Study Approach

Maican, C. C. (P.Eng., Ontario, Canada)

ISBN 978-1-57146-262-6

Paperback reissue 2013.
Previously published in 2005 under ISBN 1-57146-158-2 (paperback).

Printed in the United States of America.

TABLE OF CONTENTS

Chapter 2
APPLICATIONS OF THE GAMMA AND BETA FUNCTIONS

Chapter 3
COMPLETE ELLIPTIC INTEGRALS $K(k)$ AND $E(k)$

Chapter 4
APPLICATIONS OF ELLIPTIC INTEGRALS

APPENDIXES

FOREWORD

This book is intended to provide to the students of engineering and science faculties and to the engineers and scientists an alternative tool for integral evaluation. Using only the classical methods, many times, the work is difficult, long and because of this, prone to errors. In addition, many integrals can not be evaluated in terms of the elementary functions, but can be evaluated in terms of gamma function and elliptic integrals. This book will fill a gap that has existed for many years in engineering and science fields.

For many modern advanced technology areas, the evaluation of the integrals using the gamma and beta functions is of prime importance. It is easy to apply this method, to generalize the solutions and at the same time to obtain the convergence conditions of given integrals. The complete elliptic integrals are also important in engineering.

All chapters of this book are very lucidly treated and are important to the author's goal of making engineers and scientists effective and competitive in their fields. There are many worked examples and generalized solutions. The book is very well coordinated and I am confident that it will be very well received by a large number of engineers, scientists and students worldwide.

The book will also encourage the educators to include gamma and beta functions and complete elliptic integrals in the engineering and science curricula at universities and colleges.

University of British Columbia
Vancouver, B.C., Canada

K.D. Srivastava
Professor of Electrical Engineering

LIST OF SYMBOLS, NOTATIONS AND FORMULAS

$a \in A$ a is an element of the set A, or the point a belongs to the set A.

$a \notin A$ a is not an element of the set A.

$N = \{1, 2, 3, \ldots\}$ is the set of natural numbers (positive integers).

$Z = \{0, \pm 1, \pm 2, \ldots\}$ is the set of integers.

Q is the set of rational numbers and they are real numbers of the form m/n where m and n are integers and $n \neq 0$.

– The irrational numbers are real numbers that can not be expressed in the form m/n where m and n are integers and $n \neq 0$.

R is the set of real numbers $= R^+ \cup R^- \cup \{0\}$.

$R^+ = \{x \in R \mid x > 0\}$ is the set of positive real numbers.

$R^- = \{x \in R \mid -x \in R^+\}$ is the set of negative real numbers.

$(a, b) = \{x \in R \mid a < x < b\}$ is an open interval.

$[a, b) = \{x \in R \mid a \leq x < b\}$ is an interval closed at left and open at right.

$(a, b] = \{x \in R \mid a < x \leq b\}$ is an interval open at left and closed at right.

$[a, b] = \{x \in R \mid a \leq x \leq b\}$ is a closed interval.

R^n-space, $n \in N$, is a real n-dimensional Euclidean Space.

$n! = 1 \cdot 2 \cdot 3 \cdots n, \quad 0! = 1.$

$$(2n+1)!! = 1 \cdot 3 \cdots (2n+1) = \frac{(2n+1)!}{n!2^n}$$

$(2n)!! = 2 \cdot 4 \cdots (2n) = n! 2^n$

$(n-1)! 2^{n-1} = (2n-2)!!$

$\Gamma(p)$ is the complete gamma function

$B(p, q)$ is the complete beta function

$\gamma(p, x)$, $\Gamma(p, x)$ are the incomplete gamma functions.

$B_x(p, q)$ is the incomplete beta function

$\psi(x)$ is the digamma function

$\beta(x)$, see Eq.(1.171).

$P(x^2/v)$ is the probability integral of x^2-distribution function.

$Q(x^2/v)$ is the percentage points of x^2-distribution function.

$K(k)$ is the complete elliptic integral of the first kind.

$E(k)$ is the complete elliptic integral of the second kind.

a. *The order of the function* $f(x)$ with respect to another function $g(x)$. Suppose that the point x approaches a point x_0. We say that $f(x)$ is of the order of $g(x)$, if there exists a number $M \geq 0$ such that

$$|f(x)| \leq M |g(x)|$$

in some sufficiently small neighbourhood of the point x_0. We write

$$f(x) = \underset{x=x_0}{O} [g(x)] .$$

In particular, if

$$\lim_{x=x_0} \left| \frac{f(x)}{g(x)} \right|$$

exists and is finite, then $f(x)$ is of the order of $g(x)$ in the neighbourhood of x_0.

b. A function $f(x)$ integrable over the interval $(-a, a)$ and satisfying the relation $f(-x) = f(x)$ on that interval is called an even function and then

$$\int_{-a}^{a} f(x)\,dx = 2\int_{0}^{a} f(x)\,dx, \quad a > 0.$$

c. A function $f(x)$ integrable over the integral $(-a, a)$ and satisfying the relation $f(-x) = -f(x)$ on that interval is called an odd function and then

$$\int_{-a}^{a} f(x)\,dx = 0, \quad a > 0.$$

d. Integrating by parts, with u and v functions of x, we have

$$\int_{a}^{b} uv'\,dx = uv\Big|_{a}^{b} - \int_{a}^{b} vu'\,dx$$

or

$$\int_{a}^{b} u\frac{dv}{dx}\,dx = uv\Big|_{a}^{b} - \int_{a}^{b} v\frac{du}{dx}\,dx$$

or

$$\int_{a}^{b} u\,dv = uv\Big|_{a}^{b} - \int_{a}^{b} v\,du.$$

For applications see Secs. 1.9.2, 4.2.7 and 4.2.15.

e. For other formulas, see Appendix A6.

INTRODUCTION

For the evaluation of many integrals, the Euler's gamma and beta functions and the complete elliptic integrals are among the most useful functions in engineering, physics and probability. The gamma and beta functions are also used for the generalization of many integrals and in the definition of other special functions, e.g., the Bessel, Legendre, hypergeometric functions, and so on.

Using the properties of these functions, readers will find in the book how to use them for integral evaluations. In addition, for many integrals, using these functions we will try to eliminate the long and tedious traditional methods of integral evaluations. Note that there are many integrals that cannot be evaluated in terms of the elementary functions, but can be evaluated in terms of the gamma function or in terms of the complete elliptic integrals; e.g.,

$$\int_0^1 \frac{dx}{\sqrt{1+x^4}} = \frac{1}{8\sqrt{\pi}}\left[\Gamma\left(\frac{1}{4}\right)\right]^2 = \frac{1}{2}K\left(\frac{1}{\sqrt{2}}\right), \qquad \text{Secs.2.2.4a and 4.2.19}$$

$$\int_0^{\frac{\pi}{2}} \sqrt{\sin x}\, dx = \frac{4\pi\sqrt{2\pi}}{\left[\Gamma\left(\frac{1}{4}\right)\right]^2} = \sqrt{2}\left[2E\left(\frac{1}{\sqrt{2}}\right) - K\left(\frac{1}{\sqrt{2}}\right)\right]. \qquad \text{Sec.4.3.5}$$

This book can be used as a self-study book, textbook, supplemental textbook or reference book. The prerequisite could be any classical course of integral calculus given in college and university. In a step-by-step procedure, this book will show how to use the gamma and beta functions and complete elliptic integrals to evaluate, easily and with a high degree of accuracy, many integrals used in engineering work.

The gamma, beta, digamma functions and the function $\beta(x)$ may be used to simplify the evaluation of many definite integrals, line integrals, double, multiple and surface integrals. Using only the traditional methods, the evaluation of many integrals in engineering is very tedious. Such long calculations are prone to errors. Examples of the integrals evaluated by different methods illustrate that many times it is simpler to use the gamma, beta, digamma functions and the function $\beta(x)$. Examples

$$\int_{-\infty}^{+\infty} \frac{dx}{1+x^4} = \frac{1}{2}B\left(\frac{3}{4},\frac{1}{4}\right) = \frac{1}{2}\Gamma\left(\frac{3}{4}\right)\Gamma\left(\frac{1}{4}\right) = \frac{\pi}{\sqrt{2}}, \qquad \text{Sec.2.3.14}$$

$$\int_0^{\infty} \frac{dx}{1+x^n} = \frac{1}{n}B\left(1-\frac{1}{n},\frac{1}{n}\right) = \frac{\pi}{n\sin\dfrac{\pi}{n}}, \quad n>1, \qquad \text{Sec.2.3.16}$$

$$\int_0^1 \frac{dx}{1+x^3} = \frac{1}{3}\beta\left(\frac{1}{3}\right). \qquad \text{Sec.2.8.4}$$

Moreover, there are integrals which cannot be expressed in terms of the elementary functions, but can be in terms of the gamma function or elliptic integrals. Examples

$$\int_0^1 (1-x^2)^p\, dx = \frac{p}{1+2p}B\left(\frac{1}{2},p\right) = \frac{p\sqrt{\pi}}{1+2p}\frac{\Gamma(p)}{\Gamma\left(\dfrac{1}{2}+p\right)}, \quad p>0, \qquad \text{Sec.2.3.8}$$

$$\int_0^{\phi} \frac{d\theta}{\sqrt{\cos\theta-\cos\phi}} = \sqrt{2}K\left(\sin\frac{\phi}{2}\right), \quad \phi\in[0,\pi). \qquad \text{Sec.4.2.11}$$

There are applications where the double and triple integrals are expressed in terms of the elementary functions, but when they are generalized in the R^n-space, the integrals cannot be expressed in terms of the elementary functions, but can be in terms of the gamma and beta functions (see the surface area of a sphere in the R^3-space and then of a sphere in the R^n-space, Sec.2.5.3), the digamma function, function $\beta(p)$ and elliptic integrals.

Also, for many improper generalized integrals (integrals depending on one or more parameters), the convergence conditions of the integrals expressed in terms of the gamma and beta functions result at the end of the evaluation from the properties of the gamma and beta functions. Example

$$\int_0^{\infty} \frac{x^{p-1}\,dx}{(1+x^q)^2} = \frac{1}{q} B\left(2-\frac{p}{q}, \frac{p}{q}\right), \quad 0 < p < 2q. \qquad \text{Sec.2.3.25}$$

Many existing tables of integrals list only the indefinite and definite integrals; they do not list the double, triple, multiple, line and surface integrals which are used in engineering.

The existing tables give the gamma function for $p \in [1, 2]$, but we also need tables for $p \in (0, 1]$. Our tables are for $p \in (0, 2]$, and in addition we give $[\Gamma(p)]^2$ and the graphic of $1/\Gamma(p)$, which are important in applications. Also, the tables of complete elliptic integrals, $E(k)$ and $K(k)$, are given in terms of k^2 and the angle psi, $k = \sin\psi$. The graphs of these functions are given with enough accuracy to be usable in engineering work. The computer programs are in Microsoft QuickBASIC. With little modification, the programs can be used for other computer languages.

The book contains a valuable range of practical applications and examples. The applications are selected in such a way as to be important to any engineering branch. At the end of some applications we ask the reader to evaluate the integrals using other methods. Doing these exercises, the reader can compare different methods of evaluation for the same integral to convince himself that the gamma and beta functions method is a simple and important method of integral evaluation. This book illustrates the fact that the methods of integral evaluations using the gamma, beta and elliptic integral functions are very powerful and useful for solving practical problems in different branches of engineering and science.

In this book, the theory has been limited to only the engineering needs. Thus, the book has several limitations:

(a) The gamma function $\Gamma(z)$, where z is a complex number, is not studied. For this theory, one could refer to those books that study the functions of a complex variable.
(b) Stirling's asymptotic series and formula are given without proof; these are used in statistical mechanics and probability theory, which deal with large factorials.
(c) Hölder's theorem is not discussed.

I predict that one day $\Gamma(p)$, $\psi(x)$, $K(k)$ and $E(k)$ will be regarded as elementary functions and will be included in the scientific calculator. This publication may encourage colleges and engineering faculties to include these functions and their applications as an undergraduate or graduate term course. This book should be used as a tool for solving practical problems. Taking into consideration the many exercises and applications evaluated by a step-by-step approach, the book may be recommended to students and engineers as a self-study book.

For easy reference, a list of the most important formulas with the gamma function, beta function and complete elliptic integrals can be found in Appendix A6.

I am greatly indebted to the late Dr. H.H.E. Leipholz, former Dean of Engineering, University of Waterloo, Ontario, Canada, and Dr. K.D. Srivastava, Professor of Electrical Engineering, University of British Columbia, Vancouver, for the encouragement given to me to write this book.

C.C. Maican

CHAPTER 1

The Gamma and Beta Functions

1.1. Definitions of the Gamma Function

a. The gamma function, $\Gamma(p)$, is the Euler's integral of the second kind, defined by

$$\Gamma(p) = \int_0^\infty x^{p-1} e^{-x}\, dx, \quad p > 0, \tag{1.1}$$

which is the integral representation of $\Gamma(p)$. This is the definition of the complete gamma function. Note that the gamma function is an improper integral with the integrand not defined for $x = 0$ and $x = \infty$.

Test for convergence. According to Appendix A1.2, the integral $\Gamma(p)$ converges at the lower integration limit, since there is a number $\alpha < 1$ (condition a) for which

$$f(x) = x^{p-1} e^{-x} = \underset{x=0+}{O}\left[\frac{1}{x^{1-p}}\right], \quad \alpha = 1 - p, \quad \text{(condition b)}$$

and hence, from both conditions, we have

$$1 - p = \alpha < 1 \quad \text{or} \quad p > 0. \tag{1.2}$$

According to Appendix A1.1, the integral $\Gamma(p)$ converges at the upper integration limit, since there is a number $\alpha > 1$ (condition a) for which the limit

$$\lim_{x\to\infty} x^\alpha \frac{x^{p-1}}{e^x} = \lim_{x\to\infty} \frac{x^{\alpha+p-1}}{e^x}$$

does exist for any number $\alpha + p - 1$ (condition b). Thus, the integral $\Gamma(p)$ converges for $p > 0$. For the extension of the definition of the gamma function to the negative numbers, see Sec.1.5.2.

In particular, for $p = 1$, from Eq.(1.1), we have

$$\Gamma(1) = \int_0^\infty e^{-x}\,dx = -e^{-x}\Big|_0^\infty$$

and

$$\Gamma(1) = 1. \tag{1.3}$$

b. Gauss defined the gamma function by the infinite-product representation

$$\Gamma(x) = \lim_{n\to\infty} \frac{n!\,n^x}{x(x+1)\cdots(x+n)}, \qquad x \neq 0, -1, -2, \ldots, \tag{1.4}$$

or

$$\Gamma(x) = \lim_{n\to\infty} \frac{n^x}{x} \prod_{k=1}^{n} \frac{k}{x+k}, \qquad x \neq 0, -1, -2, \ldots, \tag{1.5}$$

and it is often used as a fundamental definition of the gamma function.

By setting $x = 1$ in Eq.(1.4), we have

$$\Gamma(1) = \lim_{n\to\infty} \frac{n!\,n}{1 \times 2 \times 3 \times \cdots \times n(n+1)} = \lim_{n\to\infty} \frac{n}{n+1} = 1.$$

See another application in Sec.1.5.2.

c. Weierstrass defined the gamma function by the infinite-product representation

$$\Gamma(x) = e^{-\gamma x} \frac{1}{x} \prod_{n=1}^{\infty} \frac{e^{\frac{x}{n}}}{1 + \dfrac{x}{n}}, \tag{1.6}$$

where $\gamma = 0.57721566\ldots$ is the Euler's constant. For application of this formula, see Secs.1.5.3 and 1.8.

In the following, we will discuss the complete gamma function defined by Eq.(1.1), and the word "complete" will be omitted unless otherwise specified.

However, the definition of incomplete gamma functions, $\gamma(p,x)$ and $\Gamma(p,x)$, will be given in Sec.1.16.

The gamma function $\Gamma(z)$, where z is a complex number, is not studied. For this, refer to books that study the functions of a complex variable. However, some of the integrals have been evaluated using Cauchy's residue theorem only to compare the complexity of this method with the evaluation of the integral using the gamma and beta functions method. See Secs.2.3.13 and 2.3.14.

d. Stirling's asymptotic series. The Stirling's asymptotic series for the gamma function is

$$\Gamma(x+1) = \sqrt{2\pi x}\, x^x e^{-x}\left(1 + \frac{1}{12x} + \frac{1}{288x^2} - \cdots\right), \quad x > 0. \tag{1.7}$$

For $x = n$, $n \in N$, and retaining only the first term in parentheses of Eq.(1.7), we obtain the Stirling's formula,

$$n! = \Gamma(n+1) \approx \sqrt{2\pi n}\, n^n e^{-n}, \tag{1.8}$$

for n sufficiently large (e.g., $n > 10$).

For a more accurate Stirling's formula than Eq.(1.8), retaining the first two terms in parentheses of Eq.(1.7), for $n \in N$, we have

$$n! = \Gamma(n+1) \approx \sqrt{2\pi n}\, n^n e^{-n}\left(1 + \frac{1}{12n}\right). \tag{1.8a}$$

For n a large positive integer, the results are adequate for many applications. With a program on a PC, with $\pi = 3.141593$, using Eq.(1.8) and Eq.(1.8a), we obtained the following results:

$n!$	From Eq.(1.8)	Error %	From Eq.(1.8a)	Error %
$3! = 6$	5.83621	2.72983	5.998327	2.788×10^{-2}
$4! = 24$	23.50618	2.05758	23.99589	1.7125×10^{-2}
$5! = 120$	118.0192	1.65067	119.9862	1.15×10^{-2}
$6! = 720$	710.0782	1.378	719.9404	8.28×10^{-3}
$7! = 5040$	4980.396	1.1826	5039.687	6.2×10^{-3}
$8! = 40\,320$	39 902.39	1.0357	40 318.05	4.83×10^{-3}
$9! = 362\,880$	359 536.9	0.921	362 865.9	3.885×10^{-3}
$10! = 3\,628\,800$	3 598 696	0.829	3 628 685	3.169×10^{-3}

Example with Eq.(1.8a): $6! = 720$ and the error is

$$\text{error} = \frac{720 - 719.9404}{720} \times 100 = 8.28 \times 10^{-3}\%.$$

From these results, we conclude that the errors are small enough for many engineering applications.

1.2. Definition of Beta Function

The beta function, $B(p, q)$, is the Euler's integral of the first kind, defined by

$$B(p, q) = \int_0^1 x^{p-1}(1-x)^{q-1}\,dx, \quad p > 0, \quad q > 0, \tag{1.9}$$

which is the integral representation of $B(p, q)$. This is the definition of the complete beta function. Note that beta function is an integral with two paremeters, p and q. The beta function is an improper integral of the second kind with the integrand not defined for $x = 0$ and $x = 1$.

Test for convergence. According to Appendix A1.2, the integral $B(p, q)$ converges for $x = 0$, since there is a number $\alpha < 1$ (condition a) for which

$$f(x) = x^{p-1}(1-x)^{q-1} = \underset{x=0+}{O}\left[\frac{1}{x^{1-p}}\right], \quad \alpha = 1 - p, \quad \text{(condition b)}$$

and hence, the integral converges for

$$1 - p = \alpha < 1 \quad \text{or} \quad p > 0. \tag{1.10}$$

Similarly, the integral $B(p,q)$ converges for $x = 1$, since there is a number $\alpha < 1$ (condition a) for which

$$x^{p-1}(1-x)^{q-1} = \underset{x=1-}{O}\left[\frac{1}{(1-x)^{1-q}}\right], \quad \alpha = 1 - q \quad \text{(condition b)}, \tag{1.11}$$

and hence,

$$1 - q = \alpha < 1 \quad \text{or} \quad q > 0.$$

Thus, the function $B(p,q)$ converges for $p, q > 0$.

a. In particular, for $q = 1$, from Eq.(1.9), we have

$$B(p,1) = \int_0^1 x^{p-1}\,dx = \frac{1}{p}x^p\Big|_0^1 = \frac{1}{p}, \quad p > 0.$$

b. Letting, in Eq.(1.9),

$$x = \sin^2\theta, \quad dx = 2\sin\theta\cos\theta\,d\theta,$$

we have

$$B(p,q) = 2\int_0^{\frac{\pi}{2}} \sin^{2p-1}\theta\cos^{2q-1}\theta\,d\theta, \quad p, q \in R^+, \tag{1.12}$$

which can be considered the *second definition of beta function.* Eq.(1.12) can be written in another form to be used for evaluating trigonometric integrals:

$$\int_0^{\frac{\pi}{2}} \sin^{2p-1}\theta\cos^{2q-1}\theta\,d\theta = \frac{1}{2}B(p,q), \quad p, q > 0. \tag{1.13}$$

For details, see Sec.1.12.

The complete beta function will be discussed in this book and the word "complete" will be omitted unless otherwise specified. However, the definition of the incomplete beta function, $B_x(p,q)$, is given in Sec.1.17.

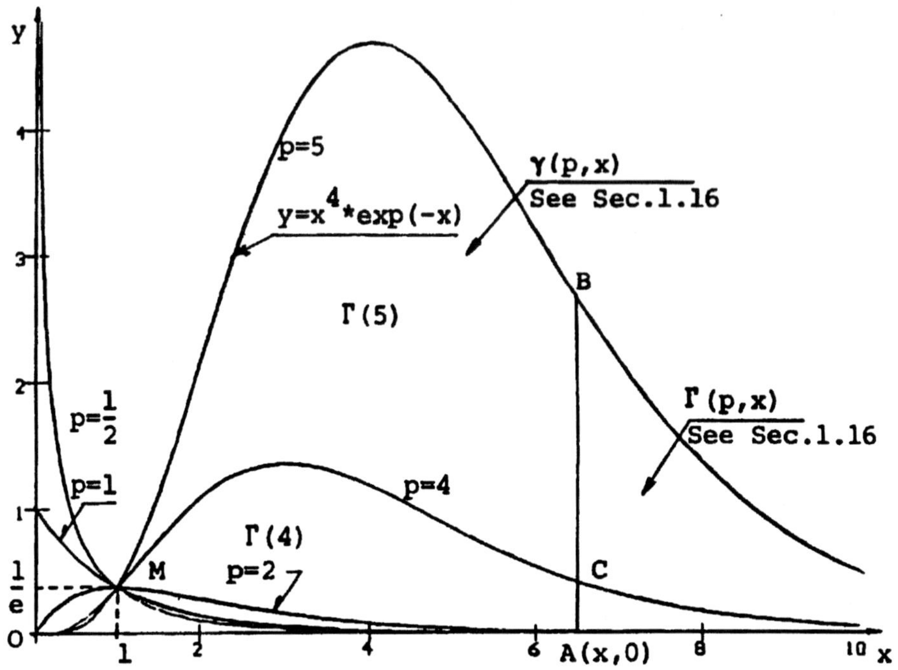

Fig.1.1

1.3. Graph of the Gamma Function

The gamma function,

$$\Gamma(p) = \int_0^\infty x^{p-1} e^{-x}\, dx, \quad p > 0, \tag{1.14}$$

for $p = 5$, is represented by the area between the integrand and the x axis. The curve lies above the x axis. See Fig.1.1. For representation of integrand

$$y = x^{p-1} e^{-x} \tag{1.15}$$

for $p = 1, 2, 4, 5$ and $1/2$, Maple V software has been used.

The graph is also important for visualizing the incomplete gamma functions, e.g., for $p = 5$ areas $OABMO$ and $AxBA$. For the incomplete gamma function definitions, see Sec.1.16.

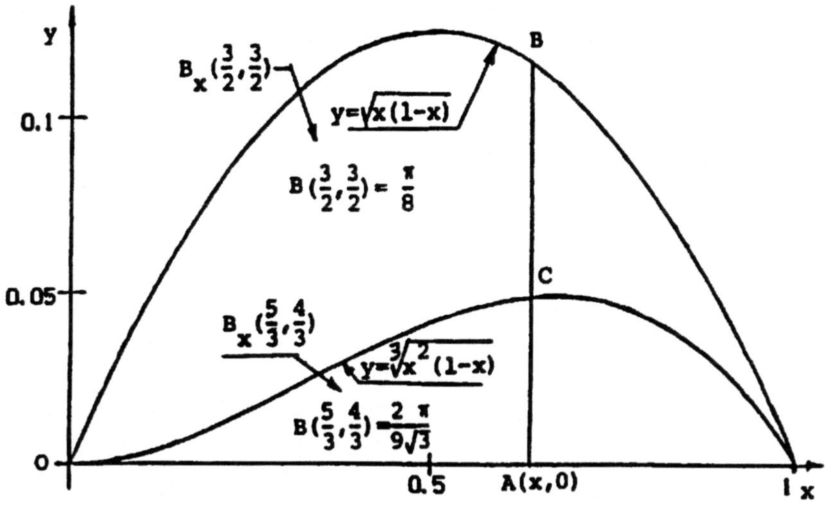

Fig.1.2

1.4. Graph of Beta Function

Let

$$B(p,q) = \int_0^1 x^{p-1}(1-x)^{q-1}\,dx, \quad p,q>0. \tag{1.16}$$

For $p = q = 3/2$, the integrand is

$$y = \sqrt{x(1-x)}. \tag{1.17}$$

The graph of Eq.(1.17), for $x \in [0,1]$, is given in Fig.1.2. The beta function for this particular case is represented by the area between the curve and the x axis. The calculated value of $B(\frac{3}{2},\frac{3}{2}) = \frac{\pi}{8}$ square units. For representation of the integrand of $B(p,q)$ for different values of p and q, Maple V software has been used. The graphs are important also for visualizing the incomplete beta function, e.g., area $OABO$. For the incomplete beta function definition, see Sec.1.17. It is noted that not all beta integrand graphs have the same shape; see Fig.1.2.

1.5. Properties of the Gamma Function

1.5.1. In Eq.(1.1), replacing x by x^2, we have a new definition of the gamma function:

$$\Gamma(p) = 2\int_0^\infty x^{2p-1}e^{-x^2}\,dx, \quad p \in R^+. \tag{1.18}$$

For applications, see Sec.1.11 and 1.14.

1.5.2. The Recurrence Formula of the Gamma Function. Integrating by parts in Eq.(1.1),

$$\begin{aligned}\Gamma(p) &= \frac{1}{p}\int_0^\infty e^{-x}\,dx^p = \frac{1}{p}x^p e^{-x}\Big|_0^\infty + \frac{1}{p}\int_0^\infty x^p e^{-x}\,dx \\ &= \frac{1}{p}\int_0^\infty x^{(p+1)-1}e^{-x}\,dx = \frac{1}{p}\Gamma(p+1),\end{aligned} \tag{1.19}$$

we get the *recurrence formula*,

$$\Gamma(p+1) = p\Gamma(p), \quad p > 0, \tag{1.20}$$

which is *the first most important functional equation for the gamma function.*

a. Using Eqs.(1.20) and (1.37), we can calculate

$$\Gamma\left(\frac{3}{2}\right) = \Gamma\left(1+\frac{1}{2}\right) = \frac{1}{2}\Gamma\left(\frac{1}{2}\right) = \frac{\sqrt{\pi}}{2}, \tag{1.21}$$

$$\Gamma\left(\frac{5}{2}\right) = \Gamma\left(1+\frac{3}{2}\right) = \frac{3}{2}\Gamma\left(\frac{3}{2}\right) = \frac{3\sqrt{\pi}}{4}, \tag{1.22}$$

$$\Gamma\left(\frac{7}{2}\right) = \Gamma\left(1+\frac{5}{2}\right) = \frac{5}{2}\Gamma\left(\frac{5}{2}\right) = \frac{5}{2}\cdot\frac{3\sqrt{\pi}}{4} = \frac{15\sqrt{\pi}}{8}, \tag{1.23}$$

$$\Gamma\left(\frac{9}{2}\right) = \Gamma\left(1+\frac{7}{2}\right) = \frac{7}{2}\Gamma\left(\frac{7}{2}\right) = \frac{7}{2}\cdot\frac{15\sqrt{\pi}}{8} = \frac{105}{16}\sqrt{\pi}. \tag{1.23a}$$

Using a scientific calculator with the gamma function built-in, the numerical values of $\Gamma(\frac{3}{2})$, $\Gamma(\frac{5}{2})$, $\Gamma(\frac{7}{2})$ and $\Gamma(\frac{9}{2})$ can be obtained straightforwardly without using Eq.(1.20).

b. The recurrence formula (1.20) can also be deduced from Gauss' infinite-product representation, Eq.(1.4). Substituting $x+1$ for x in Eq.(1.4),

$$\begin{aligned}\Gamma(x+1) &= \lim_{n\to\infty} \frac{n!\,n^{x+1}}{(x+1)(x+2)\cdots(x+n)(x+n+1)} \\ &= \lim_{n\to\infty} \frac{nx}{x+n+1} \cdot \lim_{n\to\infty} \frac{n!\,n^{x}}{x(x+1)\cdots(x+n)} = x\Gamma(x).\end{aligned}$$

Repeating Eq.(1.20) for $p = n \in N$, and knowing that $\Gamma(1) = 1$, see Eq.(1.3), we have

$$\Gamma(n+1) = n!\,, \quad n = 0, 1, 2, \ldots, \tag{1.24}$$

and

$$\Gamma(n) = (n-1)!\,, \quad n = 1, 2, 3, \ldots. \tag{1.25}$$

If $p \in (n, n+1)$, $n \in N$, and using Eq.(1.20),

$$\begin{aligned}\Gamma(p) &= (p-1)\Gamma(p-1) = (p-1)(p-2)\Gamma(p-2) = \ldots \\ &= (p-1)(p-2)\cdots(p-n)\Gamma(p-n),\end{aligned} \tag{1.26}$$

and substituting $p+n$ for p,

$$\begin{aligned}\Gamma(p+n) &= (p+n-1)(p+n-2)\cdots(p+n-n)\Gamma(p+n-n) \\ &= p(p+1)\cdots(p+n-1)\Gamma(p).\end{aligned} \tag{1.27}$$

The integral in Eq.(1.1) converges for $p > 0$. Suppose $p \in (-n, -n+1)$, $n \in N$. In this case, we define $\Gamma(p)$ by Eq.(1.27) solved for $\Gamma(p)$:

$$\Gamma(p) = \frac{\Gamma(p+n)}{p(p+1)\cdots(p+n-1)}, \quad n = 1, 2, 3, \ldots, \tag{1.28}$$

and this defines $\Gamma(p)$ over the interval $p > -n$ and $p \neq 0, -1, -2, \ldots, (-n+1)$.

c. Particular cases of Eq.(1.27):

$$\Gamma\left(n+\frac{1}{3}\right)=\frac{1}{3}\left(\frac{1}{3}+1\right)\left(\frac{1}{3}+2\right)\cdots\left(\frac{1}{3}+n-1\right)\Gamma\left(\frac{1}{3}\right)$$
$$=\frac{1\times 4\times 7\times\cdots\times(3n-2)}{3^n}\Gamma\left(\frac{1}{3}\right). \tag{1.29}$$

$$\Gamma\left(n+\frac{2}{3}\right)=\frac{2}{3}\left(\frac{2}{3}+1\right)\left(\frac{2}{3}+2\right)\cdots\left(\frac{2}{3}+n-1\right)\Gamma\left(\frac{2}{3}\right)$$
$$=\frac{2\times 5\times 8\times\cdots\times(3n-1)}{3^n}\Gamma\left(\frac{2}{3}\right). \tag{1.30}$$

$$\Gamma\left(n+\frac{1}{4}\right)=\frac{1}{4}\left(\frac{1}{4}+1\right)\left(\frac{1}{4}+2\right)\cdots\left(\frac{1}{4}+n-1\right)\Gamma\left(\frac{1}{4}\right)$$
$$=\frac{1\times 5\times 9\times\cdots\times(4n-3)}{4^n}\Gamma\left(\frac{1}{4}\right). \tag{1.31}$$

d. The gamma function—extension to negative numbers. For $p<0$, $\Gamma(p)$ is defined by the recursion relation Eq.(1.20) solved for $\Gamma(p)$,

$$\Gamma(p)=\frac{\Gamma(p+1)}{p}. \tag{1.32}$$

Since $\Gamma(1)=1$, see Eq.(1.3),

$$\lim_{p\to 0+}\Gamma(p)=\lim_{p\to 0+}\frac{\Gamma(p+1)}{p}=+\infty.$$

From successive use of Eq.(1.32), $\Gamma(p)$ will be $+\infty$ or $-\infty$ at all negative integers. The process of successive use of Eq.(1.32) is called analytic continuation. See Fig.1.6.

- Let $p=-2.4$ and, applying Eq.(1.32), we have

$$\Gamma(-2.4)=\frac{\Gamma(-2.4+1)}{-2.4}=\frac{\Gamma(-1.4)}{-2.4}=\frac{1}{-2.4}\,\frac{\Gamma(-1.4+1)}{-1.4}$$
$$=\frac{\Gamma(-0.4)}{2.4\times 1.4}=\frac{1}{2.4\times 1.4}\,\frac{\Gamma(-0.4+1)}{-0.4}$$
$$=-\frac{\Gamma(0.6)}{2.4\times 1.4\times 0.4}=-1.108.$$

- Let $p = -2.4$ and, applying Eq.(1.28), we have $n = 3$, $p + n - 1 = -2.4 + 3 - 1 = -0.4$ and

$$\Gamma(-2.4) = \frac{\Gamma(-2.4+3)}{-2.4(-2.4+1)(-2.4+3-1)} = \frac{\Gamma(0.6)}{-2.4(-1.4)(-0.4)} = -1.108.$$

1.5.3. The Complement Formula of the Gamma Function. Using Weierstrass' definition of gamma function, see Eq.(1.6), for x and $-x$, and multiplying them,

$$\Gamma(x)\Gamma(-x) = -\frac{1}{x^2}\prod_{n=1}^{\infty}\frac{e^{\frac{x}{n}}}{1+\frac{x}{n}}\frac{e^{-\frac{x}{n}}}{1-\frac{x}{n}} = -\frac{1}{x^2}\frac{1}{\prod\limits_{n=1}^{\infty}\left(1-\frac{x^2}{n^2}\right)}. \tag{1.33}$$

Knowing that the representation of $\sin x$ in the form of an infinite product is

$$\sin x = x\prod_{n=1}^{\infty}\left(1-\frac{x^2}{n^2\pi^2}\right)$$

and substituting x by $x\pi$, we have

$$\sin(x\pi) = x\pi\prod_{n=1}^{\infty}\left(1-\frac{x^2}{n^2}\right)$$

and Eq.(1.33) becomes

$$\Gamma(x)\Gamma(-x) = -\frac{\pi}{x\sin x\pi}. \tag{1.34}$$

Letting $p = -x$ in the recurrence formula (1.20), we have

$$\Gamma(-x) = -\frac{\Gamma(1-x)}{x},$$

and substituting in Eq.(1.34) yields the *complement formula* of the gamma function,

$$\Gamma(x)\Gamma(1-x) = \frac{\pi}{\sin x\pi}, \quad 0 < x < 1, \tag{1.35}$$

which is the *second most important functional equation verified by the gamma function.*

Utilizing Eq.(1.35), the values of $\Gamma(x)$ for $x \in (\frac{1}{2}, 1)$ can be calculated in terms of the values of $\Gamma(x)$ for $x \in (0, \frac{1}{2})$.

Particular values of $\Gamma(x)$ obtained using Eq.(1.35).

a. Letting $x = 1/2$ in Eq.(1.35), we have

$$\Gamma\left(\frac{1}{2}\right) \cdot \Gamma\left(\frac{1}{2}\right) = \frac{\pi}{\sin\frac{\pi}{2}} = \pi. \tag{1.36}$$

Thus,

$$\Gamma\left(\frac{1}{2}\right) = \sqrt{\pi}. \tag{1.37}$$

b. Letting $x = 1/4$ in Eq.(1.35), we have

$$\Gamma\left(\frac{1}{4}\right)\Gamma\left(\frac{3}{4}\right) = \frac{\pi}{\sin\frac{\pi}{4}} = \pi\sqrt{2}$$

and

$$\Gamma\left(\frac{3}{4}\right) = \frac{\pi\sqrt{2}}{\Gamma\left(\frac{1}{4}\right)}. \tag{1.38}$$

1.5.4. a. Evaluate the expression

$$E = \Gamma\left(\frac{1}{2}+p\right)\Gamma\left(\frac{1}{2}-p\right).$$

Letting $x = \frac{1}{2} + p$ in Eq.(1.35), we have

$$\Gamma\left(\frac{1}{2}+p\right)\Gamma\left(\frac{1}{2}-p\right) = \frac{\pi}{\sin\left(\frac{1}{2}+p\right)\pi} = \frac{\pi}{\sin\left(\frac{\pi}{2}+p\pi\right)}$$

$$\Gamma\left(\frac{1}{2}+p\right)\Gamma\left(\frac{1}{2}-p\right) = \frac{\pi}{\cos p\pi}, \quad p \neq n + \frac{1}{2}, \quad n = 0, 1, 2, \ldots. \tag{1.39}$$

b. Let $\frac{1}{2} + p = x + 1$ in Eq.(1.39),

$$\Gamma(x+1)\Gamma(-x) = \frac{\pi}{\cos\left(x+\frac{1}{2}\right)\pi},$$

but from Eq.(1.20),

$$\Gamma(-x) = \frac{\Gamma(1-x)}{-x}$$

and

$$\Gamma(x+1)\frac{\Gamma(1-x)}{-x} = \frac{\pi}{-\sin x\pi},$$

$$\Gamma(1+x)\Gamma(1-x) = \frac{x\pi}{\sin x\pi}, \quad 0 < x < 1. \tag{1.40}$$

c. Let $x = 1 + p$ in Eq.(1.35),

$$\Gamma(1+p)\cdot\Gamma(-p) = \frac{\pi}{\sin(1+p)\pi} = \frac{\pi}{-\sin p\pi}, \tag{α}$$

and substituting p by $-p$ in Eq.(1.20), from (α) we have

$$\Gamma(1+p)\frac{\Gamma(1-p)}{-p} = \frac{\pi}{-\sin p\pi}$$

or

$$\Gamma(1+p)\Gamma(1-p) = \frac{p\pi}{\sin p\pi}, \quad 0 < p < 1. \tag{1.41}$$

1.5.5. Let $p = n \in N \cup \{0\}$, and Eq.(1.39) becomes

$$\Gamma\left(\frac{1}{2}+n\right)\cdot\Gamma\left(\frac{1}{2}-n\right) = (-1)^n\pi. \tag{1.42}$$

1.5.6. Letting $p = \dfrac{1}{2}$ in Eq.(1.27),

$$\Gamma\left(\frac{1}{2}+n\right) = \frac{1\cdot 3\cdots(2n-1)}{2^n}\sqrt{\pi}$$

$$= \frac{\sqrt{\pi}}{2^n}(2n-1)!!, \quad n \in N, \tag{1.43}$$

or with $(2n)!! = n!\,2^n$,

$$\Gamma\left(\frac{1}{2}+n\right) = \frac{(2n-1)!!}{(2n)!!}n!\,\sqrt{\pi}, \quad n \in N. \tag{1.44}$$

1.5.7. From Eq.(1.42), we have

$$\Gamma\left(\frac{1}{2}-n\right) = (-1)^n\frac{\pi}{\Gamma\left(\dfrac{1}{2}+n\right)},$$

and using Eq.(1.43),

$$\Gamma\left(\frac{1}{2}-n\right) = (-1)^n\frac{\sqrt{\pi}\cdot 2^n}{(2n-1)!!}, \quad n \in N. \tag{1.45}$$

1.6. The Gamma Function Programs and Tables

The numerical values of the gamma function have been obtained with MS QuickBASIC programs (see Figs.1.3, 1.4 and 1.5). Knowing the numerical values of $\Gamma(p)$ for $p \in [1,2]$ and using the recurrence formula (1.20), the numerical values of $\Gamma(p)$ can be calculated for $p \in (0,1]$. Knowing the values of $\Gamma(p)$ for $p \in [1,2]$ and using the recurrence formula (1.20), the values of $\Gamma(p)$ can be calculated for any $p \in R$ except $p = 0, -1, -2, -3, \ldots$.

Thus, a table with the numerical values of $\Gamma(p)$ for $p \in [1,2]$ is sufficient. However, for engineering work, Table 1.1 for $p \in (0,1]$ and Table 1.2 for $p \in [1,2]$ are given with an increment of $p = 0.01$. The computer programs are self-explanatory. The programs can be used with few modifications for other programming languages.

The gamma function tables are obtained using as algorithm a polynomial approximation, for $p \in [1,2]$, [1],

$$\Gamma(p) = A_0 p^8 + A_1 p^7 + \cdots + A_7 p + 1 + \varepsilon, \tag{1.46}$$

where

$$\begin{aligned} A_0 &= 0.035868343 & A_4 &= 0.918206857 \\ A_1 &= -0.193527818 & A_5 &= -0.897056937 \\ A_2 &= 0.482199394 & A_6 &= 0.988205891 \\ A_3 &= -0.756704078 & A_7 &= -0.577191652 \end{aligned} \tag{1.47}$$

and $|\varepsilon| \leq 3 \times 10^{-7}$. With this polynomial approximation, $\Gamma(p)$ is calculated for $p \in [1,2]$ as shown in Table 1.2. With the same polynomial approximation and the recurrence formula (1.20), solved for $\Gamma(p)$,

$$\Gamma(p) = \frac{\Gamma(p+1)}{p}, \quad p > -1, \quad p \neq 0, \tag{1.48}$$

we get the values of $\Gamma(p)$ for $p \in (0,1]$ as given in Table 1.1. The error in the tables is $\leq 10^{-5}$, and this is adequate for most engineering applications. With minor changes in the program, the reader can get a smaller error. Note that for $p = 0$ and negative integers, the gamma function is undefined (see Sec. 1.1 and Fig. 1.6). Also similar tables with the values of the gamma function can be found in [4].

```
'CM11GB4 Program in QuickBASIC, to tabulate the gamma
'function and gamma square in the range of 0<P<=1,
'increment 0.01, using an approximating polynomial.
'The error is less than 10^-6. Refer to Table 1.1.
CLS
DIM I(100) AS INTEGER
DIM P(100), R(100), T(100), V(100)
DIM GAMMA(100), GAMMA1(100), GAMMA2(100)
FOR I = 1 TO 100
      P(I) = I / 100
      A0 = .035868343#
      A1 = -.193527818#
      A2 = .482199394#
      A3 = -.7567040779999999#
      A4 = .918206857#
      A5 = -.897056937#
      A6 = .9882058891#
      A7 = -.577191652#
      R(I) = P(I) ^ 6 * (A0 * P(I) ^ 2 + A1 * P(I) + A2)
      T(I) = P(I) ^ 3 * (A3 * P(I) ^ 2 + A4 * P(I) + A5)
      V(I) = A6 * P(I) ^ 2 + A7 * P(I) + 1!
      GAMMA(I) = R(I) + T(I) + V(I)
      GAMMA1(I) = GAMMA(I) / P(I)
      GAMMA2(I) = GAMMA1(I) * GAMMA1(I)
NEXT I
LPRINT CHR$(12)
LPRINT
LPRINT TAB(28); "TABLE 1.1.  GAMMA FUNCTION"
LPRINT TAB(28); "--------------------------"
LPRINT
LPRINT TAB(10); "P"; SPC(8); "GAMMA"; SPC(8); "GAMMA^2";
LPRINT SPC(9); "P"; SPC(6); "GAMMA"; SPC(6); "GAMMA^2";
LPRINT TAB(8); "----"; SPC(6); "--------"; SPC(6); "--------";
LPRINT SPC(7); "----"; SPC(4); "--------"; SPC(3); "--------";
LPRINT
LPRINT TAB(8); "0.00"; SPC(5); "UNDEFINED";
FOR I = 1 TO 50
      LPRINT TAB(8);
      LPRINT USING "#.##"; P(I);
      LPRINT TAB(17);
      LPRINT USING "##.######"; GAMMA1(I);
      LPRINT TAB(29);
      LPRINT USING "####.######"; GAMMA2(I);
      LPRINT TAB(47);
      LPRINT USING "#.##"; P(I + 50);
      LPRINT TAB(55);
      LPRINT USING "#.######"; GAMMA1(I + 50);
      LPRINT TAB(66);
      LPRINT USING "#.######"; GAMMA2(I + 50);
NEXT I
END
```

Fig.1.3

TABLE 1.1. GAMMA FUNCTION

P	GAMMA	GAMMA^2	P	GAMMA	GAMMA^2
0.00	UNDEFINED				
0.01	99.432602	9886.842773	0.51	1.738415	3.022088
0.02	49.442223	2444.533447	0.52	1.705844	2.909904
0.03	32.785004	1074.856445	0.53	1.674656	2.804474
0.04	24.460958	598.338501	0.54	1.644774	2.705281
0.05	19.470085	379.084229	0.55	1.616125	2.611859
0.06	16.145727	260.684509	0.56	1.588642	2.523783
0.07	13.773599	189.712021	0.57	1.562263	2.440666
0.08	11.996565	143.917572	0.58	1.536931	2.362156
0.09	10.616214	112.703995	0.59	1.512591	2.287930
0.10	9.513506	90.506798	0.60	1.489192	2.217694
0.11	8.612684	74.178329	0.61	1.466690	2.151179
0.12	7.863250	61.830704	0.62	1.445038	2.088136
0.13	7.230241	52.276382	0.63	1.424197	2.028338
0.14	6.688685	44.738506	0.64	1.404128	1.971576
0.15	6.220272	38.691784	0.65	1.384795	1.917658
0.16	5.811269	33.770851	0.66	1.366164	1.866405
0.17	5.451174	29.715302	0.67	1.348204	1.817653
0.18	5.131821	26.335588	0.68	1.330884	1.771252
0.19	4.846764	23.491121	0.69	1.314176	1.727060
0.20	4.590845	21.075855	0.70	1.298055	1.684947
0.21	4.359889	19.008633	0.71	1.282495	1.644794
0.22	4.150483	17.226507	0.72	1.267473	1.606487
0.23	3.959805	15.680052	0.73	1.252966	1.569924
0.24	3.785506	14.330052	0.74	1.238954	1.535007
0.25	3.625611	13.145054	0.75	1.225417	1.501646
0.26	3.478451	12.099623	0.76	1.212335	1.469756
0.27	3.342604	11.173004	0.77	1.199692	1.439261
0.28	3.216852	10.348137	0.78	1.187471	1.410087
0.29	3.100144	9.610890	0.79	1.175655	1.382164
0.30	2.991569	8.949486	0.80	1.164230	1.355431
0.31	2.890336	8.354042	0.81	1.153181	1.329826
0.32	2.795752	7.816227	0.82	1.142494	1.305293
0.33	2.707206	7.328963	0.83	1.132157	1.281780
0.34	2.624163	6.886231	0.84	1.122158	1.259238
0.35	2.546146	6.482862	0.85	1.112484	1.237621
0.36	2.472734	6.114415	0.86	1.103125	1.216884
0.37	2.403549	5.777050	0.87	1.094069	1.196988
0.38	2.338255	5.467437	0.88	1.085308	1.177894
0.39	2.276549	5.182675	0.89	1.076831	1.159565
0.40	2.218159	4.920229	0.90	1.068629	1.141968
0.41	2.162840	4.677877	0.91	1.060693	1.125070
0.42	2.110370	4.453663	0.92	1.053016	1.108842
0.43	2.060549	4.245861	0.93	1.045588	1.093255
0.44	2.013193	4.052947	0.94	1.038403	1.078281
0.45	1.968136	3.873560	0.95	1.031453	1.063896
0.46	1.925227	3.706497	0.96	1.024732	1.050075
0.47	1.884326	3.550683	0.97	1.018232	1.036796
0.48	1.845306	3.405155	0.98	1.011947	1.024037
0.49	1.808051	3.269050	0.99	1.005872	1.011778
0.50	1.772454	3.141593	1.00	1.000000	1.000000

```
'CM11GB3 Program in QuickBASIC, to tabulate the gamma
'function and gamma square in the range of 1<= P<=2,
'increment 0.01, using an approximating polynomial.
'The error is less than 10^-6. Refer to Table 1.2.
CLS
DIM I(100) AS INTEGER
DIM P(100), S(100), R(100), T(100), V(100)
DIM GAMMA(100), GAMMA2(100)
FOR I = 1 TO 100
      P(I) = I / 100
      S(I) = 1 + P(I)
      A0 = .035868343#
      A1 = -.193527818#
      A2 = .482199394#
      A3 = -.7567040779999999#
      A4 = .918206857#
      A5 = -.897056937#
      A6 = .988205891#
      A7 = -.577191652#
      R(I) = P(I) ^ 6 * (A0 * P(I) ^ 2 + A1 * P(I) + A2)
      T(I) = P(I) ^ 3 * (A3 * P(I) ^ 2 + A4 * P(I) + A5)
      V(I) = A6 * P(I) ^ 2 + A7 * P(I) + 1!
      GAMMA(I) = R(I) + T(I) + V(I)
      GAMMA2(I) = GAMMA(I) * GAMMA(I)
NEXT I
LPRINT CHR$(12)
LPRINT
LPRINT TAB(28); "TABLE 1.2.  GAMMA FUNCTION"
LPRINT TAB(28); "--------------------------"
LPRINT
LPRINT TAB(10); "P"; SPC(8); "GAMMA"; SPC(8); "GAMMA^2";
LPRINT SPC(9); "P"; SPC(6); "GAMMA"; SPC(6); "GAMMA^2";
LPRINT TAB(8); "----"; SPC(6); "--------"; SPC(6); "--------";
LPRINT SPC(7); "----"; SPC(4); "--------"; SPC(3); "--------";
LPRINT TAB(8); "1.00"; SPC(6); "1.000000";
FOR I = 1 TO 50
      LPRINT TAB(8);
      LPRINT USING "#.##"; S(I);
      LPRINT TAB(17);
      LPRINT USING "##.######"; GAMMA(I);
      LPRINT TAB(29);
      LPRINT USING "####.######"; GAMMA2(I);
      LPRINT TAB(47);
      LPRINT USING "#.##"; S(I + 50);
      LPRINT TAB(55);
      LPRINT USING "#.######"; GAMMA(I + 50);
      LPRINT TAB(66);
      LPRINT USING "#.######"; GAMMA2(I + 50);
NEXT I
END
```

Fig.1.4

TABLE 1.2. GAMMA FUNCTION

P	GAMMA	GAMMA^2	P	GAMMA	GAMMA^2
1.00	1.000000				
1.01	0.994326	0.988684	1.51	0.886592	0.786045
1.02	0.988844	0.977813	1.52	0.887039	0.786838
1.03	0.983550	0.967371	1.53	0.887568	0.787777
1.04	0.978438	0.957342	1.54	0.888178	0.788860
1.05	0.973504	0.947711	1.55	0.888869	0.790087
1.06	0.968744	0.938464	1.56	0.889639	0.791458
1.07	0.964152	0.929589	1.57	0.890490	0.792972
1.08	0.959725	0.921072	1.58	0.891420	0.794629
1.09	0.955459	0.912902	1.59	0.892428	0.796428
1.10	0.951351	0.905068	1.60	0.893516	0.798370
1.11	0.947395	0.897558	1.61	0.894681	0.800454
1.12	0.943590	0.890362	1.62	0.895924	0.802680
1.13	0.939931	0.883471	1.63	0.897244	0.805047
1.14	0.936416	0.876875	1.64	0.898642	0.807558
1.15	0.933041	0.870565	1.65	0.900117	0.810210
1.16	0.929803	0.864534	1.66	0.901668	0.813006
1.17	0.926700	0.858772	1.67	0.903296	0.815944
1.18	0.923728	0.853273	1.68	0.905001	0.819027
1.19	0.920885	0.848029	1.69	0.906782	0.822253
1.20	0.918169	0.843034	1.70	0.908639	0.825624
1.21	0.915577	0.838281	1.71	0.910572	0.829140
1.22	0.913106	0.833763	1.72	0.912580	0.832803
1.23	0.910755	0.829475	1.73	0.914665	0.836612
1.24	0.908521	0.825411	1.74	0.916826	0.840570
1.25	0.906403	0.821566	1.75	0.919062	0.844676
1.26	0.904397	0.817935	1.76	0.921375	0.848931
1.27	0.902503	0.814512	1.77	0.923763	0.853338
1.28	0.900719	0.811294	1.78	0.926227	0.857897
1.29	0.899042	0.808276	1.79	0.928767	0.862609
1.30	0.897471	0.805454	1.80	0.931384	0.867476
1.31	0.896004	0.802824	1.81	0.934076	0.872498
1.32	0.894640	0.800382	1.82	0.936845	0.877679
1.33	0.893378	0.798124	1.83	0.939690	0.883018
1.34	0.892215	0.796048	1.84	0.942612	0.888518
1.35	0.891151	0.794151	1.85	0.945611	0.894181
1.36	0.890184	0.792428	1.86	0.948687	0.900007
1.37	0.889313	0.790878	1.87	0.951840	0.906000
1.38	0.888537	0.789498	1.88	0.955071	0.912161
1.39	0.887854	0.788285	1.89	0.958380	0.918491
1.40	0.887264	0.787237	1.90	0.961766	0.924994
1.41	0.886764	0.786351	1.91	0.965231	0.931671
1.42	0.886356	0.785626	1.92	0.968774	0.938524
1.43	0.886036	0.785060	1.93	0.972397	0.945556
1.44	0.885805	0.784650	1.94	0.976099	0.952769
1.45	0.885661	0.784396	1.95	0.979881	0.960166
1.46	0.885604	0.784295	1.96	0.983742	0.967749
1.47	0.885633	0.784346	1.97	0.987685	0.975521
1.48	0.885747	0.784548	1.98	0.991708	0.983485
1.49	0.885945	0.784899	1.99	0.995813	0.991644
1.50	0.886227	0.785398	2.00	1.000000	1.000000

1.7. The Gamma Function Third Program

Using the algorithm given in Sec. 1.6, we will give a program to calculate the gamma function for $p \in (0, 1]$ using the PC Microsoft QuickBASIC programming language. The error is approximately 10^{-5}, and this is adequate for most engineering applications. With this program, the gamma function can be calculated with a high degree of accuracy even for values of p which are not given in Tables 1.1 and 1.2; e.g.,

$$\Gamma(0.3752) = 2.369131,$$
$$[\Gamma(0.3752)]^2 = 5.61278.$$

Using Eqs. (1.20) and (1.35), the gamma function can be evaluated for any real numbers, except zero and negative integer numbers. The program given in Fig. 1.5 can be easily modified for other PC computer programming languages.

```
'CMGB1, program in QuickBASIC, to calculate
'the gamma and gamma square functions
'for 0<p<=1 using an approximating
'polynomial. The error is less than 10^-5.

INPUT p
DO WHILE 0 < p <= 1
  LET A0 = .035868343#
      A1 = -.193527818#
      A2 = .482199394#
      A3 = -.7567040779999999#
      A4 = .918206857#
      A5 = -.897056937#
      A6 = .988205891#
      A7 = -.577191652#
  S1 = p ^ 5 * (A0 * p ^ 3 + A1 * p ^ 2)
  S2 = p ^ 5 * (A2 * p + A3)
  S3 = p * (A4 * p ^ 3 + A5 * p ^ 2)
  S4 = p * (A6 * p + A7) + 1!
    G1(p) = (S1 + S2 + S3 + S4) / p
    G2(p) = G1(p) ^ 2
  PRINT p, G1(p), G2(p)
  INPUT p
LOOP
END
```

```
C:\QB45>qb
? .33
 .33            2.707206        7.328964
? .7835
 .7835          1.18329         1.400175
? .2543
 .2543          3.56086         12.67972
?
```

Fig.1.5

1.8. Graphs of $\Gamma(p)$ and $\dfrac{1}{\Gamma(p)}$

With the recurrence formula of the gamma function

$$\Gamma(x+1) = x\Gamma(x), \tag{1.20}$$

the value of $\Gamma(x)$ can be calculated for any $x \in R$ except 0 and the negative integers, if the values for $x \in (0, 1]$ are known. As explained in Sec.1.5.2, the values of $\Gamma(x)$, for $x < 0$, can be calculated successively using Eq.(1.20) solved for $\Gamma(x)$,

$$\Gamma(x) = \frac{\Gamma(x+1)}{x}, \quad x > -1, \quad x \neq 0. \tag{1.48}$$

Weierstrass' definition of the gamma function, see Eq.(1.6), is

$$\Gamma(x) = e^{-\gamma x}\frac{1}{x}\prod_{n=1}^{\infty}\frac{e^{\frac{x}{n}}}{1+\frac{x}{n}}. \tag{1.49}$$

Since $\Gamma(x) > 0$ for $x > 0$, therefore $\ln\Gamma(x)$ is defined and, from Eq.(1.49),

$$\ln\Gamma(x) = -\gamma x - \ln x + \sum_{n=1}^{\infty}\left[\frac{x}{n} - \ln\left(1+\frac{x}{n}\right)\right]. \tag{1.50}$$

We can differentiate Eq.(1.50):

$$\frac{d}{dx}\ln\Gamma(x) = \frac{\Gamma'(x)}{\Gamma(x)} = -\gamma - \frac{1}{x} + \sum_{n=1}^{\infty}\left(\frac{1}{n} - \frac{1}{x+n}\right). \tag{1.51}$$

A new differentiation of Eq.(1.51) will give

$$\frac{d}{dx}\left[\frac{\Gamma'(x)}{\Gamma(x)}\right] = \sum_{n=0}^{\infty}\frac{1}{(x+n)^2}. \tag{1.52}$$

The right side of Eq.(1.52) is positive and we can write

$$\frac{\Gamma''(x)\cdot\Gamma(x) - [\Gamma'(x)]^2}{[\Gamma(x)]^2} > 0, \tag{1.53}$$

from where

$$\Gamma''(x)\cdot\Gamma(x) - [\Gamma'(x)]^2 > 0. \tag{1.54}$$

Therefore

$$\Gamma''(x) > 0, \quad \text{for} \quad x > 0,$$

and hence, for $x > 0$, $\Gamma(x)$ is convex upward and has a minimum in the interval $[1, 2]$. Eq.(1.54) shows that $\Gamma(x)$ and $\Gamma''(x)$ are both positive or both negative for each value of x. From Eq.(1.28), it follows that on the interval $(-n, -n+1)$, $\Gamma(x)$ has the sign of $(-1)^n$. The lines $x = 0$, $x = -1$, $x = -2$, ... are asymptotes to the curve. In addition, $\Gamma(x)$ has no zeros.

For the graph of the gamma function shown in Fig.1.6, we used Tables 1.1, 1.2 and the recursion relations (1.20) and (1.32). The graph of function $\frac{1}{\Gamma(x)}$ is also shown on Fig.1.6. These graphs can be used in engineering work, if a high accuracy is not required. For engineering applications, we use $\Gamma(p)$ for $p > 0$ (see the definition in Sec.1.1 and Fig.1.6).

For numerical evaluation of the gamma function, there are five possibilities:

- Using Tables 1.1, 1.2 and the recurrence formula (1.20)
- Using the computer program given in Fig.1.5
- Using the graph of $\Gamma(p)$ given in Fig.1.6
- Using the graph of $\frac{1}{\Gamma(p)}$ given in Fig.1.6
- Using a scientific calculator with the gamma function.

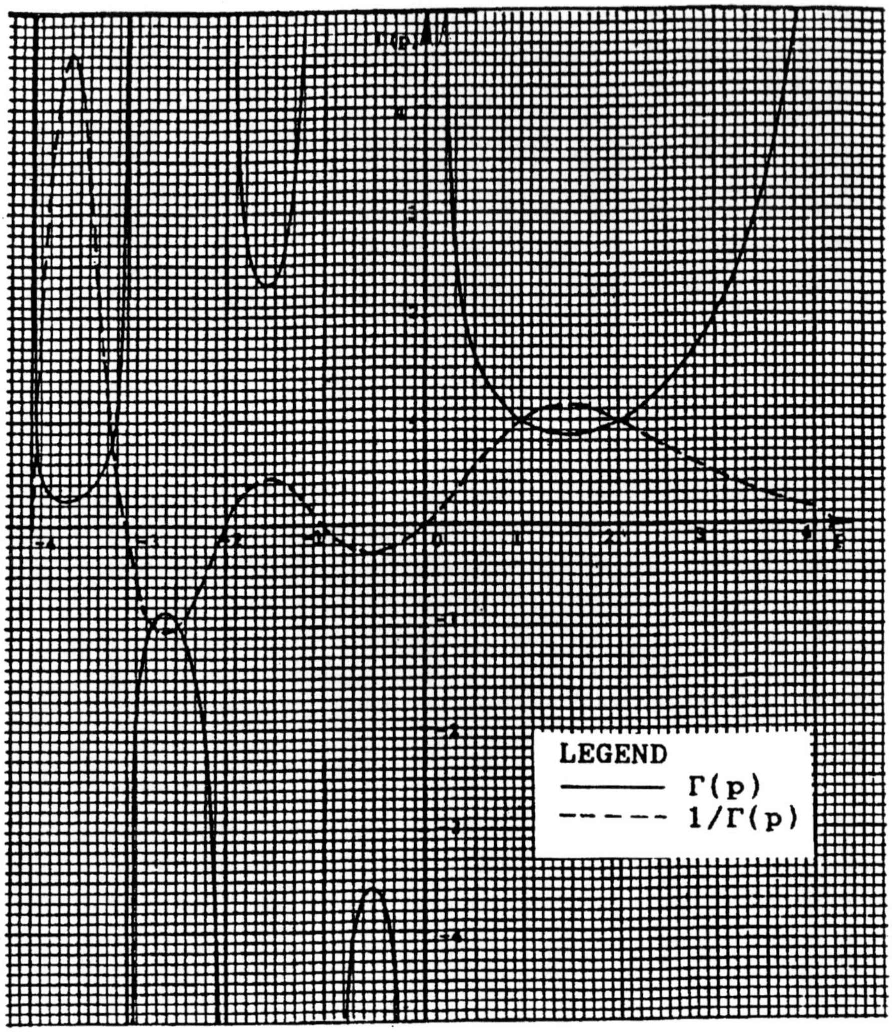

Fig.1.6 Graphs of $\Gamma(p)$ and $\dfrac{1}{\Gamma(p)}$.

1.9. Properties of Beta Function

1.9.1. Writing $1 - x$ for x in Eq.(1.9), we obtain the *symmetrical relation of beta function*

$$B(p, q) = B(q, p), \quad p, q \in R^+. \tag{1.55}$$

1.9.2. Integrating by parts in Eq.(1.9), we have

$$\begin{aligned}
B(p,q) &= \int_0^1 x^{p-1}(1-x)^{q-1}\,dx = \frac{1}{p}\int_0^1 (1-x)^{q-1}\,dx^p \\
&= \frac{1}{p}x^p(1-x)^{q-1}\Big|_0^1 + \frac{q-1}{p}\int_0^1 x^p(1-x)^{q-2}\,dx \\
&= \frac{q-1}{p}\int_0^1 [1-(1-x)]\,(1-x)^{q-2}x^{p-1}\,dx \\
&= \frac{q-1}{p}\left[\int_0^1 x^{p-1}(1-x)^{q-2}\,dx - \int_0^1 x^{p-1}(1-x)^{q-1}\,dx\right] \\
&= \frac{q-1}{p}\left[B(p,q-1) - B(p,q)\right],
\end{aligned}$$

and from this,

$$B(p,q) = \frac{q-1}{p+q-1}B(p,q-1), \quad p>0, \quad q>1; \tag{1.56}$$

and with the symmetrical property Eq.(1.55), we have

$$B(p,q) = \frac{p-1}{p+q-1}B(p-1,q), \quad p>1, \quad q>0. \tag{1.57}$$

Writing $q+1$ for q and $p+1$ for p in Eqs.(1.56) and (1.57) respectively, these formulas can be written in the forms

$$B(p,q+1) = \frac{q}{p+q}B(p,q), \quad p,q>0, \tag{1.58}$$

$$B(p+1,q) = \frac{p}{p+q}B(p,q), \quad p,q>0. \tag{1.59}$$

See also Sec. 1.11.3. Adding Eqs.(1.58) and (1.59), we have

$$B(p,q) = B(p+1,q) + B(p,q+1), \quad p,q>0. \tag{1.60}$$

1.9.3. Integrating by parts the integral of $B(p,q+1)$,

$$\begin{aligned}
B(p,q+1) &= \int_0^1 x^{p-1}(1-x)^q\,dx = \frac{1}{p}\int_0^1 (1-x)^q\,dx^p \\
&= \frac{1}{p}x^p(1-x)^q\Big|_0^1 + \frac{q}{p}\int_0^1 x^p(1-x)^{q-1}\,dx \\
&= \frac{q}{p}B(p+1,q), \quad p,q>0.
\end{aligned} \tag{1.61}$$

1.9.4. With the substitution $x = \dfrac{1-t}{t}$, $dx = -\dfrac{1}{t^2}\,dt$, in the next integral, we get a new integral representation of beta function

$$\int_0^\infty \frac{x^{q-1}\,dx}{(1+x)^{p+q}} = \int_0^1 t^{p-1}(1-t)^{q-1}\,dt = B(p,q), \quad p,q>0. \tag{1.62}$$

1.9.5. Also using the same substitution,

$$\int_0^\infty \frac{x^{p-1}\,dx}{(1+x)^{p+q}} = \int_0^1 t^{q-1}(1-t)^{p-1}\,dt = B(q,p), \quad p,q>0. \tag{1.63}$$

For application, see Sec.2.3.23.

1.9.6. Adding Eqs.(1.62) and (1.63), we have

$$\begin{aligned} 2B(p,q) &= \int_0^\infty \frac{x^{p-1}+x^{q-1}}{(1+x)^{p+q}}\,dx \\ &= \int_0^1 \frac{x^{p-1}+x^{q-1}}{(1+x)^{p+q}}\,dx + \int_1^\infty \frac{x^{p-1}+x^{q-1}}{(1+x)^{p+q}}\,dx. \end{aligned}$$

Replacing x by $\frac{1}{x}$ in the last integral, we have

$$2B(p,q) = \int_0^1 \frac{x^{p-1}+x^{q-1}}{(1+x)^{p+q}}\,dx + \int_0^1 \frac{x^{p-1}+x^{q-1}}{(1+x)^{p+q}}\,dx$$

and the beta function can be defined also by the integral

$$B(p,q) = \int_0^1 \frac{x^{p-1}+x^{q-1}}{(1+x)^{p+q}}\,dx, \quad p,q>0, \tag{1.64}$$

and for $q=p$,

$$B(p,p) = 2\int_0^1 \frac{x^{p-1}\,dx}{(1+x)^{2p}}, \quad p>0. \tag{1.64a}$$

For application, see Sec.2.2.4a.

1.9.7. Replacing x by x^2 in Eqs.(1.62) and (1.63), we have other definitions of the beta function,

$$B(p,q) = 2\int_0^{\infty} \frac{x^{2q-1}\,dx}{(1+x^2)^{p+q}}, \quad p,q>0, \tag{1.65}$$

$$B(q,p) = 2\int_0^{\infty} \frac{x^{2p-1}\,dx}{(1+x^2)^{p+q}}, \quad p,q>0. \tag{1.66}$$

1.9.8. Replacing x by x^2 in Eq.(1.9), we have another definition of the beta function,

$$B(p,q) = 2\int_0^1 x^{2p-1}(1-x^2)^{q-1}\,dx, \quad p,q>0. \tag{1.67}$$

1.9.9. With the substitution $x = at$, $a > 0$, we evaluate the integral,

$$\begin{aligned}\int_0^a x^{p-1}(a-x)^{q-1}\,dx &= a^{p+q-1}\int_0^1 t^{p-1}(1-t)^{q-1}\,dt \\ &= a^{p+q-1}B(p,q), \quad a>0, \quad p,q>0. \end{aligned} \tag{1.68}$$

1.10. Exponential Functions

1.10.1. Evaluate the integral

$$I = \int_0^{\infty} e^{-x^2}\,dx. \tag{1.69}$$

Let

$$I_a = \int_0^a e^{-x^2}\,dx = \int_0^a e^{-y^2}\,dy \tag{1.70}$$

and

$$\lim_{a\to\infty} I_a = I.$$

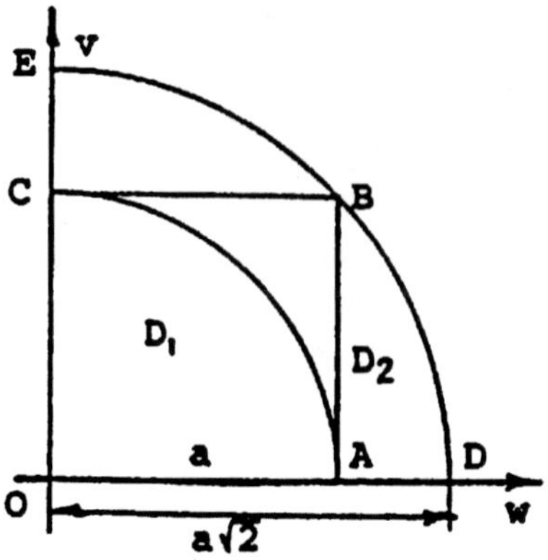

Fig.1.7

Then,

$$I_a^2 = \left(\int_0^a e^{-x^2}\,dx\right)\left(\int_0^a e^{-y^2}\,dy\right) = \int_0^a\int_0^a e^{-(x^2+y^2)}\,dx\,dy$$
$$= \iint_D e^{-(x^2+y^2)}\,dx\,dy, \tag{1.71}$$

where D is the $OABC$-square of side a, Fig.1.7. The integrand is positive and then we have

$$\iint_{D_1} e^{-(x^2+y^2)}\,dx\,dy \le I_a^2 \le \iint_{D_2} e^{-(x^2+y^2)}\,dx\,dy, \tag{1.72}$$

where D_1 and D_2 are the regions in the first quadrant bounded by the circles having radii a and $a\sqrt{2}$ respectively. Changing to polar coordinates,

$$x = r\cos\theta, \quad y = r\sin\theta, \quad J = r, \tag{1.73}$$

$$\int_0^{\frac{\pi}{2}} d\theta \int_0^a e^{-r^2} r\,dr \le I_a^2 \le \int_0^{\frac{\pi}{2}} d\theta \int_0^{a\sqrt{2}} e^{-r^2} r\,dr \tag{1.74}$$

and with $\int_0^{\frac{\pi}{2}} d\theta = \frac{\pi}{2}$, Eq.(1.74) becomes

$$\frac{\pi}{2}\int_0^a e^{-r^2} r\,dr \le I_a^2 \le \frac{\pi}{2}\int_0^{a\sqrt{2}} e^{-r^2} r\,dr \tag{1.75}$$

and with $r^2 = t$, $2r\,dr = dt$, $r\,dr = \frac{1}{2}\,dt$ for $r = a \Rightarrow t = a^2$ and for $r = 0 \Rightarrow t = 0$ we have

$$\begin{aligned}\int_0^a e^{-r^2} r\,dr &= \int_0^{a^2} e^{-t}\frac{dt}{2} = \frac{1}{2}\int_0^{a^2} e^{-t}\,dt \\ &= \frac{1}{2}\frac{e^{-t}}{-1}\Big|_0^{a^2} = -\frac{1}{2}e^{-t}\Big|_0^{a^2} \\ &= -\frac{1}{2}\left(e^{-a^2} - e^0\right) = \frac{1}{2}\left(1 - e^{-a^2}\right)\end{aligned}$$

and from Eq.(1.75)

$$\begin{aligned}\frac{\pi}{2}\cdot\frac{1}{2}\left(1 - e^{-a^2}\right) \le I_a^2 \le \frac{\pi}{2}\cdot\frac{1}{2}\left(1 - e^{-2a^2}\right) \\ \frac{\pi}{4}\left(1 - e^{-a^2}\right) \le I_a^2 \le \frac{\pi}{4}\left(1 - e^{-2a^2}\right).\end{aligned}$$

Taking the limit,

$$\lim_{a\to\infty} I_a^2 = I^2 = \frac{\pi}{4}$$

and

$$I = \int_0^\infty e^{-x^2}\,dx = \frac{\sqrt{\pi}}{2}. \tag{1.76}$$

For other solutions, see Sec.2.3.31 and Sec.2.7.1.

1.10.2. Other Formulas. a. For $p = 1/2$ in Eq.(1.1) and using Eq.(1.37) we have

$$\int_0^\infty x^{\frac{1}{2}-1}e^{-x}\,dx = \Gamma\left(\frac{1}{2}\right) = \sqrt{\pi}. \tag{1.77}$$

b. If in the integral (1.76) we make the change of variable $x = au$, $dx = a\,du$, then

$$\int_0^\infty e^{-x^2}\,dx = a\int_0^\infty e^{-a^2u^2}\,du = \frac{\sqrt{\pi}}{2} \tag{1.78}$$

or

$$\int_0^\infty e^{-a^2u^2}\,du = \frac{\sqrt{\pi}}{2a}, \quad a > 0. \tag{1.79}$$

c. Evaluate the integral

$$I = \int_0^\infty e^{-x^2 - \frac{a^2}{x^2}}\, dx, \quad a > 0. \tag{1.80}$$

Differentiating with respect to a, we have

$$\frac{dI}{da} = -2\int_0^\infty \frac{a}{x^2} e^{-x^2 - \frac{a^2}{x^2}}\, dx, \tag{1.81}$$

and changing the variable of integration,

$$x = \frac{a}{y}, \quad dx = -\frac{a}{y^2}\, dy, \quad x = 0, \quad y = \infty, \quad x = \infty, \quad y = 0, \tag{1.82}$$

Eq.(1.81) becomes

$$\frac{dI}{da} = 2\int_0^\infty \frac{ay^2}{a^2} e^{-\frac{a^2}{y^2} - \frac{y^2 a^2}{a^2}} \left(-\frac{a}{y^2}\, dy\right)$$

$$= -2\int_0^\infty e^{-\frac{a^2}{y^2} - y^2}\, dy = -2I \tag{1.83}$$

or

$$\frac{dI}{I} = -2\, da \quad \text{and} \quad \ln I = -2a + \ln C, \quad \ln\frac{I}{C} = -2a,$$

$$\frac{I}{C} = e^{-2a}, \quad I = Ce^{-2a}$$

or

$$\int_0^\infty e^{-x^2 - \frac{a^2}{x^2}}\, dx = Ce^{-2a}. \tag{1.84}$$

To determine the integration constant C, for $a = 0$ in Eq.(1.84) we have

$$\int_0^\infty e^{-x^2}\, dx = C,$$

but from Eq.(1.76), $\int_0^\infty e^{-x^2}\, dx = \frac{\sqrt{\pi}}{2}$ and $C = \frac{\sqrt{\pi}}{2}$ and from Eq.(1.84) we have

$$\int_0^\infty e^{-x^2 - \frac{a^2}{x^2}}\, dx = \frac{\sqrt{\pi}}{2} e^{-2a} = \frac{\sqrt{\pi}}{2e^{2a}}, \quad a > 0. \tag{1.85}$$

d. Letting $2p = m$ in Eq.(1.18), we have

$$\Gamma\left(\frac{m}{2}\right) = 2\int_0^\infty e^{-x^2} x^{m-1}\, dx, \quad m > 0, \tag{1.86}$$

or

$$\int_0^\infty e^{-x^2} x^{m-1}\, dx = \frac{1}{2}\Gamma\left(\frac{m}{2}\right), \quad m > 0. \tag{1.87}$$

e. Replacing x by ax^2 and p by n, Eq.(1.1) becomes

$$\begin{aligned}\Gamma(n) &= \int_0^\infty (ax^2)^{n-1} e^{-ax^2} \cdot 2ax\, dx \\ &= 2a^n \int_0^\infty x^{2n-1} e^{-ax^2}\, dx,\end{aligned} \tag{1.88}$$

or

$$\int_0^\infty x^{2n-1} e^{-ax^2}\, dx = \frac{\Gamma(n)}{2a^n}, \quad n \in N, \quad a > 0. \tag{1.89}$$

f. Replacing $(2n-1)$ by n, from Eq.(1.89) we have

$$I_n = \int_0^\infty x^n e^{-ax^2}\, dx = \frac{1}{2a^{\frac{n+1}{2}}}\Gamma\left(\frac{n+1}{2}\right), \quad n \in N \cup \{0\}, \quad a > 0. \tag{1.90}$$

Particular cases:

$$\text{For } n = 0, \quad I_0 = \int_0^\infty e^{-ax^2}\, dx = \frac{1}{2a^{\frac{1}{2}}}\Gamma\left(\frac{1}{2}\right) = \frac{1}{2}\left(\frac{\pi}{a}\right)^{\frac{1}{2}},$$

$$\text{For } n = 1, \quad I_1 = \int_0^\infty x e^{-ax^2}\, dx = \frac{1}{2a},$$

$$\text{For } n = 2, \quad I_2 = \int_0^\infty x^2 e^{-ax^2}\, dx = \frac{1}{2a^{\frac{3}{2}}}\Gamma\left(\frac{3}{2}\right) = \frac{\sqrt{\pi}}{4a^{\frac{3}{2}}}.$$

g. Replacing n by $(2n)$ in Eq.(1.90), the integrand is an even function and

$$\int_{-\infty}^{+\infty} x^{2n} e^{-ax^2}\, dx = 2\int_0^{+\infty} x^{2n} e^{-ax^2}\, dx$$
$$= \frac{1}{a^{n+\frac{1}{2}}}\Gamma\left(n + \frac{1}{2}\right), \quad n \in N \cup \{0\}. \tag{1.91}$$

h. The integrand in Eq.(1.89) is an odd function and then,

$$\int_{-\infty}^{+\infty} x^{2n-1} e^{-ax^2}\, dx = 0, \quad a > 0, \quad n \in N. \tag{1.92}$$

i. Replacing n by 0 in Eq.(1.91), we obtain

$$\int_{-\infty}^{+\infty} e^{-ax^2}\, dx = \sqrt{\frac{\pi}{a}}, \quad a > 0,$$

and replacing x by $(x - x_0)$ and a by $\frac{1}{2\sigma^2}$, we have

$$\int_{-\infty}^{+\infty} e^{-\frac{(x-x_0)^2}{2\sigma^2}}\, dx = \sigma\sqrt{2\pi},$$

or

$$\frac{1}{\sigma\sqrt{2\pi}}\int_{-\infty}^{+\infty} e^{-\frac{(x-x_0)^2}{2\sigma^2}}\, dx = 1, \tag{1.93}$$

which is used to define the Gaussian probability function.

j. Letting $x = at$, $a > 0$, in Eq.(1.1), we have

$$\int_0^\infty t^{p-1}e^{-at}\,dt = \frac{\Gamma(p)}{a^p}, \qquad a, p \in R^+. \tag{1.94}$$

1.11. Beta Function in Terms of the Gamma Function

1.11.1. Using the definition of the gamma function (1.1) for $p = p$ and $p = q$, and multiplying,

$$\Gamma(p)\cdot\Gamma(q) = \int_0^\infty e^{-x}x^{p-1}\,dx \int_0^\infty e^{-y}y^{q-1}\,dy, \quad p, q \in R^+, \tag{1.95}$$

then substituting x^2 for x and y^2 for y,

$$\begin{aligned}\Gamma(p)\Gamma(q) &= \int_0^\infty e^{-x^2}x^{2(p-1)}2x\,dx \int_0^\infty e^{-y^2}y^{2(q-1)}2y\,dy\\ &= 4\int_0^\infty\int_0^\infty e^{-(x^2+y^2)}x^{2(p-1)+1}y^{2(q-1)+1}\,dx\,dy\\ &= 4\int_0^\infty\int_0^\infty e^{-(x^2+y^2)}x^{2p-1}y^{2q-1}\,dx\,dy,\end{aligned}$$

and changing to polar coordinates,

$$x = r\cos\theta, \quad y = r\sin\theta, \quad dx\,dy = r\,dr\,d\theta, \tag{1.96}$$

we have

$$\Gamma(p)\cdot\Gamma(q) = 4\int_0^\infty e^{-r^2}r^{2(p+q)-1}\,dr \int_0^{\frac{\pi}{2}} \sin^{2p-1}\theta\cdot\cos^{2q-1}\theta\,d\theta. \tag{1.97}$$

From the definition of the gamma function, Eq.(1.18), the first integral is

$$\int_0^\infty e^{-r^2}r^{2(p+q)-1}\,dr = \frac{1}{2}\Gamma(p+q),$$

and using the second definition of the beta function, Eq.(1.12), from Eq.(1.97) we have

$$\Gamma(p)\Gamma(q) = \Gamma(p+q)\cdot B(p,q), \tag{1.98}$$

and thus, we have the formula for calculating the beta function in terms of the gamma functions,

$$B(p,q) = \frac{\Gamma(p)\cdot\Gamma(q)}{\Gamma(p+q)}, \quad p,q \in R^+. \tag{1.99}$$

In particular, for $p = q = 1/2$ and Eq.(1.37),

$$B\left(\frac{1}{2},\frac{1}{2}\right) = \left[\Gamma\left(\frac{1}{2}\right)\right]^2 = \pi. \tag{1.100}$$

Using the transformation from rectangular to polar coordinates, the integration on area A_2 has been neglected (see Fig.1.8). For other demonstrations of Eq.(1.99), see Secs.1.11.2 and 2.7.2.

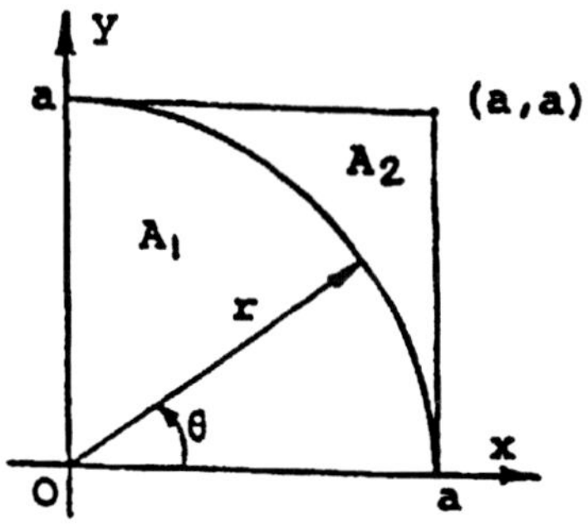

Fig.1.8

1.11.2. Another proof of Eq.(1.99). Let

$$I = \int_0^\infty \int_0^\infty e^{-x(y+1)} x^{p+q-1} y^{q-1}\, dx\, dy, \quad p,q \in R^+. \tag{1.101}$$

Integrating first with respect to x and then with respect to y, we have

$$I = \int_0^\infty y^{q-1}\, dy \int_0^\infty e^{-x(y+1)} x^{p+q-1}\, dx.$$

Letting $x(y+1) = z,\ dx = \dfrac{dz}{y+1}$,

$$I_1 = \int_0^\infty e^{-x(y+1)} x^{p+q-1}\, dx = \int_0^\infty e^{-z} \left(\frac{z}{y+1}\right)^{p+q-1} \cdot \frac{dz}{y+1}$$
$$= \frac{1}{(y+1)^{p+q}} \int_0^\infty e^{-z} z^{p+q-1}\, dz = \frac{1}{(y+1)^{p+q}} \Gamma(p+q).$$

Thus,

$$I = \Gamma(p+q) \int_0^\infty \frac{y^{q-1}}{(y+1)^{p+q}}\, dy,$$

and with Eq.(1.63), we have

$$I = \Gamma(p+q) \cdot B(p,q). \tag{1.102}$$

Secondly, evaluate the integral Eq.(1.101), first with respect to y and then with respect to x, and we have

$$I = \int_0^\infty x^{p+q-1}\, dx \int_0^\infty e^{-x(y+1)} y^{q-1}\, dy. \tag{1.103}$$

Letting $z = xy,\ dz = x\, dy$,

$$I_3 = \int_0^\infty e^{-x(y+1)} y^{q-1}\, dy$$
$$= \int_0^\infty e^{-xy-x} y^{q-1}\, dy = \int_0^\infty e^{-z} e^{-x} y^{q-1}\, dy$$
$$= e^{-x} \int_0^\infty e^{-z} y^{q-1}\, dy = e^{-x} \int_0^\infty e^{-z} \left(\frac{z}{x}\right)^{q-1} \frac{dz}{x}$$
$$= e^{-x} \int_0^\infty e^{-z} \frac{z^{q-1}}{x^q}\, dz = e^{-x} x^{-q} \int_0^\infty e^{-z} z^{q-1}\, dz$$
$$= e^{-x} x^{-q} \cdot \Gamma(q),$$

and Eq.(1.103) becomes

$$I = \Gamma(q) \cdot \int_0^\infty x^{p+q-1} e^{-x} x^{-q}\, dx = \Gamma(q) \int_0^\infty x^{p-1} e^{-x}\, dx$$
$$= \Gamma(q)\Gamma(p). \tag{1.104}$$

From Eqs.(1.102) and (1.104), we have

$$B(p,q) = \frac{\Gamma(p)\Gamma(q)}{\Gamma(p+q)}, \quad p, q \in R^+. \tag{1.98}$$

In particular, the value of $\Gamma\left(\frac{1}{2}\right)$ can be calculated using Eqs.(1.98) and (1.9) with $p = q = 1/2$:

$$\left[\Gamma\left(\tfrac{1}{2}\right)\right]^2 = \int_0^1 x^{-\frac{1}{2}}(1-x)^{-\frac{1}{2}}\,dx = \int_0^\infty x^{\left(1-\frac{1}{2}\right)-1}(1-x)^{\left(1-\frac{1}{2}\right)-1}\,dx$$
$$= \int_0^1 x^{\frac{1}{2}-1}(1-x)^{\frac{1}{2}-1}\,dx = B\left(\frac{1}{2},\frac{1}{2}\right) = \pi.$$

Therefore

$$\Gamma\left(\frac{1}{2}\right) = \sqrt{\pi}. \tag{1.105}$$

1.11.3. Other Formulas. Using Eqs.(1.98) and (1.20), we have

$$B(p,q+1) = \frac{\Gamma(p)\Gamma(q+1)}{\Gamma(p+q+1)} = \frac{\Gamma(p)\cdot q\Gamma(q)}{(p+q)\Gamma(p+q)}$$

or

$$B(p,q+1) = \frac{q}{p+q}B(p,q), \quad p,q \in R^+, \tag{1.106}$$

and similarly,

$$B(p+1,q) = \frac{p}{p+q}B(p,q), \quad p,q \in R^+. \tag{1.107}$$

For results found with another method, see Eqs.(1.58) and (1.59) respectively.

If $q = n+1$, $n \in N$, from Eq.(1.98)

$$B(p,n+1) = \frac{n!\,\Gamma(p)}{\Gamma(p+n+1)} = \frac{n!\,\Gamma(p)}{(n+p)\Gamma(p+n)},$$

and using Eq.(1.27) for $p \in (0,1)$, we have

$$B(p,n+1) = \frac{n!\,\Gamma(p)}{(n+p)p(p+1)\cdots(p+n-1)\Gamma(p)}$$
$$= \frac{n!}{(p+n)(p+n-1)\cdots(p+1)p}, \quad p \in (0,1), \quad n \in N. \tag{1.108}$$

Also, we have

$$B(p_1,p_2)\cdot B(p_1+p_2,p_3) = \frac{\Gamma(p_1)\Gamma(p_2)\Gamma(p_3)}{\Gamma(p_1+p_2+p_3)}. \tag{1.109}$$

For application of Eq.(1.109), see Sec.1.14b. Using the circular permutations yields

$$B(p_1, p_2) \cdot B(p_1 + p_2, p_3) = B(p_2, p_3)B(p_2 + p_3, p_1) = B(p_3, p_1)B(p_3 + p_1, p_2).$$

Now it is easy to prove the formula

$$B(p_1, p_2) \cdot B(p_1 + p_2, p_3) \cdot B(p_1 + p_2 + p_3, p_4) \cdots B(p_1 + p_2 + \cdots + p_{n-1}, p_n)$$
$$= \frac{\Gamma(p_1)\Gamma(p_2)\cdots\Gamma(p_n)}{\Gamma(p_1 + p_2 + \cdots + p_n)}. \tag{1.110}$$

1.11.4. If n is an odd integer, show that

$$\Gamma\left(\frac{n}{2}\right) = \frac{\sqrt{\pi}\,(n-1)!}{2^{n-1}\left(\dfrac{n-1}{2}\right)!}. \tag{1.111}$$

Reasoning by mathematical induction, the formula is valid for $n = 1$ and $n = 3$:

$$\text{For } n = 1, \quad \Gamma\left(\frac{1}{2}\right) = \sqrt{\pi}.$$

$$\text{For } n = 3, \quad \Gamma\left(\frac{3}{2}\right) = \frac{\sqrt{\pi}\,(3-1)!}{2^{3-1}\left(\dfrac{3-1}{2}\right)!} = \frac{\sqrt{\pi}}{2}.$$

Supposing that the formula is valid for n, we should prove it for $n + 2$:

$$\Gamma\left(\frac{n+2}{2}\right) = \Gamma\left(\frac{n}{2} + 1\right) = \frac{n}{2}\Gamma\left(\frac{n}{2}\right) = \frac{n}{2}\,\frac{\sqrt{\pi}\,(n-1)!}{2^{n-1}\left(\dfrac{n-1}{2}\right)!}$$
$$= \frac{\sqrt{\pi}\cdot n!}{2^n\left(\dfrac{n-1}{2}\right)!} = \frac{\sqrt{\pi}\,(n+1)!}{2^n\left(\dfrac{n-1}{2}\right)!\,(n+1)} = \frac{\sqrt{\pi}\,(n+1)!}{2^{n+1}\left(\dfrac{n+1}{2}\right)!}.$$

Thus, the given formula is valid for any odd positive integer.

1.12. Trigonometric Integral Evaluations

1.12.1. With the change of variable $x = \sin^2\theta$ in Eq.(1.9), we have the integral

$$\int_0^{\frac{\pi}{2}} \sin^{2p-1}\theta \cdot \cos^{2q-1}\theta \, d\theta = \frac{1}{2}B(p,q), \quad p,q \in R^+. \tag{1.112}$$

a. Let $2p = m$, $2q = n$, and from Eq.(1.112) we have

$$\int_0^{\frac{\pi}{2}} \sin^{m-1}\theta \cdot \cos^{n-1}\theta \, d\theta = \frac{1}{2}B\left(\frac{m}{2}, \frac{n}{2}\right), \quad m,n \in R^+. \tag{1.113}$$

In particular, for $m = n = 1$,

$$B\left(\frac{1}{2}, \frac{1}{2}\right) = 2\int_0^{\frac{\pi}{2}} d\theta = \pi, \tag{1.114}$$

and with Eq.(1.98),

$$\Gamma\left(\frac{1}{2}\right) = \sqrt{\pi}. \tag{1.115}$$

b. Let $m - 1 = p$, $n - 1 = q$, and Eq.(1.113) becomes

$$\int_0^{\frac{\pi}{2}} \sin^p\theta \cdot \cos^q\theta \, d\theta = \frac{1}{2}B\left(\frac{p+1}{2}, \frac{q+1}{2}\right), \quad p,q > -1, \tag{1.116}$$

$$= \frac{\Gamma\left(\frac{p+1}{2}\right) \cdot \Gamma\left(\frac{q+1}{2}\right)}{(p+q)\Gamma\left(\frac{p+q}{2}\right)}, \quad p,q > -1, \quad p+q > 0. \tag{1.117}$$

1.12.2. From Eq.(1.112) with $q = 1/2$, then $p = 1/2$ and $q = p$,

$$\int_0^{\frac{\pi}{2}} \sin^{2p-1}\theta\, d\theta = \frac{1}{2}B\left(p, \frac{1}{2}\right), \quad \int_0^{\frac{\pi}{2}} \cos^{2p-1}\theta\, d\theta = \frac{1}{2}B\left(\frac{1}{2}, p\right), \quad p > 0,$$

and using Eq.(1.55),

$$\int_0^{\frac{\pi}{2}} \sin^{2p-1}\theta\, d\theta = \int_0^{\frac{\pi}{2}} \cos^{2p-1}\theta\, d\theta = \frac{1}{2}B\left(p, \frac{1}{2}\right) = \frac{1}{2}B\left(\frac{1}{2}, p\right). \qquad (1.118)$$

Also, with $2p = m$ in Eq.(1.118),

$$\int_0^{\frac{\pi}{2}} \sin^{m-1}\theta\, d\theta = \int_0^{\frac{\pi}{2}} \cos^{m-1}\theta\, d\theta = \frac{1}{2}B\left(\frac{m}{2}, \frac{1}{2}\right), \quad m > 0. \qquad (1.119)$$

If we substitute $m = p + 1$, from Eq.(1.119) we have

$$\int_0^{\frac{\pi}{2}} \sin^{p}\theta\, d\theta = \int_0^{\frac{\pi}{2}} \cos^{p}\theta\, d\theta = \frac{1}{2}B\left(\frac{p+1}{2}, \frac{1}{2}\right), \quad p > -1. \qquad (1.120)$$

1.12.3. With $q = p$, from Eq.(1.112), we have the integral

$$I = \int_0^{\frac{\pi}{2}} (\sin\theta\cos\theta)^{2p-1}\, d\theta = \frac{1}{2}B(p, p), \qquad (1.121)$$

but $\sin\theta\cos\theta = \dfrac{1}{2}\sin 2\theta$ and

$$I = \frac{1}{2^{2p-1}} \int_0^{\frac{\pi}{2}} \sin^{2p-1}(2\theta)\, d\theta,$$

and with $2\theta = t$,

$$I = \frac{1}{2^{2p}} \int_0^{\pi} \sin^{2p-1} t\, dt, \qquad (1.121a)$$

but

$$\int_0^{\pi} \sin^{2p-1} t\, dt = \int_0^{\frac{\pi}{2}} \sin^{2p-1} t\, dt + \int_{\frac{\pi}{2}}^{\pi} \sin^{2p-1} t\, dt,$$

and with $t = \dfrac{\pi}{2} + \varphi$,

$$\int_{\frac{\pi}{2}}^{\pi} \sin^{2p-1} t\, dt = \int_0^{\frac{\pi}{2}} \sin^{2p-1}\left(\frac{\pi}{2} + \varphi\right) d\varphi = \int_0^{\frac{\pi}{2}} \cos^{2p-1}\varphi\, d\varphi,$$

and with Eq.(1.118),

$$\int_0^{\pi} \sin^{2p-1} t \, dt = 2 \int_0^{\frac{\pi}{2}} \sin^{2p-1} t \, dt,$$

and from Eq.(1.121a),

$$I = \frac{1}{2^{2p-1}} \int_0^{\frac{\pi}{2}} \sin^{2p-1} t \, dt = \frac{1}{2} B(p, p), \quad p > 0,$$

and

$$\int_0^{\frac{\pi}{2}} \sin^{2p-1} t \, dt = \int_0^{\frac{\pi}{2}} \cos^{2p-1} t \, dt = 2^{2p-2} B(p, p), \quad p > 0. \tag{1.122}$$

For $2p = m$, we have the integrals

$$\int_0^{\frac{\pi}{2}} \sin^{m-1} t \, dt = \int_0^{\frac{\pi}{2}} \cos^{m-1} t \, dt = 2^{m-2} B\left(\frac{m}{2}, \frac{m}{2}\right), \quad m > 0. \tag{1.123}$$

1.12.4. With the notation $2p - 1 = n$ and $2q - 1 = -n$, Eq.(1.112) becomes

$$\int_0^{\frac{\pi}{2}} \tan^n \theta \, d\theta = \frac{1}{2} B\left(\frac{1+n}{2}, \frac{1-n}{2}\right), \quad n \in (-1, 1), \tag{1.124}$$

or using Eq.(1.40),

$$\int_0^{\frac{\pi}{2}} \tan^n \theta \, d\theta = \frac{\pi}{2 \cos \frac{n\pi}{2}} = \frac{\pi}{2} \sec \frac{n\pi}{2}, \quad n \in (-1, 1). \tag{1.125}$$

1.12.5. Evaluate the trigonometric integrals

$$I_1 = \int_0^{\frac{\pi}{2}} \sin^{2n} \theta \, d\theta, \quad I_2 = \int_0^{\frac{\pi}{2}} \cos^{2n} \theta \, d\theta, \quad n \in N. \tag{1.126}$$

Solution. With the change of variable $\sin^2\theta = t$, $\sin\theta = t^{\frac{1}{2}}$, $\cos\theta = (1-t)^{\frac{1}{2}}$, $d\theta = \dfrac{dt}{2\sin\theta\cos\theta} = \dfrac{1}{2}t^{-\frac{1}{2}}(1-t)^{-\frac{1}{2}}\,dt$ and

$$\begin{aligned} I_1 &= \int_0^1 t^n \cdot \frac{1}{2}t^{-\frac{1}{2}}(1-t)^{-\frac{1}{2}}\,dt = \frac{1}{2}\int_0^1 t^{n-\frac{1}{2}}(1-t)^{-\frac{1}{2}}\,dt \\ &= \frac{1}{2}\int_0^1 t^{\left(1+n-\frac{1}{2}\right)-1}(1-t)^{\left(1-\frac{1}{2}\right)-1}\,dt \\ &= \frac{1}{2}\int_0^1 t^{\left(\frac{1}{2}+n\right)-1}(1-t)^{\frac{1}{2}-1}\,dt = \frac{1}{2}B\left(n+\frac{1}{2},\frac{1}{2}\right). \end{aligned} \tag{1.127}$$

The second integral,

$$\begin{aligned} I_2 &= \frac{1}{2}\int_0^1 t^{-\frac{1}{2}}(1-t)^{n-\frac{1}{2}}\,dt = \frac{1}{2}\int_0^1 t^{\left(1-\frac{1}{2}\right)-1}(1-t)^{\left(1+n-\frac{1}{2}\right)-1}\,dt \\ &= \frac{1}{2}\int_0^1 t^{\frac{1}{2}-1}(1-t)^{\left(\frac{1}{2}+n\right)-1}\,dt = \frac{1}{2}B\left(\frac{1}{2},\frac{1}{2}+n\right). \end{aligned} \tag{1.128}$$

With Eq.(1.43),

$$B\left(n+\frac{1}{2},\frac{1}{2}\right) = \frac{\sqrt{\pi}}{n!}\frac{\sqrt{\pi}}{2^n}(2n-1)!! = \frac{(2n-1)!!}{(2n)!!}\pi$$

and

$$\int_0^{\frac{\pi}{2}} \sin^{2n}\theta\,d\theta = \int_0^{\frac{\pi}{2}} \cos^{2n}\theta\,d\theta = \frac{(2n-1)!!}{(2n)!!}\cdot\frac{\pi}{2}, \qquad n\in N. \tag{1.129}$$

Particular case. Let $n=3$:

$$\int_0^{\frac{\pi}{2}} \sin^6\theta\,d\theta = \int_0^{\frac{\pi}{2}} \cos^6\theta\,d\theta = \frac{5!!}{6!!}\frac{\pi}{2} = \frac{5\pi}{32}.$$

1.12.6. Evaluate the trigonometric integrals,

$$I_1 = \int_0^{\frac{\pi}{2}} \sin^{2n+1}\theta\,d\theta, \quad I_2 = \int_0^{\frac{\pi}{2}} \cos^{2n+1}\theta\,d\theta, \quad n\in N. \tag{1.130}$$

Solution. Using Eq.(1.120), for $p = 2n + 1$, we have

$$
\begin{aligned}
I_1 &= \int_0^{\frac{\pi}{2}} \sin^{2n+1}\theta \, d\theta \\
&= \frac{1}{2} B\left(n+1, \frac{1}{2}\right) = \frac{1}{2} \frac{n}{n+\frac{1}{2}} B\left(n, \frac{1}{2}\right) \\
&= \frac{n}{2n+1} B\left(n, \frac{1}{2}\right) = \frac{n}{2n+1} \frac{\Gamma(n)\sqrt{\pi}}{\Gamma\left(n+\dfrac{1}{2}\right)},
\end{aligned}
$$

and with Eq.(1.43),

$$
I_1 = \frac{n!\, 2^n}{(2n+1)!!} = \frac{(2n)!!}{(2n+1)!!}. \tag{1.131}
$$

$$
I_2 = \int_0^{\frac{\pi}{2}} \cos^{2n+1}\theta \, d\theta = \frac{1}{2} B\left(\frac{1}{2}, n+1\right),
$$

and

$$
\int_0^{\frac{\pi}{2}} \sin^{2n+1}\theta \, d\theta = \int_0^{\frac{\pi}{2}} \cos^{2n+1}\theta \, d\theta = \frac{(2n)!!}{(2n+1)!!}, \quad n \in N. \tag{1.132}
$$

Particular case. Let $n = 2$:

$$
\int_0^{\frac{\pi}{2}} \sin^5\theta \, d\theta = \int_0^{\frac{\pi}{2}} \cos^5\theta \, d\theta = \frac{4!!}{5!!} = \frac{8}{15}. \tag{1.133}
$$

Exercise. Evaluate this integral using a traditional method.

1.13. Duplication Formula for the Gamma Function

From Eqs.(1.118) and (1.122), we have

$$
B\left(p, \frac{1}{2}\right) = 2^{2p-1} B(p, p), \tag{1.134}
$$

and using Eq.(1.98) we get

$$\frac{\Gamma(p)\sqrt{\pi}}{\Gamma\left(p+\frac{1}{2}\right)} = 2^{2p-1}\frac{[\Gamma(p)]^2}{\Gamma(2p)},$$

from where the *Legendre duplication formula* for the gamma function is obtained:

$$\Gamma(2p) = \frac{2^{2p-1}}{\sqrt{\pi}}\Gamma(p)\Gamma\left(p+\frac{1}{2}\right), \quad p > 0. \tag{1.135}$$

- For $p = x/2$,

$$\Gamma(x) = \frac{2^{x-1}}{\sqrt{\pi}}\Gamma\left(\frac{x}{2}\right)\Gamma\left(\frac{x+1}{2}\right), \quad x > 0.$$

- For $p = n \in N$, from Eq.(1.135), we have

$$\Gamma\left(n+\frac{1}{2}\right) = \frac{\sqrt{\pi}}{2^{2n-1}}\frac{\Gamma(2n)}{\Gamma(n)} = \frac{\sqrt{\pi}}{2^{2n}}\frac{(2n)!}{n!}, \quad n \in N \cup 0. \tag{1.136}$$

- Knowing that $\Gamma(2n) = (2n-1)!$, from Eq.(1.136) we have

$$\Gamma\left(n+\frac{1}{2}\right) = \frac{\sqrt{\pi}}{2^{2n-1}}\frac{(2n-1)!}{(n-1)!} = \frac{\sqrt{\pi}}{2^n \cdot 2^{n-1}} \cdot \frac{(2n-1)!}{(n-1)!},$$

and using the relation $2^{n-1} \cdot (n-1)! = (2n-2)!!$, we have

$$\Gamma\left(n+\frac{1}{2}\right) = \frac{\sqrt{\pi}}{2^n}\frac{(2n-1)!}{(2n-2)!!}, \quad n \in N. \tag{1.137}$$

See also Sec.1.5.2 and 1.5.6.

Particular cases:

$$\text{For } n=1, \quad \Gamma\left(\frac{3}{2}\right)=\frac{\sqrt{\pi}}{2}.$$

$$\text{For } n=2, \quad \Gamma\left(\frac{5}{2}\right)=\frac{\sqrt{\pi}}{4}3!!=\frac{3}{4}\sqrt{\pi}.$$

$$\text{For } n=3, \quad \Gamma\left(\frac{7}{2}\right)=\frac{\sqrt{\pi}}{8}5!!=\frac{15}{8}\sqrt{\pi}.$$

1.14. Generalization of Beta Function in R^n-Space

Writing the representation of the gamma function given in Eq.(1.18) for $p=p_1,p_2,\ldots,p_n>0$, $x=x_1,x_2,\ldots,x_n$, and making the product of these relations, we have

$$\Gamma(p_1)\Gamma(p_2)\cdots\Gamma(p_n)=2^n\int\cdots\int_D e^{-(x_1^2+x_2^2+\cdots+x_n^2)}x_1^{2p_1-1}\cdots x_n^{2p_n-1}\,dx_1\cdots dx_n, \tag{1.138}$$

where

$$D\{x_1,x_2,\ldots,x_n\geq 0\}.$$

Let a hypersphere with the centre at the origin, radius r, be

$$x_1^2+x_2^2+\cdots+x_n^2=r^2,$$

and with the change of variable,

$$x_i=rt_i, \quad i=1,2,\ldots,n, \quad r\geq 0,$$

with $t_1^2+t_2^2+\cdots+t_n^2=1$ and the Jacobian of transformation, see also Appendix A5,

$$J_{n-1}=\begin{vmatrix} r & 0 & \ldots & 0 \\ 0 & r & \ldots & 0 \\ \ldots & \ldots & \ldots & \ldots \\ 0 & \ldots & \ldots & r \end{vmatrix}=r^{n-1},$$

Eq.(1.138) will be

$$\begin{aligned}\Gamma(p_1)\cdots\Gamma(p_n) &= 2^n \int_0^\infty \int\cdots\int_S e^{-r^2(t_1^2+\cdots+t_n^2)} \\ &\quad \times r^{2p_1-1}t_1^{2p_1-1}\cdots r^{2p_n-1}t_n^{2p_n-1}\cdot r^{n-1}\,dr\,dS \\ &= 2^n \int_0^\infty e^{-r^2} r^{2(p_1+p_2+\cdots+p_n)^{-n}} r^{n-1}\,dr \\ &\quad \times \int\cdots\int_S t_1^{2p_1-1}t_2^{2p_2-1}\cdots t_n^{2p_n-1}\,dS \\ &= 2^n \int_0^\infty e^{-r^2} r^{2(p_1+p_2+\cdots+p_n)^{-1}}\,dr \\ &\quad \times \int\cdots\int_S t_1^{2p_1-1}t_2^{2p_2-1}\cdots t_n^{2p_n-1}\,dS. \qquad (1.139)\end{aligned}$$

Using Eq.(1.18) with $n \in N$, we have

$$\begin{aligned}\Gamma(p_1)\cdots\Gamma(p_n) &= 2^{n-1}\Gamma(p_1+p_2+\cdots+p_n) \\ &\quad \times \int\cdots\int_S t_1^{2p_1-1}t_2^{2p_2-1}\cdots t_n^{2p_n-1}\,dS \qquad (1.140)\end{aligned}$$

or

$$\begin{aligned}\frac{\Gamma(p_1)\Gamma(p_2)\cdots\Gamma(p_n)}{\Gamma(p_1+p_2+\cdots+p_n)} &= B(p_1,p_2,\ldots,p_n) \qquad (1.141) \\ &= 2^{n-1}\int\cdots\int_S t_1^{2p_1-1}t_2^{2p_2-1}\cdots t_n^{2p_n-1}\,dS, \qquad (1.142)\end{aligned}$$

where

$$S\left\{t_1,t_2,\ldots,t_n \geq 0 \mid t_1^2+t_2^2+\cdots+t_n^2=1\right\}, \quad p_1,p_2,\ldots,p_n>0.$$

Particular cases:

a. Prove that the integral

$$I_2 = 2\iint_S t_1^{2p_1-1} t_2^{2p_2-1}\, dS, \quad p_1, p_2 \geq 0,$$
$$S\left\{t_1, t_2 \geq 0 \mid t_1^2 + t_2^2 = 1\right\},$$

is another definition of beta function $B(p_1, p_2)$. See Eq.(1.12). Changing to polar coordinates in plane

$$t_1 = \cos\theta_1, \quad t_2 = \sin\theta_1,$$

we have

$$dS = \frac{dt_1}{t_2} = -\frac{\sin\theta_1\, d\theta_1}{\sin\theta_1} = -\,d\theta_1$$

and with Eq.(1.112),

$$I_2 = 2\int_0^{\frac{\pi}{2}} (\cos\theta_1)^{2p_1-1} (\sin\theta_1)^{2p_2-1}\, d\theta_1 = B(p_1, p_2).$$

Therefore

$$2\iint_S t_1^{2p_1-1} t_2^{2p_2-1}\, dS = B(p_1, p_2). \tag{1.143}$$

As a particular case, let $p_1 = p_2 = 1/2$, and from Eq.(1.143), we get the circle length (L_c), with radius $r = 1$,

$$2 \cdot \frac{L_c}{4} = B\left(\frac{1}{2}, \frac{1}{2}\right) = \pi,$$

from where $L_c = 2\pi$.

Without changing to polar coordinates, we have

$$t_2 = \left(1 - t_1^2\right)^{\frac{1}{2}}, \quad dS = \frac{dt_1}{t_2} = \left(1 - t_1^2\right)^{-\frac{1}{2}} dt_1,$$

and the integral I_2 becomes

$$I_2 = 2\iint_S t_1^{2p_1-1} t_2^{2p_2-1}\, dS = 2\int_0^1 t_1^{2p_1-1} \left[\left(1 - t_1^2\right)^{\frac{1}{2}}\right]^{2p_2-1} \left(1 - t_1^2\right)^{-\frac{1}{2}} dt_1$$
$$= 2\int_0^1 t_1^{2p_1-1} \left(1 - t_1^2\right)^{\frac{2p_2-1}{2} - \frac{1}{2}} dt_1 = 2\int_0^1 t_1^{2p_1-1} \left(1 - t_1^2\right)^{p_2-1} dt_1$$

and using Eq.(1.67) we have

$$I_2 = B(p_1, p_2).$$

b. Let $n = 3$, and from Eq.(1.141),

$$I_3 = 2^{3-1} \iint_{\mathcal{S}} t_1^{2p_1-1} t_2^{2p_2-1} t_3^{2p_3-1}\, dS = 4 \iint_{\mathcal{S}} t_1^{2p_1-1} t_2^{2p_2-1} t_3^{2p_3-1} \frac{dt_1\, dt_2}{t_3}, \qquad (1.144)$$

$$S\left\{t_1, t_2, t_3 \geq 0, t_1^2 + t_2^2 + t_3^2 = 1\right\}.$$

Using the transformation given in Appendix A3.4, for a sphere with $r = 1$,

$$t_1 = \cos\theta_1, \quad t_2 = \sin\theta_1 \cos\theta_2, \quad t_3 = \sin\theta_1 \sin\theta_2, \qquad (1.145)$$

with the Jacobian

$$J_2 = \sin^2\theta_1 \sin\theta_2 \qquad (1.146)$$

and

$$dt_1\, dt_2 = |J_2|\, d\theta_1\, d\theta_2 = \sin^2\theta_1 \sin\theta_2 \cdot d\theta_1\, d\theta_2, \qquad (1.147)$$

substituting in Eq.(1.144), knowing that $dS = \dfrac{dt_1\, dt_2}{t_3}$, yields

$$\begin{aligned} I_3 &= 4 \iint_{\mathcal{S}'} (\cos\theta_1)^{2p_1-1} (\sin\theta_1 \cos\theta_2)^{2p_2-1} (\sin\theta_1 \sin\theta_2)^{2p_3-1} \frac{\sin^2\theta_1 \sin\theta_2}{\sin\theta_1 \sin\theta_2}\, d\theta_1\, d\theta_2 \\ &= 4 \int_0^{\frac{\pi}{2}} (\sin\theta_1)^{2p_2+2p_3-1} (\cos\theta_1)^{2p_1-1}\, d\theta_1 \int_0^{\frac{\pi}{2}} (\sin\theta_2)^{2p_3-1} (\cos\theta_2)^{2p_2-1}\, d\theta_2, \end{aligned}$$

and with Eq.(1.112) and Eq.(1.109) or Eq.(1.110),

$$I_3 = 4 \cdot \frac{1}{2} B\left(p_2 + p_3, p_1\right) \cdot \frac{1}{2} B(p_2, p_3) = \frac{\Gamma(p_1)\Gamma(p_2)\Gamma(p_3)}{\Gamma(p_1 + p_2 + p_3)}$$

or

$$4 \iint_{\mathcal{S}} t_1^{2p_1-1} t_2^{2p_2-1} t_3^{2p_3-1}\, dS = B(p_1, p_2, p_3) = \frac{\Gamma(p_1)\Gamma(p_2)\Gamma(p_3)}{\Gamma(p_1 + p_2 + p_3)}. \qquad (1.148)$$

Particular case. Let $p_1 = p_2 = p_3 = 1/2$, and from Eqs.(1.144) and (1.148), we get the area of a sphere with radius $r = 1$,

$$4\frac{A_s}{8} = B\left(\frac{1}{2}, \frac{1}{2}, \frac{1}{2}\right) = \frac{\left[\Gamma\left(\frac{1}{2}\right)\right]^3}{\Gamma\left(\frac{3}{2}\right)} = 2\pi$$

or

$$A_s = 4\pi.$$

c. Let $p_1 = p_2 = p_3 = \cdots = p_n = 1/2$, and the integral in Eq.(1.141) is the area of the portion of the hypersphere with $r = 1,\ t_1, t_2, \ldots, t_n \geq 0$,

$$2^{n-1}\frac{A_n}{2^n} = B\left(\frac{1}{2}, \frac{1}{2}, \ldots, \frac{1}{2}\right) = \frac{\left[\Gamma\left(\frac{1}{2}\right)\right]^n}{\Gamma\left(\frac{n}{2}\right)}$$

or for all sphere,

$$A_n = 2\frac{\left[\Gamma\left(\frac{1}{2}\right)\right]^n}{\Gamma\left(\frac{n}{2}\right)} = \frac{2\pi^{\frac{n}{2}}}{\Gamma\left(\frac{n}{2}\right)}, \quad n \in N, \tag{1.149}$$

a result also found in Sec.2.5.3 and 2.6.6.

1.15. Laplace Transform and the Gamma Function

1.15.1. Given a function $F(t)$ defined for $t \geq 0$, if the integral

$$\int_0^\infty e^{-st} F(t)\, dt$$

exists, the integral is called the Laplace transform of the *original function* $F(t)$. The Laplace transform is denoted by $\mathcal{L}\{F(t)\}$ and determines a function $f(s)$ at those values of s for which the integral exists:

$$\mathcal{L}\{F(t)\} = \int_0^\infty e^{-st} F(t)\, dt = f(s). \tag{1.150}$$

1.15.2. The Laplace transform of $F(t) = t^x$. From Eq.(1.150), letting $st = u$, $s > 0$, we have

$$\begin{aligned}\mathcal{L}\{t^x\} &= \int_0^\infty e^{-st} t^{tsx}\, dt = \int_0^\infty e^{-u} \left(\frac{u}{s}\right)^x \frac{1}{s}\, du = \frac{1}{s^{x+1}} \int_0^\infty e^{-u} u^x\, du \\ &= \frac{1}{s^{x+1}} \int_0^\infty e^{-u} u^{(1+x)-1}\, du = \frac{1}{s^{x+1}} \Gamma(1+x), \quad x > -1, \quad s > 0. \end{aligned} \tag{1.151}$$

Also,

$$\mathcal{L}\left\{\frac{t^x}{\Gamma(x+1)}\right\} = \frac{1}{\Gamma(x+1)} \int_0^\infty e^{-st} t^x\, dt,$$

and letting $st = u$, $s\, dt = du$,

$$\begin{aligned}\mathcal{L}\left\{\frac{t^x}{\Gamma(x+1)}\right\} &= \frac{1}{\Gamma(x+1)} \int_0^\infty e^{-u} \left(\frac{u}{s}\right)^x \frac{du}{s} = \frac{1}{\Gamma(x+1)} \frac{1}{s^{x+1}} \int_0^\infty e^{-u} u^x\, du \\ &= \frac{1}{s^{x+1}\Gamma(x+1)} \int_0^\infty e^{-u} u^{(1+x)-1}\, du = \frac{\Gamma(1+x)}{s^{x+1}\Gamma(x+1)} \\ &= \frac{1}{s^{x+1}}. \end{aligned} \tag{1.151a}$$

Particular cases.

a. Let $x = n$, $n \in N$:

$$\mathcal{L}\{t^n\} = \frac{\Gamma(n+1)}{s^{n+1}} = \frac{n!}{s^{n+1}}, \quad s > 0. \tag{1.152}$$

b. Let $x = n + \frac{1}{2}$, $n \in N$, and using Eqs.(1.151), (1.20) and (1.137),

$$\mathcal{L}\left\{t^{n+\frac{1}{2}}\right\} = \frac{\Gamma\left[\left(n+\frac{1}{2}\right)+1\right]}{s^{n+\frac{3}{2}}} = \frac{1}{s^{n+\frac{3}{2}}}\frac{2n+1}{2}\Gamma\left(n+\frac{1}{2}\right)$$

$$= \frac{1}{s^{n+\frac{3}{2}}}\frac{2n+1}{2}\frac{\sqrt{\pi}}{2^n}(2n-1)!! = \frac{\sqrt{\pi}}{s^{n+\frac{3}{2}}}\frac{(2n+1)!!}{2^{n+1}}. \qquad (1.153)$$

1.15.3. Find $\mathcal{L}\{\sin at\}$ and $\mathcal{L}\{\cos at\}$, where $a \in R^+$. According to Eq.(1.150),

$$\mathcal{L}\{\sin at\} = \int_0^\infty e^{-st} \cdot \sin at\, dt = -\frac{1}{s}\int_0^\infty \sin at\, d(e^{-st})$$

$$= -\frac{1}{s}e^{-st}\sin at\Big|_0^\infty + \frac{a}{s}\int_0^\infty e^{-st}\cos at\, dt$$

$$= \frac{a}{s}\int_0^\infty e^{-st}\cos at\, dt = \frac{a}{s}\mathcal{L}\{\cos at\},$$

$$\mathcal{L}\{\cos at\} = \int_0^\infty e^{-st} \cdot \cos at\, dt = -\frac{1}{s}\int_0^\infty \cos at\, d(e^{-st})$$

$$= -\frac{1}{s}\cos at\, e^{-st}\Big|_0^\infty - \frac{a}{s}\int_0^\infty e^{-st}\sin at\, dt$$

$$= \frac{1}{s} - \frac{a}{s}\mathcal{L}\{\sin at\}.$$

We have the following system of equations:

$$\begin{cases} \mathcal{L}\{\sin at\} = \frac{a}{s}\mathcal{L}\{\cos at\} \\ \mathcal{L}\{\cos at\} = \frac{1}{s} - \frac{a}{s}\mathcal{L}\{\sin at\}. \end{cases}$$

Solving this system of equations yields

$$\mathcal{L}\{\sin at\} = \int_0^\infty e^{-st}\sin at\, dt = \frac{a}{s^2+a^2}, \quad s > 0, \quad a \in R^+, \qquad (1.154)$$

$$\mathcal{L}\{\cos at\} = \int_0^\infty e^{-st}\cos at\, dt = \frac{s}{s^2+a^2}, \quad s > 0, \quad a \in R^+. \qquad (1.155)$$

1.15.4. Inverse Laplace Transform. Given a function $F(t)$ with $\mathcal{L}\{F(t)\} = f(s)$, the function $F(t)$ is defined as the inverse Laplace transform of $f(s)$, and it is denoted by $\mathcal{L}^{-1}$. We can write

$$\mathcal{L}^{-1}\{f(s)\} = F(t), \tag{1.156}$$

and it is called the inverse operator of the Laplace transform.

Examples. From Eq.(1.151), we have

$$\mathcal{L}^{-1}\left\{\frac{\Gamma(x+1)}{s^{x+1}}\right\} = t^x$$

$$\mathcal{L}^{-1}\left\{\frac{1}{s^{x+1}}\right\} = \frac{t^x}{\Gamma(x+1)}, \quad x > -1. \tag{1.157}$$

1.15.5. The Convolution Theorem. If $\mathcal{L}^{-1}\{f(s)\} = F(t)$ and $\mathcal{L}^{-1}\{g(s)\} = G(t)$, then

$$\mathcal{L}^{-1}\{f(s)g(s)\} = \int_0^t F(u)G(t-u)\,du = F * G. \tag{1.158}$$

For applications of the Laplace transform related to the gamma and beta functions, see Sec.2.7.

1.16. Incomplete Gamma Functions

The *incomplete gamma functions are defined by the integral*

$$\gamma(p, x) = \int_0^x t^{p-1}e^{-t}\,dt, \quad p > 0, \tag{1.159}$$

and the *complementary incomplete gamma function*

$$\Gamma(p, x) = \int_x^\infty t^{p-1}e^{-t}\,dt, \quad p > 0. \tag{1.160}$$

These two functions are related by the equation

$$\gamma(p, x) + \Gamma(p, x) = \Gamma(p). \tag{1.161}$$

See also Sec.1.3, and for applications, see Sec.2.3.37.

1.17. Incomplete Beta Function

The incomplete beta function is defined by the integral

$$B_x(p,q) = \int_0^x t^{p-1}(1-t)^{q-1}\,dt, \quad x \in [0,1], \quad p,q > 0. \tag{1.162}$$

If $x = 1$, $B_x(p,q)$ will become $B(p,q)$. See also Sec.1.4.

1.18. Digamma Function and the Function $\beta(x)$

The psi function (Digamma function), $\psi(x)$, is defined by

$$\psi(x) = \frac{d}{dx}\ln\Gamma(x) = \frac{\Gamma'(x)}{\Gamma(x)}, \quad x \neq 0, -1, -2, -3, \ldots \tag{1.163}$$

From Eqs.(1.51) and (1.163), we have

$$\psi(x) = \frac{\Gamma'(x)}{\Gamma(x)} = -\gamma - \frac{1}{x} + \sum_{n=1}^{\infty}\left(\frac{1}{n} - \frac{1}{x+n}\right),$$

and let $x = 1$,

$$\psi(1) = \frac{\Gamma'(1)}{\Gamma(1)} = \Gamma'(1) = -\gamma, \tag{1.164}$$

where γ is the Euler constant $\gamma = 0.5772157\ldots$.

Taking the logarithmic derivative of recurrence formula Eq.(1.20), we have

$$\psi(x+1) = \psi(x) + \frac{1}{x}, \quad x \neq 0, -1, -2, -3, \ldots, \tag{1.165}$$

which for applications could be written

$$\psi(x) = \psi(x+1) - \frac{1}{x}. \tag{1.166}$$

From Eq.(1.165) could be obtained particular values of $\psi(x)$,

$$\begin{aligned} \psi(2) &= -\gamma + 1 = -0.5772 + 1 = 0.4228, \\ \psi(3) &= \psi(2) + \frac{1}{2} = 0.4228 + \frac{1}{2} = 0.9228, \\ \psi(4) &= \psi(3) + \frac{1}{3} = 0.9228 + \frac{1}{3} = 1.2561, \\ \psi(5) &= \psi(4) + \frac{1}{4} = 1.2561 + \frac{1}{4} = 1.5061, \\ &- - - - - \\ \psi(n+1) &= -\gamma + 1 + \frac{1}{2} + \frac{1}{3} + \cdots + \frac{1}{n}, \end{aligned} \tag{1.167}$$

and also

$$\psi(n+x) = \frac{1}{(n-1)+x} + \frac{1}{(n-2)+x} + \cdots + \frac{1}{2+x} + \frac{1}{1+x} + \psi(1+x), \quad n \in N. \tag{1.168}$$

Taking the logarithmic derivative of the complement formula Eq.(1.35) and knowing that $[\ln \Gamma(u)]' = \psi(u) \cdot u'$ leads to

$$\psi(1-x) - \psi(x) = \pi \cot \pi x, \tag{1.169}$$

and the logarithmic derivative of Eq.(1.135) leads to

$$2\psi(2x) = \psi(x) + \psi\left(x + \frac{1}{2}\right) + 2\ln 2. \tag{1.170}$$

The definition of the function $\beta(x)$ is

$$\beta(x) = \frac{1}{2}\left[\psi\left(\frac{x+1}{2}\right) - \psi\left(\frac{x}{2}\right)\right], \quad x > 0. \tag{1.171}$$

Some formulas for the integral evaluations are

$$\int_0^1 \frac{x^{p-1}}{1+x}\,dx = \beta(p), \quad p > 0, \tag{1.172}$$

$$\int_0^1 \frac{x^{p-1}}{1+x^q}\,dx = \frac{1}{q}\beta\left(\frac{p}{q}\right), \quad p, q > 0, \tag{1.173}$$

$$\int_0^1 \frac{x^p\,dx}{1+x^2} = \frac{1}{2}\beta\left(\frac{p+1}{2}\right), \quad p > -1, \tag{1.174}$$

$$\int_0^1 \frac{x^p\,dx}{(1+x^2)^2} = \frac{p-1}{4}\beta\left(\frac{p-1}{2}\right) - \frac{1}{4}, \quad p > 1, \tag{1.175}$$

$$\int_0^1 \frac{x^{m-1}\,dx}{1+x^n} = \frac{1}{n}\left[B\left(1-\frac{m}{n}, \frac{m}{n}\right) - \beta\left(\frac{m}{n}\right)\right], \quad 0 < m < n, \tag{1.176}$$

and for applications, see Sec.2.8.

The evaluation of the function $\psi(x)$ for $x \in [1,2]$ can be obtained from Table 6.1 in [1]. Knowing the values of $\psi(x)$ for $x \in [1,2]$, the evaluation of $\psi(x)$ for $x \in (0,1]$ and $x > 2$ can be obtained using Eqs.(1.165), (1.167) and (1.168). Table 6.3 in [1] can also be used. A general graph of $y = \psi(x)$ is given in Fig.1.9. The evaluation of the function $\beta(x)$ can be obtained with Eq.(1.171), which gives the values of $\beta(x)$ in terms of $\psi(x)$. The evaluation of integrals, using the function $\beta(x)$, can be obtained with Eqs.(1.172) to (1.176).

a. Evaluate $\psi(5.47)$. Using Eq.(1.168), we have

$$\psi(5.47) = \psi(5+0.47) = \frac{1}{4.47} + \frac{1}{3.47} + \frac{1}{2.47} + \frac{1}{1.47} + \psi(1.47).$$

From tables, $\psi(1.47) = 0.00806$ and then $\psi(5.47) = 1.605088$.

b. Evaluate $\beta(4.2)$. Using Eq.(1.171), we have

$$\beta(4.2) = \frac{1}{2}\left[\psi\left(\frac{4.2+1}{2}\right) - \psi\left(\frac{4.2}{2}\right)\right] = \frac{1}{2}\left[\psi(2.6) - \psi(2.1)\right].$$

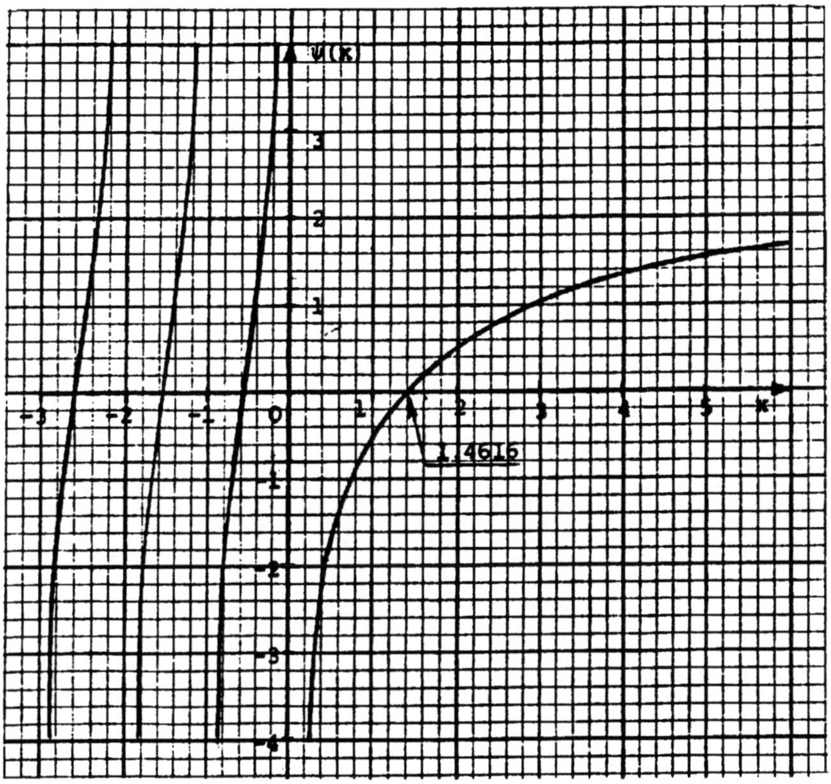

Fig.1.9

But, using Eq.(1.165), we have

$$\psi(2.6) = \frac{1}{1.6} + \psi(1.6) = \frac{1}{1.6} + 0.12604 = 0.75104,$$
$$\psi(2.1) = \frac{1}{1.1} + \psi(1.1) = \frac{1}{1.1} - 0.42375 = 0.48534,$$

and

$$\beta(4.2) = \frac{1}{2}(0.75104 - 0.48534) = 0.13285.$$

c. Verify Eq.(1.170) for $x = 0.75$. Let

$$
\begin{aligned}
A &= \psi(x) + \psi\left(x + \frac{1}{2}\right) + 2\ln 2 \\
&= \psi(0.75) + \psi(0.75 + 0.5) + 2\ln 2 \\
&= \psi(0.75) + \psi(1.25) + 2\ln 2.
\end{aligned}
$$

But, according to Eq.(1.166),

$$
\psi(0.75) = \psi(1.75) - \frac{1}{0.75} = 0.24747 - \frac{1}{0.75} = -1.08586,
$$

$$
A = -1.08586 - 0.22745 + 2\ln 2 = 0.0729.
$$

Let $B = 2\psi(2x) = 2\psi(2 \times 0.75) = 2 \times 0.03648 = 0.0729$, and Eq.(1.170) has been verified for $x = 0.75$.

Examples with the evaluation of integrals using the functions $\psi(x)$ and $\beta(x)$ are given in Sec.2.8.

CHAPTER 2

Applications of the Gamma and Beta Functions

2.1. Introduction

For application of the gamma and beta functions for the integral evaluations, the most important formulas and the integral representations of these functions will be given here and written as follows:

$$\int_0^\infty x^{p-1} e^{-x}\, dx = \Gamma(p), \quad p > 0. \tag{1.1}$$

$$\int_0^1 x^{p-1}(1-x)^{q-1}\, dx = B(p,q), \quad p,q > 0. \tag{1.9}$$

$$B(p,q) = \frac{\Gamma(p)\Gamma(q)}{\Gamma(p+q)}, \quad p,q > 0. \qquad \text{((1.99) and Sec.2.7.2))}$$

$$\Gamma(p+1) = p \cdot \Gamma(p), \quad p > 0. \tag{1.20}$$

$$\Gamma(p)\cdot\Gamma(1-p) = \frac{\pi}{\sin p\pi}, \quad 0 < p < 1. \tag{1.35}$$

$$\Gamma(1+p)\cdot\Gamma(1-p) = \frac{p\pi}{\sin p\pi}, \quad 0 < p < 1. \tag{1.41}$$

$$\int_0^{\frac{\pi}{2}} \sin^p\theta \cdot \cos^q\theta\, d\theta = \frac{1}{2} B\left(\frac{p+1}{2}, \frac{q+1}{2}\right), \quad p,q > -1. \tag{1.116}$$

$$\int_0^\infty \frac{x^{m+1}}{1+x}\, dx = B(1-m, m) = \frac{\pi}{\sin m\pi}, \quad 0 < m < 1. \qquad \text{(Sec.2.3.13)}$$

$$\int_0^\infty \frac{x^{q-1}}{(1+x)^{p+q}}\, dx = B(p,q), \quad p,q > 0. \tag{1.62}$$

$$\int_1^\infty \frac{x^{m-1}}{1+x^n}\, dx = \frac{1}{n}\left[B\left(1-\frac{m}{n}, \frac{m}{n}\right) - \beta\left(\frac{m}{n}\right)\right], \quad 0 < m < n. \qquad \text{(Sec.2.8.8)}$$

Also, Appendix A6 can be used. Many definite integrals, line integrals, double, multiple and surface integrals from any engineering branch can be evaluated using the properties of the gamma and beta functions. Note that for the convergence test of the improper integrals: - if the improper integral does not depend on parameters, the test for convergence should be done at the beginning of the integral evaluation, using Appendixes A1.1 and A1.2; - if the improper integral depends on parameters, the convergence conditions of the integral expressed in terms of the $\Gamma(p)$ and $B(p,q)$ functions can be done at the end of the evaluation, from the properties of the $\Gamma(p)$ and $B(p,q)$ functions.

Using substitutions and algebraic manipulations, many integrals can be expressed in terms of the $\Gamma(p)$ and $B(p,q)$ functions. Once the integral is expressed in terms of the $\Gamma(p)$ and $B(p,q)$ functions, then using Eq.(1.98), which gives the beta function in terms of the gamma function, the integral will be expressed in terms of the gamma function only. Then, using a scientific calculator with the gamma function built-in, or Table 1.1, Table 1.2, or the program given in Fig.1.5 or the program given in Fig.1.4, the integral can be easily evaluated.

2.2. Definite Integrals

2.2.1. Evaluate the definite integral

$$I = \int_0^1 x^3(1-x)^5\,dx. \tag{1}$$

Solution 1. The gamma and beta method. Using Eqs.(1.9), (1.98) and (1.24), with $p-1=3$, $q-1=5$ or $p=4$, $q=6$, we have

$$I = \int_0^1 x^{4-1}(1-x)^{6-1}\,dx = B(4,6) = \frac{\Gamma(4)\Gamma(6)}{\Gamma(10)} = \frac{3!\,5!}{9!} = \frac{1}{504}. \tag{2}$$

See the integrand graphic in Fig.2.1.

Solution 2. The traditional method:

$$\begin{aligned}\int_0^1 x^3(1-x)^5\,dx &= \int_0^1 x^3(1-5x+10x^2-10x^3+5x^4-x^5)\,dx \\ &= \int_0^1 (x^3-5x^4+10x^5-10x^6+5x^7-x^8)\,dx \\ &= \left(\frac{1}{4}x^4-\frac{5}{5}x^5+\frac{10}{6}x^6-\frac{10}{7}x^7+\frac{5}{8}x^8-\frac{1}{9}x^9\right)\Big|_0^1 \\ &= \frac{1}{4}-1+\frac{5}{3}-\frac{10}{7}+\frac{5}{8}-\frac{1}{9}=\frac{1}{504}. \qquad (3)\end{aligned}$$

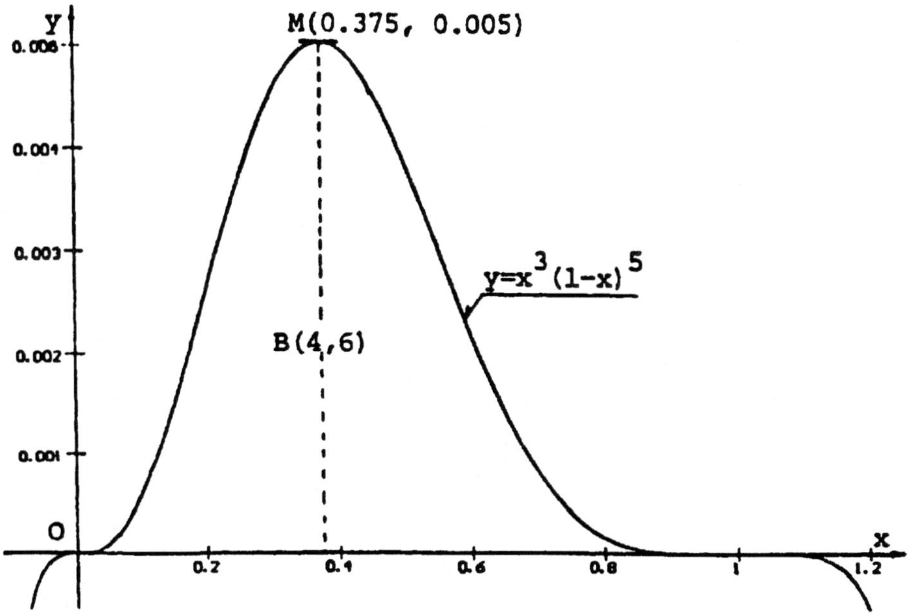

Fig.2.1

2.2.2. a. Prove Eq.(1.35),

$$\Gamma(p)\cdot\Gamma(1-p)=\frac{\pi}{\sin p\pi} \quad (p \text{ nonintegral}), \qquad (1)$$

for $p=0.6$.

Solution. We should prove the equality

$$\Gamma(0.6)\Gamma(1-0.6) = \frac{\pi}{\sin 0.6\pi}. \tag{2}$$

We have

$$\Gamma(0.6)\Gamma(0.4) = 1.489192 \times 2.218159 = 3.30326$$

and

$$\frac{\pi}{\sin p\pi} = \frac{\pi}{\sin 0.6\pi} = 3.30326$$

and (1) has been proved for $p = 0.6$.

b. Prove Eq.(1.40),

$$\Gamma(1+p)\Gamma(1-p) = \frac{p\pi}{\sin p\pi}, \quad (p \text{ nonintegral}) \tag{3}$$

for $p = 0.8$.

Solution. We should prove the equality

$$\Gamma(1+0.8)\Gamma(1-0.8) = \frac{0.8\pi}{\sin 0.8\pi}. \tag{4}$$

We have

$$\Gamma(1+0.8) = \Gamma(1.8) = 0.931384,$$
$$\Gamma(1-0.8) = \Gamma(0.2) = 4.590844$$

and

$$\Gamma(1+0.8)\Gamma(1-0.8) = 0.931384 \times 4.590844 = 4.2758$$

and

$$\frac{0.8\pi}{\sin 0.8\pi} = 4.2758$$

and Eq.(3) has been proved for $p = 0.8$.

2.2.3. Evaluate the definite integral

$$I = \int_0^1 \sqrt{x - x^2}\, dx. \tag{1}$$

Solution 1. Using the gamma and beta functions, we have

$$\begin{aligned} I &= \int_0^1 x^{\frac{1}{2}} (1-x)^{\frac{1}{2}}\, dx = \int_0^1 x^{\left(1+\frac{1}{2}\right)-1} (1-x)^{\left(1+\frac{1}{2}\right)-1}\, dx \\ &= \int_0^1 x^{\frac{3}{2}-1} (1-x)^{\frac{3}{2}-1}\, dx \end{aligned} \tag{2}$$

and with Eqs.(1.9), (1.98) and (1.20), we have

$$I = B\left(\frac{3}{2}, \frac{3}{2}\right) = \frac{\Gamma\left(\frac{3}{2}\right)\Gamma\left(\frac{3}{2}\right)}{\Gamma(3)} = \frac{1}{2}\left[\Gamma\left(1+\frac{1}{2}\right)\right]^2 = \frac{1}{2}\left[\frac{1}{2}\Gamma\left(\frac{1}{2}\right)\right]^2 = \frac{\pi}{8}. \tag{3}$$

Solution 2. Let

$$\sqrt{x - x^2} = tx, \quad x = \frac{1}{1+t^2}, \quad dx = -2\frac{t\, dt}{(1+t^2)^2} \tag{4}$$

and the integral becomes

$$I = 2\int_0^\infty \frac{t^2\, dt}{(1+t^2)^3}. \tag{5}$$

Using Eq.(1.66) with $2p - 1 = 2$ and $p + q = 3$ from where $p = q = 3/2$, we have

$$I = 2\frac{1}{2}B\left(\frac{3}{2}, \frac{3}{2}\right) = \frac{\Gamma\left(\frac{3}{2}\right)\Gamma\left(\frac{3}{2}\right)}{\Gamma(3)} = \frac{\pi}{8}. \tag{6}$$

Solution 3. Using a traditional method, from (1), we have

$$I = \int_0^1 \sqrt{x - x^2}\, dx = \int_0^1 \sqrt{\frac{1}{4} - \left(x - \frac{1}{2}\right)^2}\, dx \tag{7}$$

and with the change of variable,

$$x - \frac{1}{2} = \frac{1}{2}\sin\theta, \quad dx = \frac{1}{2}\cos\theta\, d\theta,$$

from (7) we have

$$\int \sqrt{\frac{1}{4} - \left(x - \frac{1}{2}\right)^2}\, dx = \frac{1}{2}\int \sqrt{\frac{1}{4} - \frac{1}{4}\sin^2\theta} \cdot \cos\theta\, d\theta$$
$$= \frac{1}{4}\int \cos^2\theta\, d\theta$$

and (7) becomes

$$I = \frac{1}{4}\int_{-\frac{\pi}{2}}^{\frac{\pi}{2}} \cos^2\theta\, d\theta = \frac{1}{2}\left[\frac{\cos\theta\sin\theta}{2}\bigg|_0^{\frac{\pi}{2}} + \frac{1}{2}\int_0^{\frac{\pi}{2}} \cos^0\theta\, d\theta\right] = \frac{1}{4}\theta\bigg|_0^{\frac{\pi}{2}} = \frac{\pi}{8}.$$

2.2.4. Evaluate the definite integral

$$I = \int_0^1 \sqrt{1 - x^2}\, dx.$$

Solution. With the substitution $x^2 = t$, $dx = \frac{1}{2}t^{-\frac{1}{2}}\, dt$ the integral becomes

$$I = \frac{1}{2}\int_0^1 t^{-\frac{1}{2}}(1-t)^{\frac{1}{2}}\, dt = \frac{1}{2}\int_0^1 t^{\left(1-\frac{1}{2}\right)-1}(1-t)^{\left(1+\frac{1}{2}\right)-1}\, dt$$
$$= \frac{1}{2}\int_0^1 t^{\frac{1}{2}-1}(1-t)^{\frac{3}{2}-1}\, dt = \frac{1}{2}B\left(\frac{1}{2}, \frac{3}{2}\right) = \frac{1}{2}\Gamma\left(\frac{1}{2}\right) \cdot \Gamma\left(\frac{3}{2}\right)$$
$$= \frac{1}{4}\left[\Gamma\left(\frac{1}{2}\right)\right]^2 = \frac{\pi}{4}.$$

One can also use the substitution $x = \sin\theta$ and then Eq.(1.120).

2.2.4. a. Evaluate the integral

$$I = \int_0^1 \frac{dx}{ts\sqrt{1 + x^4}}. \tag{1}$$

Solution. Let $x^4 = t$, $4x^3\,dx = dt$, $dx = \frac{1}{4}t^{-\frac{3}{4}}\,dt$ and (1) becomes

$$I = \frac{1}{4}\int_0^1 \frac{t^{-\frac{3}{4}}\,dt}{\sqrt{1+t}} \tag{2}$$

and using Eq.(1.64a) for $p - 1 = -3/4$ or $p = 1/4$, (2) becomes

$$\begin{aligned} I &= \frac{1}{4}\int_0^1 \frac{t^{-\frac{3}{4}}\,dt}{(1+t)^{\frac{1}{2}}} = \frac{1}{8}B\left(\frac{1}{4}, \frac{1}{4}\right) \\ &= \frac{1}{8}\frac{\left[\Gamma\left(\frac{1}{4}\right)\right]^2}{\Gamma\left(\frac{1}{2}\right)} = \frac{1}{8\sqrt{\pi}}\left[\Gamma\left(\frac{1}{4}\right)\right]^2 = 0.927. \end{aligned}$$

See also Sec.4.2.19 with the integral evaluation in terms of the elliptic integrals

$$\int_0^1 \frac{dx}{\sqrt{1+x^4}} = \frac{1}{2}K\left(\frac{1}{\sqrt{2}}\right) = \frac{1}{2} \times 1.854 = 0.927.$$

2.2.5. Evaluate the definite integral

$$I = \int_0^1 x^4 \sqrt[3]{1-x^3}\,dx. \tag{1}$$

Solution. With the substitution $x^3 = t$, $dx = \frac{1}{3}t^{-\frac{2}{3}}\,dt$, we have

$$\begin{aligned} I &= \frac{1}{3}\int_0^1 t^{\frac{2}{3}}(1-t)^{\frac{1}{3}}\,dt = \frac{1}{3}\int_0^1 t^{\left(1+\frac{2}{3}\right)-1}(1-t)^{\left(1+\frac{1}{3}\right)-1}\,dt \\ &= \frac{1}{3}\int_0^1 t^{\frac{5}{3}-1}(1-t)^{\frac{4}{3}-1}\,dt = \frac{1}{3}B\left(\frac{5}{3}, \frac{4}{3}\right) = \frac{1}{27}\Gamma\left(\frac{2}{3}\right)\Gamma\left(\frac{1}{3}\right) \end{aligned}$$

and with the complement formula (1.35), we have

$$\int_0^1 x^4 \sqrt[3]{1-x^3}\,dx = \frac{1}{27}\frac{\pi}{\sin\frac{\pi}{3}} = \frac{2\pi}{27\sqrt{3}} = 0.134. \tag{2}$$

2.2.5. a. Evaluate the integral

$$I = \int_0^1 \sqrt{1 - x^4}\, dx. \tag{1}$$

Solution. With the substitution $x^4 = y$, we have

$$x = y^{\frac{1}{4}}, \quad 4x^3\, dx = dy, \quad 4y^{\frac{3}{4}}\, dx = dy, \quad dx = \frac{dy}{4y^{\frac{3}{4}}} = \frac{1}{4} y^{-\frac{3}{4}}\, dy,$$

$$I = \frac{1}{4}\int_0^1 \sqrt{1-y}\cdot y^{-\frac{3}{4}}\, dy = \frac{1}{4}\int_0^1 y^{\left(1-\frac{3}{4}\right)-1}(1-y)^{\left(1+\frac{1}{2}\right)-1}\, dy \tag{2}$$

$$= \frac{1}{4}\int_0^1 y^{\frac{1}{4}-1}(1-y)^{\frac{3}{2}-1}\, dy = \frac{1}{4} B\left(\frac{1}{4}, \frac{3}{2}\right)$$

$$= \frac{1}{4}\frac{\Gamma\left(\frac{1}{4}\right)\Gamma\left(\frac{3}{2}\right)}{\Gamma\left(\frac{1}{4}+\frac{3}{2}\right)} = \frac{1}{4}\frac{\Gamma\left(\frac{1}{4}\right)\Gamma\left(\frac{3}{2}\right)}{\Gamma\left(\frac{7}{4}\right)},$$

but with Eq.(1.20),

$$\Gamma\left(\frac{7}{4}\right) = \Gamma\left(\frac{3}{4}+1\right) = \frac{3}{4}\Gamma\left(\frac{3}{4}\right)$$

and

$$I = \frac{\sqrt{\pi}}{6}\frac{\Gamma\left(\frac{1}{4}\right)}{\Gamma\left(\frac{3}{4}\right)} = \frac{\sqrt{\pi}}{6}\frac{\Gamma\left(\frac{1}{4}\right)}{\dfrac{\pi\sqrt{2}}{\Gamma\left(\frac{1}{4}\right)}}$$

$$= \frac{1}{6\sqrt{2\pi}}\left[\Gamma\left(\frac{1}{4}\right)\right]^2 = 0.874. \tag{3}$$

2.2.6. Evaluate the definite integral

$$i = \int_0^{\frac{\pi}{2}} \sin^2 x \cos^4 x\, dx. \tag{1}$$

Solution. Using the substitution

$$\sin^2 x = t, \quad dx = \frac{dt}{2\sqrt{t(1-t)}} \tag{2}$$

the integral becomes

$$\begin{aligned} I &= \frac{1}{2}\int_0^1 t^{\frac{1}{2}}(1-t)^{\frac{3}{2}}\,dt = \frac{1}{2}\int_0^1 t^{(1+\frac{1}{2})-1}(1-t)^{\left(1+\frac{3}{2}\right)-1}\,dt \\ &= \frac{1}{2}\int_0^1 t^{\frac{3}{2}-1}(1-t)^{\frac{5}{2}-1}\,dt = \frac{1}{2}B\left(\frac{3}{2},\frac{5}{2}\right) = \frac{1}{2}\frac{\Gamma\left(\frac{3}{2}\right)\Gamma\left(\frac{5}{2}\right)}{\Gamma(4)} \\ &= \frac{1}{8}\left[\Gamma\left(\frac{3}{2}\right)\right]^2 = \frac{\pi}{32}. \end{aligned} \tag{3}$$

Exercise. Evaluate the integral using Eqs.(1.12) or (1.116).

2.2.7. Evaluate the integral

$$I = \int_0^{\frac{\pi}{2}} \sin^8\theta\cos^6\theta\,d\theta. \tag{1}$$

Solution. Using Eq.(1.116), we have

$$\begin{aligned} I &= \int_0^{\frac{\pi}{2}} \sin^8\theta\cos^6\theta\,d\theta = \frac{1}{2}B\left(\frac{9}{2},\frac{7}{2}\right) \\ &= \frac{\Gamma\left(\frac{9}{2}\right)\Gamma\left(\frac{7}{2}\right)}{2\Gamma(8)} = \frac{1}{7!\times 2}\Gamma\left(\frac{9}{2}\right)\Gamma\left(\frac{7}{2}\right) = 0.003835. \end{aligned} \tag{2}$$

Also, using Eq.(1.43),

$$\Gamma\left(\frac{9}{2}\right) = \Gamma\left(\frac{1}{2}+4\right) = \frac{\sqrt{\pi}}{2^4}7!!, \quad \Gamma\left(\frac{7}{2}\right) = \Gamma\left(\frac{1}{2}+3\right) = \frac{\sqrt{\pi}}{2^3}5!!,$$

we have

$$I = \int_0^{\frac{\pi}{2}} \sin^8\theta\cos^6\theta\,d\theta = \frac{1}{7!\times 2}\cdot\frac{\sqrt{\pi}}{2^4}7!!\frac{\sqrt{\pi}}{2^3}5!! = \frac{5\pi}{2^{12}}. \tag{3}$$

Exercise. Evaluate the integral using a traditional method. Also evaluate the integral using Eq.(1.12).

2.2.8. Evaluate the definite integral,

$$I = \int_0^{\frac{\pi}{2}} \sin^4\theta\, d\theta. \tag{1}$$

Solution 1. The gamma and beta method. Let

$$\sin^2\theta = x, \quad d\theta = \frac{dx}{2x^{\frac{1}{2}}\sqrt{1-x}}. \tag{2}$$

The integral becomes

$$I = \frac{1}{2}\int_0^1 x^{\frac{3}{2}}(1-x)^{-\frac{1}{2}}\, dx = \frac{1}{2}\int_0^1 x^{\left(1+\frac{3}{2}\right)-1}(1-x)^{\left(1-\frac{1}{2}\right)-1}\, dx$$

$$= \frac{1}{2}\int_0^1 x^{\frac{5}{2}-1}(1-x)^{\frac{1}{2}-1}\, dx = \frac{1}{2}B\left(\frac{5}{2},\frac{1}{2}\right) = \frac{1}{2}\frac{\Gamma\left(\frac{5}{2}\right)\Gamma\left(\frac{1}{2}\right)}{\Gamma(3)} = \frac{\sqrt{\pi}}{4}\Gamma\left(\frac{5}{2}\right)$$

and by Eq.(1.22),

$$\int_0^{\frac{\pi}{2}} \sin^4\theta\, d\theta = \frac{3\pi}{16} = 0.589. \tag{3}$$

- Also, using Eq.(1.120), for $p = 4$, the integral becomes

$$\int_0^{\frac{\pi}{2}} \sin^4\theta\, d\theta = \frac{1}{2}B\left(\frac{5}{2},\frac{1}{2}\right). \tag{4}$$

- Also, using Eq.(1.129), for $n = 2$, the integral becomes

$$\int_0^{\frac{\pi}{2}} \sin^4\theta\, d\theta = \frac{3!!}{4!!}\frac{\pi}{2} = \frac{3\pi}{16}. \tag{5}$$

- For generalization, see Sec.1.12.

Solution 2. Using a traditional method. See Fig.2.2. We have

$$\begin{aligned}\int_0^{\frac{\pi}{2}} \sin^4\theta\,d\theta &= \frac{1}{4}\int_0^{\frac{\pi}{2}} (1-\cos 2\theta)^2\,d\theta \\ &= \frac{1}{4}\int_0^{\frac{\pi}{2}} (1-2\cos 2\theta+\cos^2 2\theta)\,d\theta \\ &= \frac{1}{4}\int_0^{\frac{\pi}{2}} \left[1-2\cos 2\theta+\frac{1}{2}(1+\cos 4\theta)\right] d\theta \\ &= \frac{1}{4}\int_0^{\frac{\pi}{2}} \left(\frac{3}{2}-2\cos 2\theta+\frac{1}{2}\cos 4\theta\right) d\theta \\ &= \frac{1}{4}\int_0^{\frac{\pi}{2}} \left(\frac{3}{2}\theta-\sin 2\theta+\frac{1}{8}\sin 4\theta\right)\Big|_0^{\frac{\pi}{2}} = \frac{3\pi}{16}.\end{aligned}$$

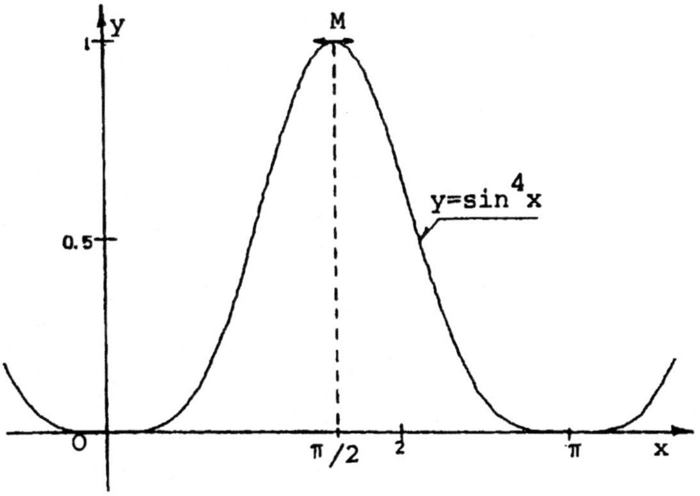

Fig.2.2

2.2.9. Evaluate the integrals

$$\int_0^{\frac{\pi}{2}} \sin^{\frac{3}{2}} x\,dx, \qquad \int_0^{\frac{\pi}{2}} \cos^{\frac{3}{2}} x\,dx. \tag{1}$$

Solution. Using Eq.(1.120), we have

$$\int_0^{\frac{\pi}{2}} \sin^{\frac{3}{2}} x\,dx = \int_0^{\frac{\pi}{2}} \cos^{\frac{3}{2}} x\,dx = \frac{1}{2} B\left(\frac{5}{4}, \frac{1}{2}\right) = \frac{\sqrt{\pi}}{6} \frac{\Gamma\left(\frac{1}{4}\right)}{\Gamma\left(\frac{3}{4}\right)}$$

and using Eq.(1.38), we have

$$\int_0^{\frac{\pi}{2}} \sin^{\frac{3}{2}} x\,dx = \int_0^{\frac{\pi}{2}} \cos^{\frac{3}{2}} x\,dx = \frac{1}{6\sqrt{2\pi}} \left[\Gamma\left(\frac{1}{4}\right)\right]^2 = 0.874. \quad (2)$$

Notes.

a. These integrals can also be evaluated using the substitution

$$\sin^2 x = t, \quad dx = \frac{1}{2} t^{-\frac{1}{2}} (1-t)^{-\frac{1}{2}}\,dt$$

and then using the definition of the beta function, Eq.(1.9).

b. From Sec.4.3.1, $[\Gamma(1/4)]^2 = 4\sqrt{\pi} K(1/\sqrt{2})$ and

$$\int_0^{\frac{\pi}{2}} \sin^{\frac{3}{2}} x\,dx = \int_0^{\frac{\pi}{2}} \cos^{\frac{3}{2}} x\,dx = \frac{\sqrt{2}}{3} K\left(\frac{1}{\sqrt{2}}\right) = 0.874. \quad (3)$$

2.2.10. Evaluate the area bounded by the curve

$$x^n + y^n = a^n, \quad x, y \geq 0, \quad a, n \in R^+. \quad (1)$$

Solution. The area A can be evaluated by the integral

$$A = \int_0^a y\,dx = \int_0^a (a^n - x^n)^{\frac{1}{n}}\,dx. \quad (2)$$

Using the transformation

$$x = a t^{\frac{1}{n}}, \quad dx = \frac{a}{n} t^{\frac{1}{n}-1}\,dt, \quad (3)$$

the area is

$$
\begin{aligned}
A &= \frac{a^2}{n}\int_0^1 t^{\frac{1}{n}-1}(1-t)^{\frac{1}{n}}\,dt = \frac{a^2}{n}\int_0^1 t^{\frac{1}{n}-1}(1-t)^{\left(\frac{1}{n}+1\right)-1}\,dt \\
&= \frac{a^2}{n}B\left(\frac{1}{n},\frac{1}{n}+1\right) \qquad (4)
\end{aligned}
$$

and using Eq.(1.58) yields

$$
A = \frac{a^2}{2n}B\left(\frac{1}{n},\frac{1}{n}\right) = \frac{a^2}{2n}\frac{\left[\Gamma\left(\frac{1}{n}\right)\right]^2}{\Gamma\left(\frac{2}{n}\right)}, \quad x,y \geq 0, \quad a > 0, \qquad (5)
$$

and the integral converges for $n \in R^+$.

Particular cases:

a. Let $n = 2$ and the curve is $x^2 + y^2 = a^2$ (circle) with the area of a quarter circle

$$
A = \frac{a^2}{4}B\left(\frac{1}{2},\frac{1}{2}\right) = \frac{\pi a^2}{4}, \quad x,y \geq 0, \quad a > 0. \qquad (6)
$$

b. Let $n = 4$ and the curve is $x^4 + y^4 = a^4$ with the area

$$
\begin{aligned}
A &= \frac{a^2}{8}B\left(\frac{1}{4},\frac{1}{4}\right) = \frac{a^2}{8}\frac{\Gamma\left(\frac{1}{4}\right)\Gamma\left(\frac{1}{4}\right)}{\Gamma\left(\frac{1}{2}\right)} \\
&= \frac{a^2}{8\sqrt{\pi}}\left[\Gamma\left(\frac{1}{4}\right)\right]^2 = 0.927 \times a^2. \qquad (7)
\end{aligned}
$$

c. Let $n = 2/3$ and the curve is $x^{\frac{2}{3}} + y^{\frac{2}{3}} = a^{\frac{2}{3}}$, with the area

$$A = \frac{3a^2}{4} B\left(\frac{3}{2}, \frac{3}{2}\right)$$

$$= \frac{3a^2}{8}\left[\Gamma\left(\frac{3}{2}\right)\right]^2 = \frac{3a^2}{32}\left[\Gamma\left(\frac{1}{2}\right)\right]^2 = \frac{3\pi}{32}a^2. \qquad (8)$$

2.2.11. Evaluate the area bounded by the curve

$$\left(\frac{x}{a}\right)^n + \left(\frac{y}{b}\right)^n = 1, \quad x, y \geq 0, \quad a, b, n \in R^+. \qquad (1)$$

Solution. The area A can be calculated by the integral

$$A = \int_0^a y\, dx = b\int_0^a \left[1 - \left(\frac{x}{a}\right)^n\right]^{\frac{1}{n}} dx, \quad x, y \geq 0.$$

Using the transformation

$$\left(\frac{x}{a}\right)^n = t, \quad dx = \frac{a}{n} t^{\frac{1}{n}-1}\, dt \qquad (2)$$

the area becomes

$$A = \frac{ab}{n}\int_0^1 t^{\frac{1}{n}-1}(1-t)^{\frac{1}{n}}\, dt = \frac{ab}{n}\int_0^1 t^{\frac{1}{n}-1}(1-t)^{\left(\frac{1}{n}+1\right)-1}\, dt$$

$$= \frac{ab}{n} B\left(\frac{1}{n}, \frac{1}{n} + 1\right). \qquad (3)$$

With Eq.(1.58), the area is

$$A = \frac{ab}{2n} B\left(\frac{1}{n}, \frac{1}{n}\right), \quad n > 0, \quad a, b > 0. \qquad (4)$$

See also Sec.2.2.16.

Particular cases:

a. Let $b = a$ and (4) becomes

$$A = \frac{a^2}{2n} B\left(\frac{1}{n}, \frac{1}{n}\right), \quad n > 0. \qquad (5)$$

See also Sec.2.2.10.

b. Let $n = 2$, $a \neq b$, and the curve is an ellipse

$$\frac{x^2}{a^2} + \frac{y^2}{b^2} = 1$$

with the total area

$$A_T = abB\left(\frac{1}{2}, \frac{1}{2}\right) = \pi ab.$$

Exercise. Evaluate the above areas using traditional methods.

2.2.12. Evaluate the total area bounded by the curve

$$r^{2n} = a^{2n} \sin m\theta, \quad a = R^+, \quad m, n \in N. \tag{1}$$

Solution. We have

$$r = a \sqrt[2n]{\sin m\theta}, \quad r^2 = a^2 \sin^{\frac{1}{n}} m\theta. \tag{2}$$

The curve is symmetric about the y axis. The regions where the function r is a real number are given by the inequality $\sin m\theta \geq 0$ or

$$2k\pi \leq m\theta \leq \pi + 2k\pi \quad \text{or} \quad \frac{2k\pi}{m} \leq \theta \leq \frac{(1+2k)\pi}{m}, \quad k \in I. \tag{3}$$

The maximum values of r are given by the equation $\sin m\theta = 1$, or

$$m\theta = \frac{\pi}{2} + 2k\pi, \quad \text{or} \quad \theta = \frac{(1+4k)\pi}{2m}, \quad k = 0, 1, 2, \ldots, \tag{4}$$

and the curve has m loops.

The tangents in origin to the curve are given by the equation $\sin m\theta = 0$ or

$$m\theta = k\pi, \quad \text{or} \quad \theta = \frac{k\pi}{m}, \quad k = 0, 1, 2, \ldots. \tag{5}$$

The total area bounded by the curve (1) is given by the integral

$$A = 2m \int_0^{\frac{\pi}{2m}} \frac{1}{2} r^2 \, d\theta = ma^2 \int_0^{\frac{\pi}{2m}} \sin^{\frac{1}{n}} m\theta \, d\theta. \tag{6}$$

Let $m\theta = t$ and with Eq.(1.120), we have

$$A = a^2 \int_0^{\frac{\pi}{2}} \sin^{\frac{1}{n}} t\, dt = \frac{a^2}{2} B\left(\frac{n+1}{2n}, \frac{1}{2}\right), \quad n \in N. \tag{7}$$

Let $n = 2$, $m = 3$, $a > 0$ and the curve has three loops (see Fig.2.3), and

$$r^4 = a^4 \sin 3\theta, \quad r = \sqrt[4]{\sin 3\theta} \tag{8}$$

and from (7) we have

$$A = \frac{a^2}{2} B\left(\frac{3}{4}, \frac{1}{2}\right) = \frac{a^2}{2} \frac{\Gamma\left(\frac{3}{4}\right)\sqrt{\pi}}{\Gamma\left(\frac{5}{4}\right)} = \frac{2\pi\sqrt{2\pi}}{\left[\Gamma\left(\frac{1}{4}\right)\right]^2} a^2 = 1.198 \times a^2. \tag{9}$$

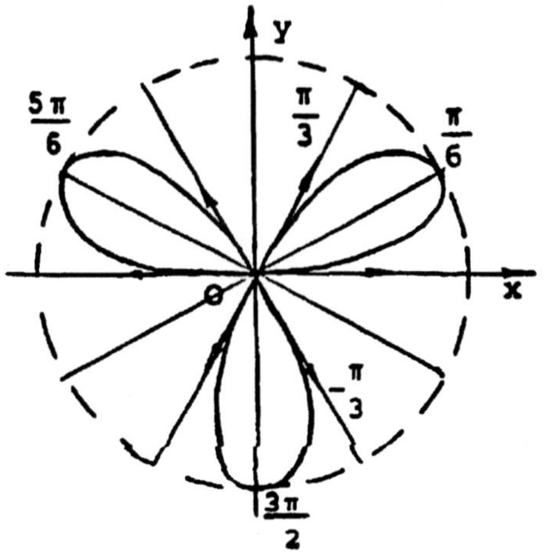

Fig.2.3

- Using (7) calculate the area of the lemniscate $r^2 = a^2 \sin 2\theta$. $A = a^2$. For the graph and length of the lemniscate, see Sec.4.2.25.

2.2.13. Evaluate the total area bounded by the curve

$$r^{2n} = a^{2n} \cos m\theta, \quad a \in R^+, \quad m \in N, \quad n \in N. \tag{1}$$

Solution. We have

$$r = a\sqrt[2n]{\cos m\theta}, \quad r^2 = a^2 \cos^{\frac{1}{n}} m\theta. \tag{2}$$

The curve is symmetric about the x axis. The regions where the function r is a real number are given by the inequality $\cos m\theta \geq 0$ or

$$-\frac{\pi}{2} + 2k\pi \leq m\theta \leq \frac{\pi}{2} + 2k\pi,$$

or

$$-\frac{\pi}{2m} + \frac{2k\pi}{m} \leq \theta \leq \frac{\pi}{2m} + \frac{2k\pi}{m}, \quad k \in I. \tag{3}$$

The maximum values of r are given by the equation $\cos m\theta = 1$, or

$$m\theta = 2k\pi \quad \text{or} \quad \theta\frac{2k\pi}{m}, \quad k = 0, 1, 2, \ldots, \tag{4}$$

and the curve has m loops. The tangents in origin to the curve (1) are given by the equation $\cos m\theta = 0$, or

$$m\theta = \frac{\pi}{2} + k\pi, \quad \text{or} \quad \theta = \frac{\pi}{2m} + \frac{k\pi}{m}, \quad k = 0, 1, 2, \ldots. \tag{5}$$

The total area bounded by the curve (1) is given by the integral

$$A = 2m \int_0^{\frac{\pi}{2m}} \frac{1}{2} r^2 \, d\theta = ma^2 \int_0^{\frac{\pi}{2m}} \cos^{\frac{1}{n}} m\theta \, d\theta. \tag{6}$$

Let $m\theta = t$ and with Eq.(1.120), we have

$$A = a^2 \int_0^{\frac{\pi}{2}} \cos^{\frac{1}{n}} t \, dt = \frac{a^2}{2} B\left(\frac{n+1}{2n}, \frac{1}{2}\right), \quad n \in N. \tag{7}$$

- Let $n = 2$, $m = 3$, $a > 0$ and the curve has three loops (see Fig.2.4), and we have

$$r^4 = a^4 \cos 3\theta, \quad r = \sqrt[4]{\cos 3\theta}, \tag{8}$$

and from (7),

$$A = \frac{a^2}{2} B\left(\frac{3}{4}, \frac{1}{2}\right) = \frac{a^2}{2} \frac{\Gamma\left(\frac{3}{4}\right)\sqrt{\pi}}{\Gamma\left(\frac{5}{4}\right)} = \frac{2\pi\sqrt{2\pi}}{\left[\Gamma\left(\frac{1}{4}\right)\right]^2} a^2 = 1.198 \times a^2. \tag{9}$$

- Using (7) calculate the area of the lemniscate $r^2 = a^2 \cos 2\theta$. $A = a^2$.

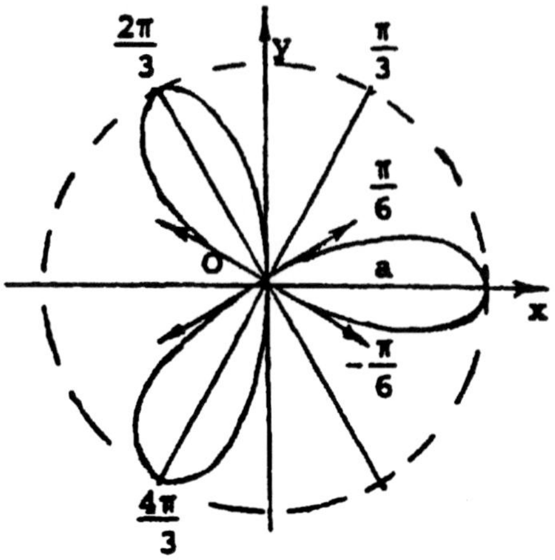

Fig.2.4

2.2.14. Evaluate the total area bounded by the curve

$$r^{2n+1} = a^{2n+1} \sin m\theta, \quad a \in R^+, \quad m \in N, \quad n \in N \cup \{0\}. \tag{1}$$

Solution. We have

$$r = a \sqrt[2n+1]{\sin m\theta}, \quad r^2 = a^2 \sin^{\frac{2}{2n+1}} m\theta. \tag{2}$$

The curve is symmetric about the x axis, y axis and the pole. The function r is a real number for any value of θ. The maximum values of r are given by the equation $\sin m\theta = \pm 1$, or

$$m\theta = \frac{\pi}{2} + k\pi \quad \text{or} \quad \theta = \frac{\pi}{2m} + \frac{k\pi}{m}, \quad k \in I, \tag{3}$$

and the curve has $2m$ loops. The tangents in origin to the curve are given by the equation $\sin m\theta = 0$, or

$$m\theta = k\pi \quad \text{or} \quad \theta = \frac{k\pi}{m}, \quad k \in N \cup \{0\}. \tag{4}$$

The total area bounded by the curve (1) is given by the integral

$$A = 4m \int_0^{\frac{\pi}{2m}} \frac{1}{2} r^2 \, d\theta = 2m^a \int_0^{\frac{\pi}{2m}} \sin^{\frac{2}{2n+1}} m\theta \, d\theta. \tag{5}$$

Let $m\theta = t$ and with Eq.(1.120), we have

$$A = 2a^2 \int_0^{\frac{\pi}{2}} \sin^{\frac{2}{2n+1}} t\, dt = a^2 B\left(\frac{2n+3}{2(2n+1)}, \frac{1}{2}\right), \quad n \in N. \tag{6}$$

a. Let $n = m = 2$ and the curve has four loops (see Fig.2.5), and

$$r^5 = a^5 \sin 2\theta, \quad A = a^2 B\left(\frac{7}{10}, \frac{1}{2}\right) = a^2\sqrt{\pi}\frac{\Gamma\left(\frac{7}{10}\right)}{\Gamma\left(\frac{6}{5}\right)} = 2.506 \times a^2. \tag{7}$$

b. Let $n = 0$, $m = 3$ and the curve has six loops (see Fig.2.6), and

$$r = a\sin 3\theta, \quad A = a^2 B\left(\frac{3}{2}, \frac{1}{2}\right) = \frac{\pi}{2}a^2. \tag{8}$$

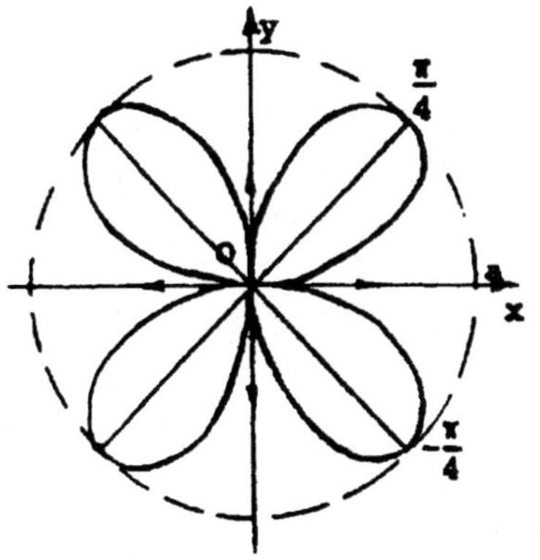

Fig.2.5

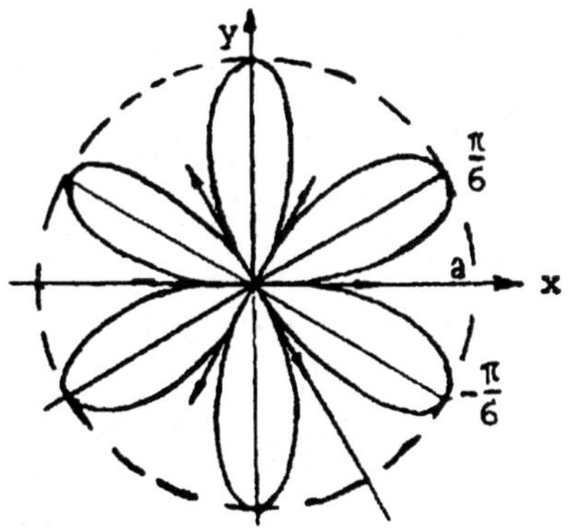

Fig.2.6

2.2.15. Evaluate the total area bounded by the curve

$$r^{2n+1} = a^{2n+1}\cos m\theta, \quad a \in R^+, \quad m \in N, \quad n \in N \cup \{0\}. \tag{1}$$

Solution. We have

$$r = a\sqrt[2n+1]{\cos m\theta}, \quad r^2 = a^2 \cos^{\frac{2}{2n+1}} m\theta. \tag{2}$$

The curve is symmetric about the x axis and the y axis. The function r is a real number for any value of θ. The maximum values of r are given by the equation $\cos m\theta = \pm 1$, or

$$m\theta = k\pi \quad \text{or} \quad \theta = \frac{k\pi}{m}, \qquad k \in N \cup \{0\}, \tag{3}$$

and the curve has $2m$ loops. The tangents in origin to the curve (1) are given by the equation $\cos m\theta = 0$, or

$$m\theta = \frac{\pi}{2} + k\pi \quad \text{or} \quad \theta = \frac{\pi}{2m} + \frac{k\pi}{m}, \qquad k \in N \cup \{0\}. \tag{4}$$

The total area bounded by the curve (1) is given by the integral

$$A = 4m \int_0^{\frac{\pi}{2m}} \frac{1}{2} r^2 \, d\theta = 2ma^2 \int_0^{\frac{\pi}{2m}} \cos^{\frac{2}{2n+1}} m\theta \, d\theta. \tag{5}$$

Let $m\theta = t$ and with Eq.(1.120), we have

$$A = 2a^2 \int_0^{\frac{\pi}{2}} \cos^{\frac{2}{2n+1}} t \, dt = a^2 B\left(\frac{2n+3}{2(2n+1)}, \frac{1}{2}\right), \qquad n \in N. \tag{6}$$

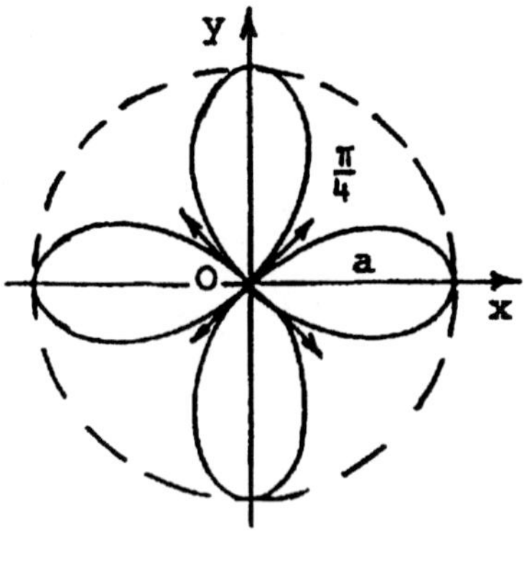

Fig.2.7

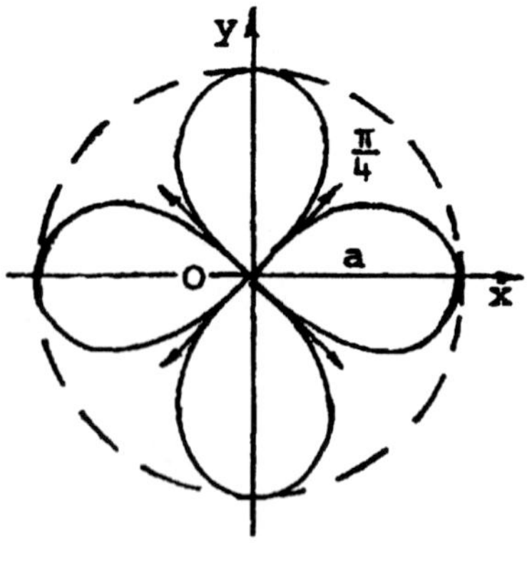

Fig.2.8

a. Let $n = 0$, $m = 2$ and the curve has four loops (see Fig.2.7), and from (1) and (6) we have

$$r = a\cos 2\theta \quad \text{and} \quad A = a^2 B\left(\frac{3}{2}, \frac{1}{2}\right) = \frac{\pi}{2}a^2. \tag{7}$$

b. Let $n = m = 2$ and the curve has four loops (see Fig.2.8), and from (1) and (6) we have

$$r^5 = a^5 \cos 2\theta \quad \text{and} \quad A = a^2 B\left(\frac{7}{10}, \frac{1}{2}\right) = a^2\sqrt{\pi}\frac{\Gamma\left(\frac{7}{10}\right)}{\Gamma\left(\frac{6}{5}\right)} = 2.506 \times a^2. \tag{8}$$

Notes.

a. The total areas bounded by the curves, in Sec.2.2.12 to 2.2.15, do not depend on m (the number of loops).

b. The areas A_{2n} and A_{2n+1} increase with n, and for $n \to \infty$ we have

$$\lim_{n\to\infty} A_{2n} = \frac{a^2}{2}\lim_{n\to\infty} B\left(\frac{n+1}{2n}, \frac{1}{2}\right) = \frac{a^2}{2}B\left(\frac{1}{2}, \frac{1}{2}\right) = \frac{\pi a^2}{2},$$

which is half of the area of the circle with the a radius; also,

$$\lim_{n\to\infty} A_{2n+1} = a^2 \lim_{n\to\infty} B\left[\frac{2n+3}{2(2n+1)}, \frac{1}{2}\right] = a^2 B\left(\frac{1}{2}, \frac{1}{2}\right) = \pi a^2,$$

which is the area of the circle with the a radius.

2.2.16. Find the volume generated by rotating the curve

$$\left(\frac{x}{a}\right)^n + \left(\frac{y}{b}\right)^n = 1, \quad a, b, n > 0, \tag{1}$$

about the x axis.

Solution. The volume is given by the integral

$$V = 2\pi \int_0^a y^2\,dx = 2\pi b^2 \int_0^a \left(1 - \frac{x^n}{a^n}\right)^{\frac{2}{n}} dx. \tag{2}$$

Letting

$$x = at^{\frac{1}{n}}, \qquad dx = \frac{a}{n} t^{\frac{1}{n}-1}\,dt, \tag{3}$$

the volume (2) becomes

$$\begin{aligned} V &= \frac{2\pi ab^2}{n} \int_0^1 t^{\frac{1}{n}-1}(1-t)^{\frac{2}{n}}\,dt \\ &= \frac{2\pi ab^2}{n} \int_0^1 t^{\frac{1}{n}-1}(1-t)^{\left(\frac{2}{n}+1\right)-1}\,dt \\ &= \frac{2\pi ab^2}{n} B\left(\frac{1}{n}, \frac{2}{n}+1\right) \end{aligned} \tag{4}$$

and using Eq.(1.58) we have

$$V = \frac{4\pi ab^2}{3n} B\left(\frac{1}{n}, \frac{2}{n}\right), \qquad a, b, n > 0. \tag{5}$$

Particular cases:

a. Let $n = 2$ and the curve (1) becomes an ellipse

$$\frac{x^2}{a^2} + \frac{y^2}{b^2} = 1 \tag{6}$$

and from (5) we have the volume of an ellipsoid of revolution with the semi-axes a and b,

$$\begin{aligned} V &= \frac{4\pi ab^2}{6} B\left(\frac{1}{2}, 1\right) = \frac{2\pi ab^2}{3} \frac{\Gamma\left(\frac{1}{2}\right)}{\Gamma\left(\frac{3}{2}\right)} \\ &= \frac{2\pi ab^2}{3} \frac{\Gamma\left(\frac{1}{2}\right)}{\frac{1}{2}\Gamma\left(\frac{1}{2}\right)} = \frac{4\pi ab^2}{3}. \end{aligned} \tag{7}$$

b. Let $n = 2/3$, $b = a$ and (1) becomes the equation of the four-cusped hypocycloid, see Fig.2.9,

$$x^{\frac{2}{3}} + y^{\frac{2}{3}} = a^{\frac{2}{3}}, \quad a > 0. \tag{8}$$

The volume generated by rotating the curve (8) about the x axis, using formula (5), is

$$V = 2\pi a^3 B\left(\frac{3}{2}, 3\right). \tag{9}$$

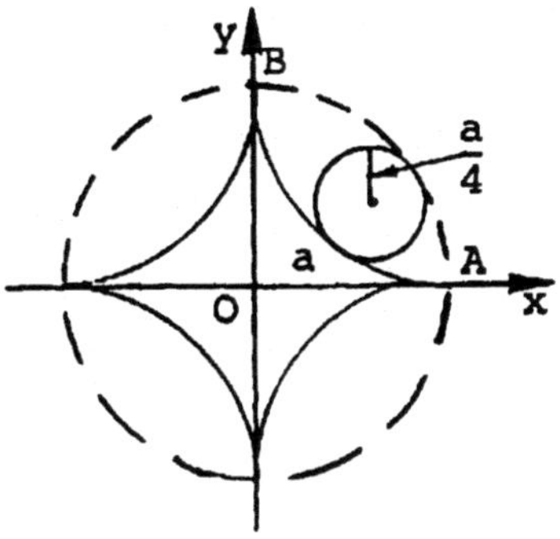

Fig.2.9

Using Eqs.(1.59) and (1.58), we have

$$\begin{aligned} B\left(\frac{3}{2}, 3\right) &= \frac{1}{7} B\left(\frac{1}{2}, 3\right) \\ &= \frac{4}{35} B\left(\frac{1}{2}, 2\right) = \frac{8}{105} B\left(\frac{1}{2}, 1\right) \\ &= \frac{8}{105} \frac{\Gamma\left(\frac{1}{2}\right)}{\Gamma\left(\frac{3}{2}\right)} = \frac{16}{105} \end{aligned}$$

and from (9),

$$V = \frac{32\pi}{105} a^3. \tag{10}$$

Exercise. Evaluate the above volumes using traditional methods.

2.3. Improper Integrals

2.3.1. Study the improper integral

$$I = \int_0^1 \frac{dx}{\sqrt[5]{1-x}}. \tag{1}$$

Test for convergence. According to Appendix A1.2, the integral converges at the upper limit, because we can write

$$f(x) = \frac{1}{\sqrt[5]{1-x}} \underset{x=1}{=} O\left[\frac{1}{(1-x)^{\frac{1}{5}}}\right] \tag{2}$$

and hence $\alpha = 1/5 < 1$.

Integral evaluation. The integral can be written

$$I = \int_0^1 (1-x)^{-\frac{1}{5}}\,dx = \int_0^1 (1-x)^{\left(1-\frac{1}{5}\right)-1}\,dx$$

$$= \int_0^1 x^{1-1}(1-x)^{\frac{4}{5}-1}\,dx = B\left(1,\frac{4}{5}\right) = \frac{\Gamma(1)\Gamma\left(\frac{4}{5}\right)}{\Gamma\left(1+\frac{4}{5}\right)} = \frac{\Gamma\left(\frac{4}{5}\right)}{\frac{4}{5}\Gamma\left(\frac{4}{5}\right)} = \frac{5}{4}. \tag{3}$$

Exercise. Evaluate the integral using the substitution $1 - x = y$.

2.3.1. **a.** Evaluate the integral

$$I = \int_{-\infty}^{+\infty} \frac{dx}{1+x^2}.$$

Solution. We have

$$I = 2\int_0^{\infty} \frac{dx}{1+x^2},$$

and let

$$x^2 = t, \quad x = t^{\frac{1}{2}}, \quad 2x\,dx = dt, \quad dx = \frac{dt}{2x} = \frac{dt}{2t^{\frac{1}{2}}},$$

$$I = 2\int_0^{\infty} \frac{dt}{2t^{\frac{1}{2}}(1+t)} = \int_0^{\infty} \frac{t^{-\frac{1}{2}}}{1+t}\,dt,$$

and using Sec.2.3.13 with $m = 1/2$,

$$I = B\left(1-\frac{1}{2},\frac{1}{2}\right) = B\left(\frac{1}{2},\frac{1}{2}\right) = \pi.$$

See also Sec.2.3.16. Evaluate the integral using a traditional method.

2.3.2. Study the improper integrals

$$I_1 = \int_0^1 \frac{dx}{\sqrt{1-x^n}}, \quad I_2 = \int_0^{\infty} \frac{dx}{\sqrt{1+x^n}}, \quad n \in R,$$

and prove the relation

$$I_1 = I_2 \cdot \cos\frac{\pi}{n}. \tag{1}$$

Solution. With the substitution

$$x^n = t, \quad x = t^{\frac{1}{n}}, \quad nx^{n-1}\,dx = dt, \quad dx = \frac{dt}{nt^{\frac{n-1}{n}}} = \frac{1}{n}t^{\frac{1}{n}-1}\,dt \tag{2}$$

the integral I_1 becomes

$$\begin{aligned} I_1 &= \frac{1}{n}\int_0^1 t^{\frac{1}{n}-1}(1-t)^{-\frac{1}{2}}\,dt = \frac{1}{n}\int_0^1 t^{\frac{1}{n}-1}(1-t)^{\left(1-\frac{1}{2}\right)-1}\,dt \\ &= \frac{1}{n}\int_0^1 t^{\frac{1}{n}-1}(1-t)^{\frac{1}{2}-1}\,dt = \frac{1}{n}B\left(\frac{1}{n},\frac{1}{2}\right) \end{aligned} \tag{3}$$

and the integral converges for $n > 0$. This result is also found in Sec.2.3.10.

With the substitution (2) and using Eq.(1.63) with $p = 1/n$, $q = \frac{1}{2} - \frac{1}{n}$, the integral I_2 is

$$\int_0^\infty \frac{dx}{\sqrt{1+x^n}} = \frac{1}{n}\int_0^\infty \frac{t^{\frac{1}{n}-1}\,dt}{\sqrt{1+t}} = \frac{1}{n}B\left(\frac{1}{n}, \frac{1}{2} - \frac{1}{n}\right) \tag{4}$$

and the integral I_2 converges for $n > 2$. This result is also found in Sec.2.3.22.

Using (3) and (4) we have

$$\frac{I_1}{I_2} = \frac{B\left(\frac{1}{n}, \frac{1}{2}\right)}{B\left(\frac{1}{n}, \frac{1}{2} - \frac{1}{n}\right)} = \frac{\pi}{\Gamma\left(\frac{1}{n} + \frac{1}{2}\right)\Gamma\left(\frac{1}{2} - \frac{1}{n}\right)} \tag{5}$$

and with Eq.(1.39) we have

$$\frac{I_1}{I_2} = \cos\frac{\pi}{n} \quad \text{or} \quad I_1 = I_2 \cos\frac{\pi}{n}, \qquad n > 2. \tag{6}$$

Find the convergence conditions using Appendixes A1.1 and A1.2.

Particular cases:

a. Let $n = 3$ in I_1,

$$\int_0^1 \frac{dx}{\sqrt{1-x^3}} = \frac{1}{3}B\left(\frac{1}{3}, \frac{1}{2}\right) = \frac{1}{3}\frac{\Gamma\left(\frac{1}{3}\right)\Gamma\left(\frac{1}{2}\right)}{\Gamma\left(\frac{5}{6}\right)}$$

$$= \frac{\sqrt{\pi}}{3}\frac{\Gamma\left(\frac{1}{3}\right)}{\Gamma\left(\frac{5}{6}\right)} = 1.402. \tag{7}$$

This result is also found in Sec.2.3.10. Also with Eq.(1.35) we have

$$I_1 = \frac{\sqrt{\pi}}{3 \times 2\pi}\Gamma\left(\frac{1}{3}\right)\Gamma\left(\frac{1}{6}\right) = \frac{1}{6\sqrt{\pi}}\Gamma\left(\frac{1}{3}\right)\Gamma\left(\frac{1}{6}\right) = 1.402.$$

b. Let $n = 3$ in I_2 and from (4) we have

$$I_2 = \int_0^\infty \frac{dx}{\sqrt{1+x^3}} = \frac{1}{3}B\left(\frac{1}{3}, \frac{1}{2} - \frac{1}{3}\right)$$
$$= \frac{1}{3}B\left(\frac{1}{3}, \frac{1}{6}\right)$$
$$= \frac{1}{3\sqrt{\pi}}\Gamma\left(\frac{1}{3}\right) \cdot \Gamma\left(\frac{1}{6}\right) = 2.804. \tag{8}$$

The evaluation of integral (8) can also be found with (1) for $n = 3$:

$$I_2 = \frac{I_1}{\cos\frac{\pi}{3}} = \frac{1.402}{\frac{1}{2}} = 2.804. \tag{9}$$

c. Let $n = 2$ in I_1,

$$I_1 = \int_0^1 \frac{dx}{\sqrt{1-x^2}} = \frac{1}{2}B\left(\frac{1}{2}, \frac{1}{2}\right) = \frac{\pi}{2}. \tag{10}$$

d. Let $n = 4$ in I_1:

$$I_1 = \int_0^1 \frac{dx}{\sqrt{1-x^4}} = \frac{1}{4}B\left(\frac{1}{4}, \frac{1}{2}\right)$$
$$= \frac{1}{4}\frac{\Gamma\left(\frac{1}{4}\right)\Gamma\left(\frac{1}{2}\right)}{\Gamma\left(\frac{3}{4}\right)} = \frac{1}{4\sqrt{2\pi}}\left[\Gamma\left(\frac{1}{4}\right)\right]^2 = \frac{1}{\sqrt{2}}K\left(\frac{1}{\sqrt{2}}\right) = 1.311. \tag{11}$$

This result is also found in Sec.2.3.10. See also Sec.4.3.6.

e. Let $n = 4$ in I_2 and we have

$$I_2 = \int_0^\infty \frac{dx}{\sqrt{1+x^4}} = \frac{1}{4}B\left(\frac{1}{4}, \frac{1}{4}\right)$$
$$= \frac{\left[\Gamma\left(\frac{1}{4}\right)\right]^2}{4\sqrt{\pi}} = 1.854. \tag{12}$$

See also the evaluation found in Sec.4.2.19 as an elliptic integral. Knowing I_1 (for $n = 4$) and using (1) we get the evaluation of the integral I_2 (for $n = 4$),

$$\int_0^\infty \frac{dx}{\sqrt{1+x^4}} = \frac{I_1}{\cos\dfrac{\pi}{4}} = \sqrt{2}I_1 = \sqrt{2} \times 1.311 = 1.854. \tag{13}$$

Exercise. Evaluate the integrals using traditional methods.

2.3.3. Study the integral

$$I_n = \int_0^1 \frac{dx}{\sqrt[n]{1-x^n}}, \qquad n \in N. \tag{1}$$

Solution. Using the substitution

$$x^n = t, \quad dx = \frac{1}{n} t^{\frac{1}{n}-1}\, dt \tag{2}$$

the integral becomes

$$\begin{aligned} I_n &= \frac{1}{n}\int_0^1 t^{\frac{1}{n}-1}(1-t)^{-\frac{1}{n}}\, dt = \frac{1}{n}\int_0^1 t^{\frac{1}{n}-1}(1-t)^{\left(1-\frac{1}{n}\right)-1}\, dt \\ &= \frac{1}{n}B\left(\frac{1}{n}, 1-\frac{1}{n}\right) = \frac{1}{n}\Gamma\left(\frac{1}{n}\right)\cdot\Gamma\left(1-\frac{1}{n}\right). \end{aligned} \tag{3}$$

Using the complement formula (1.35),

$$\int_0^1 \frac{dx}{\sqrt[n]{1-x^n}} = \frac{\pi}{n\sin\dfrac{\pi}{n}}. \tag{4}$$

From formula (3), the integral converges for $n > 0$ and $1 - \dfrac{1}{n} > 0$ from where $n > 1$.

Particular cases:

a. Let $n = 2$ and the integral becomes

$$\int_0^1 \frac{dx}{\sqrt{1-x^2}} = \frac{\pi}{2\sin\frac{\pi}{2}} = \frac{\pi}{2}. \tag{5}$$

b. Let $n = 3$ and the integral becomes

$$\int_0^1 \frac{dx}{\sqrt[3]{1-x^3}} = \frac{\pi}{3\sin\frac{\pi}{3}} = \frac{2\pi}{3\sqrt{3}} = 1.209. \tag{6}$$

2.3.4. Evaluate the definite integral

$$I = \int_{-1}^{1} (1+x)^{p-1}(1-x)^{q-1}\,dx, \quad p, q \in R. \tag{1}$$

Solution. We should find a linear substitution $x = At + B$ which will transform the integration interval $[-1, 1]$ into the interval $[0, 1]$. The simultaneous equations are

$$\begin{cases} -1 = B \\ 1 = A + B \end{cases} \Rightarrow \begin{cases} A = 2 \\ B = -1 \end{cases} \tag{2}$$

and the linear transformation is

$$x = 2t - 1, \quad dx = 2\,dt. \tag{3}$$

With (3) the integral (1) becomes

$$\begin{aligned} I &= \int_0^1 (2t)^{p-1}(1-2t+1)^{q-1} 2\,dt \\ &= 2^{p+q-1}\int_0^1 t^{p-1}(1-t)^{q-1}\,dt \\ &= 2^{p+q-1}B(p,q) \end{aligned} \tag{4}$$

and from (4), the integral converges for $p, q > 0$.

Particular cases:

a. Let $p = 2$, $q = 5$. Using Eq.(1.24) we have

$$\int_{-1}^{1} (1+x)(1-x)^4\,dx = 2^6 B(2,5) = 2^6 \frac{\Gamma(2)\Gamma(5)}{\Gamma(7)} = 2^6 \frac{4\,!}{6\,!} = \frac{32}{15}.$$

b. Let $p = 1/3$, $q = 2/3$ and the integral becomes

$$\int_{-1}^{1} (1+x)^{\frac{1}{3}-1}(1-x)^{\frac{2}{3}-1}\,dx = B\left(\frac{1}{3},\frac{2}{3}\right) = \Gamma\left(\frac{1}{3}\right)\Gamma\left(\frac{2}{3}\right) = \frac{\pi}{\sin\dfrac{\pi}{3}} = \frac{2\pi}{\sqrt{3}}.$$

2.3.5. Evaluate the integral

$$I = \int_2^5 \sqrt[3]{(5-x)(x-2)}\,dx. \tag{1}$$

Solution. Similar to Sec.2.3.4, the linear transformation $x = At + B$ which will transform the integration interval $[2,5]$ into the interval $[0,1]$ is

$$x = 3t + 2, \qquad dx = 3\,dt \tag{2}$$

and the integral (1) becomes

$$\begin{aligned}
I &= 3^{\frac{5}{3}} \int_0^1 t^{\frac{1}{3}}(1-t)^{\frac{1}{3}}\,dt = 3^{\frac{5}{3}} \int_0^1 t^{\left(1+\frac{1}{3}\right)-1}(1-t)^{\left(1+\frac{1}{3}\right)-1}\,dt \\
&= 3^{\frac{5}{3}} \int_0^1 t^{\frac{4}{3}-1}(1-t)^{\frac{4}{3}-1}\,dt \\
&= 3^{\frac{5}{3}} B\left(\frac{4}{3},\frac{4}{3}\right) = 3^{\frac{5}{3}} B\left(1+\frac{1}{3},1+\frac{1}{3}\right) \\
&= 3^{\frac{5}{3}}\frac{1}{5} B\left(\frac{1}{3},1+\frac{1}{3}\right) = 3^{\frac{5}{3}}\frac{1}{10} B\left(\frac{1}{3},\frac{1}{3}\right) \\
&= \frac{3^{\frac{5}{3}}}{10}\,\frac{\left[\Gamma\left(\frac{1}{3}\right)\right]^2}{\Gamma\left(\frac{2}{3}\right)} = 128.8.
\end{aligned} \tag{3}$$

2.3.6. Evaluate the integral

$$I = \int_{-a}^{b} \sqrt[n]{(a+x)^p (b-x)^{n-p}}\, dx, \quad n \in N, \quad a, b > 0. \tag{1}$$

Solution. Similar to Sec.2.3.4, the linear transformation $x = At + B$ which will transform the integration interval $[-a, b]$ to the interval $[0, 1]$ is $x = (a+b)t - a$ and (1) becomes

$$\begin{aligned}
I &= \int_0^1 \sqrt[n]{[(a+b)t]^p [(a+b) - (a+b)t]^{n-p}} \cdot (a+b)\, dt \\
&= (a+b)^{\frac{p}{n}+1} \int_0^1 t^{\frac{p}{n}} [(a+b)(1-t)]^{\frac{n-p}{n}}\, dt \\
&= (a+b)^2 \int_0^1 t^{\frac{p}{n}} (1-t)^{1-\frac{p}{n}}\, dt \\
&= (a+b)^2 \int_0^1 t^{\left(\frac{p}{n}+1\right)-1} (1-t)^{\left(2-\frac{p}{n}\right)-1}\, dt \\
&= (a+b)^2 B\left(\frac{p}{n}+1, 2-\frac{p}{n}\right) && (2) \\
&= (a+b)^2 B\left[1+\frac{p}{n}, 1+\left(1-\frac{p}{n}\right)\right] \\
&= \frac{(a+b)^2}{2}\left(1-\frac{p}{n}\right) B\left(1+\frac{p}{n}, 1-\frac{p}{n}\right) \\
&= \frac{(a+b)^2}{2}\left(1-\frac{p}{n}\right) \Gamma\left(1+\frac{p}{n}\right) \Gamma\left(1-\frac{p}{n}\right), \quad -1 < \frac{p}{n} < 1, && (3) \\
&= \frac{(a+b)^2}{2}\left(1-\frac{p}{n}\right) \frac{\frac{p}{n}\pi}{\sin \frac{p}{n}\pi}. && (4)
\end{aligned}$$

Particular cases:

a. Let $n = 1$ and from (4),

$$I = \int_{-a}^{b} (a+x)^p (b-x)^{1-p}\, dx = \frac{(a+b)^2}{2}(1-p)\frac{p\pi}{\sin p\pi}. \tag{5}$$

b. Let $a = 2$, $b = 4$, $n = 3$, $p = 1/2$ and from (4) we have

$$I = \int_{-2}^{4} \sqrt[3]{(2+x)^{\frac{1}{2}}(4-x)^{3-\frac{1}{2}}}\, dx = \frac{(2+4)^2}{2}\left(1-\frac{1}{6}\right)\frac{\frac{\pi}{6}}{\sin\frac{\pi}{6}} = 5\pi.$$

2.3.7. Evaluate the integral

$$I = \int_a^b \sqrt{\frac{x-a}{b-x}} \cdot x\, dx, \qquad b > a > 0. \tag{1}$$

According to Appendix A1.2, the integral converges at $x = b$ since we have

$$f(x) = \sqrt{\frac{x-a}{b-x}} \cdot x = \underset{x\to b}{O}\left[\frac{x}{(b-x)^{\frac{1}{2}}}\right]$$

and therefore $\alpha = 1/2 < 1$.

Integral evaluation. Similar to Sec.2.3.4, the linear transformation $x = At+b$ which will transform the integration interval $[a, b]$ into the interval $[0, 1]$ is

$$x = (b-a)t + a, \qquad dx = (b-a)\, dt \tag{2}$$

and the integral (1) becomes

$$
\begin{aligned}
I &= \int_0^1 \sqrt{\frac{(b-a)t}{b-(b-a)t-a}} \cdot [(b-a)t+a](b-a)\,dt \\
&= (b-a)\int_0^1 \sqrt{\frac{t}{1-t}}[(b-a)t+a]\,dt \\
&= (b-a)\left[(b-a)\int_0^1 t^{\frac{3}{2}}(1-t)^{-\frac{1}{2}}\,dt + a\int_0^1 t^{\frac{1}{2}}(1-t)^{-\frac{1}{2}}\,dt\right] \\
&= (b-a)\left[(b-a)\int_0^1 t^{1+\frac{3}{2}-1}(1-t)^{\frac{1}{2}-1}\,dt + a\int_0^1 t^{\frac{3}{2}-1}(1-t)^{\frac{1}{2}-1}\,dt\right] \\
&= (b-a)\left[(b-a)B\left(1+\frac{3}{2},\frac{1}{2}\right) + aB\left(\frac{3}{2},\frac{1}{2}\right)\right] \\
&= (b-a)\left[\frac{3}{4}(b-a)+a\right] B\left(\frac{3}{2},\frac{1}{2}\right) \\
&= (b-a)\left[\frac{3}{4}(b-a)+a\right]\cdot\frac{1}{2}\left[\Gamma\left(\frac{1}{2}\right)\right]^2 \\
&= \frac{b-a}{2}\pi\left[\frac{3}{4}(b-a)+a\right] \\
&= \frac{\pi}{8}(b-a)(a+3b). \qquad (3)
\end{aligned}
$$

Particular case. Let $a=2$, $b=4$ and from (3),

$$\int_2^4 \sqrt{\frac{x-2}{4-x}} \cdot x\,dx = \frac{7\pi}{2}.$$

2.3.8. Study the integral

$$I = \int_0^1 (1-x^2)^p\,dx, \quad p \in R. \qquad (1)$$

Solution. With the substitution $x^2 = t$ the integral becomes

$$I = \int_0^1 (1-t)^p \frac{dt}{2\sqrt{t}} = \frac{1}{2}\int_0^1 t^{-\frac{1}{2}}(1-t)^p\, dt = -\frac{1}{2}\int_0^1 t^{\left(1-\frac{1}{2}\right)-1}(1-t)^{(1+p)-1}\, dt$$

$$= \frac{1}{2}\int_0^1 t^{\frac{1}{2}-1}(1-t)^{(p+1)-1}\, dt$$

$$= \frac{1}{2}B\left(\frac{1}{2}, p+1\right), \quad n > -1. \tag{2}$$

If $p > 0$, we have

$$I = \frac{1}{2}\frac{p}{p+\dfrac{1}{2}}B\left(\frac{1}{2}, p\right) = \frac{p}{2p+1}B\left(\frac{1}{2}, p\right), \quad p > 0. \tag{3}$$

If $p = n \in N$, from (3) we have

$$I = \frac{n}{2n+1}\frac{\Gamma\left(\dfrac{1}{2}\right)\Gamma(n)}{\Gamma\left(n+\dfrac{1}{2}\right)} = \frac{\sqrt{\pi}}{2n+1}\frac{n!}{\Gamma\left(n+\dfrac{1}{2}\right)}, \tag{4}$$

and using Eq.(1.43),

$$I = \frac{\sqrt{\pi}}{2n+1}\frac{n!}{\dfrac{\sqrt{\pi}}{2^n}(2n-1)!!} = \frac{2^n}{2n+1}\frac{n!}{(2n-1)!!} \tag{5}$$

and using Eq.(1.136),

$$I = \frac{\sqrt{\pi}}{2n+1}\frac{n!}{\dfrac{\sqrt{\pi}}{2^{2n}}\cdot\dfrac{(2n)!}{n!}} = \frac{2^{2n}}{2n+1}\frac{(n!)^2}{(2n)!}$$

$$= 2^{2n}\frac{(n!)^2}{(2n+1)!}. \tag{6}$$

Particular cases:

a. Let $n = 3$ and from (5),

$$I = \int_0^1 (1-x^2)^3\, dx = \frac{2^3}{7}\frac{3!}{5!!} = \frac{16}{35}.$$

b. Let $n = 3$ and from (6),

$$I = \int_0^1 (1 - x^2)^3\, dx = 2^6 \frac{(3\,!)^2}{7\,!} = \frac{16}{35}.$$

c. Let $p = 1/2$ and from (1) and (3),

$$I = \int_0^1 (1 - x^2)^{\frac{1}{2}}\, dx = \frac{1}{4} B\left(\frac{1}{2}, \frac{1}{2}\right) = \frac{\pi}{4}.$$

See also Sec.2.2.4.

2.3.9. Study the integral

$$I = \int_0^a x^p (a^2 - x^2)^q\, dx, \quad a > 0, \quad p, q \in R. \tag{1}$$

Solution. With the substitution

$$x^2 = a^2 t, \quad dx = \frac{a}{2} t^{-\frac{1}{2}}\, dt \tag{2}$$

the integral becomes

$$\begin{aligned} I &= \int_0^1 \left(a t^{\frac{1}{2}}\right)^p [a^2(1 - t)]^q \frac{a}{2} t^{-\frac{1}{2}}\, dt \\ &= \frac{a^{p+2q+1}}{2} \int_0^1 t^{\frac{p}{2} - \frac{1}{2}} (1 - t)^q\, dt \\ &= \frac{a^{p+2q+1}}{2} \int_0^1 t^{\left(1 + \frac{p}{2} - \frac{1}{2}\right) - 1} (1 - t)^{(1+q)-1}\, dt \\ &= \frac{a^{p+2q+1}}{2} \int_0^1 t^{\frac{p+1}{2} - 1} (1 - t)^{(1+q)-1}\, dt \\ &= \frac{a^{p+2q+1}}{2} B\left(\frac{p+1}{2}, 1 + q\right) \end{aligned} \tag{3}$$

and the integral converges for $p, q > -1$. For applications of this integral see Secs.2.6.2, 2.6.5 and 2.6.10.

Particular cases:

a. Let $p = 0$, $q = 1/2$ and the integral becomes

$$\int_0^a \sqrt{a^2 - x^2}\, dx = \frac{a^2}{2} B\left(\frac{1}{2}, 1 + \frac{1}{2}\right)$$
$$= \frac{a^2}{4} B\left(\frac{1}{2}, \frac{1}{2}\right) = \frac{\pi a^2}{4}. \tag{4}$$

b. Let $p = 0$, $q = -1/2$, $a = 1$ and the integral becomes

$$\int_0^1 \frac{dx}{\sqrt{1 - x^2}} = \frac{1}{2} B\left(\frac{1}{2}, \frac{1}{2}\right) = \frac{\pi}{2}. \tag{5}$$

See also Sec.2.3.2.

c. Let $p = 4$, $q = -1/2$, $a = 1$ and the integral becomes

$$\int_0^1 \frac{x^4\, dx}{\sqrt{1 - x^2}} = \frac{1}{2} B\left(\frac{5}{2}, \frac{1}{2}\right) = \frac{\sqrt{\pi}}{4} \Gamma\left(\frac{5}{2}\right) = \frac{3\pi}{16}. \tag{6}$$

Exercise. Evaluate the integrals using traditional methods.

2.3.10. Study the improper integral

$$I = \int_0^1 \frac{x^{p-1}}{\sqrt{1 - x^q}}\, dx, \quad q > 0, \quad p \in R. \tag{1}$$

Solution. Using the substitution

$$x^q = t, \quad x = t^{\frac{1}{q}}, \quad dx = \frac{1}{q} t^{\frac{1}{q} - 1}\, dt \tag{2}$$

the equation becomes

$$I = \frac{1}{q}\int_0^1 t^{\frac{p-q}{q}}(1-t)^{-\frac{1}{2}}\,dt = \frac{1}{q}\int_0^1 t^{\frac{p}{q}-1}(1-t)^{\left(1-\frac{1}{2}\right)-1}\,dt$$

$$= \frac{1}{q}\int_0^1 t^{\frac{p}{q}-1}(1-t)^{\frac{1}{2}-1}\,dt = \frac{1}{q}B\left(\frac{p}{q},\frac{1}{2}\right) \qquad (3)$$

and the integral converges for $p, q > 0$.

Particular cases:

a. Let $p = 1$ and the integral becomes

$$\int_0^1 \frac{dx}{\sqrt{1-x^q}} = \frac{1}{q}B\left(\frac{1}{q},\frac{1}{2}\right), \quad q > 0. \qquad (4)$$

This evaluation is also found in Sec.2.3.2.

b. Let $p = 1$, $q = 4$ and the integral becomes

$$\int_0^1 \frac{dx}{\sqrt{1-x^4}} = \frac{1}{4}B\left(\frac{1}{4},\frac{1}{2}\right) = \frac{\sqrt{\pi}}{4}\frac{\Gamma\left(\frac{1}{4}\right)}{\Gamma\left(\frac{3}{4}\right)}$$

$$= \frac{1}{4\sqrt{2\pi}}\left[\Gamma\left(\frac{1}{4}\right)\right]^2 = \frac{1}{\sqrt{2}}K\left(\frac{1}{\sqrt{2}}\right) = 1.311. \qquad (5)$$

c. Let $p = 2$, $q = 4$ and the integral becomes

$$\int_0^1 \frac{x\,dx}{\sqrt{1-x^4}} = \frac{1}{4}B\left(\frac{1}{2},\frac{1}{2}\right) = \frac{1}{4}\left[\Gamma\left(\frac{1}{2}\right)\right]^2 = \frac{\pi}{4}. \qquad (6)$$

d. Let $p = 1$, $q = 3$ and the integral becomes

$$\int_0^1 \frac{dx}{\sqrt{1-x^3}} = \frac{1}{3}B\left(\frac{1}{3},\frac{1}{2}\right) = \frac{\sqrt{\pi}}{3}\frac{\Gamma\left(\frac{1}{3}\right)}{\Gamma\left(\frac{5}{6}\right)} = 1.402.$$

2.3.10. a. Evaluate the integral

$$I = \int_0^\infty \frac{x^{\frac{1}{4}}\,dx}{(1+x)^3}.$$

Test for convergence. According to Appendix A1.1, the integral converges at the upper limit, because we can write

$$\frac{x^{\frac{1}{4}}}{(1+x)^3} = \underset{x\to\infty}{O}\left[\frac{1}{x^{3-\frac{1}{4}}}\right] = \underset{x\to\infty}{O}\left[\frac{1}{x^{\frac{11}{4}}}\right]$$

and $\alpha = 11/4 > 1$.

Solution. Using Eq.(1.63) with $p-1 = 1/4$, $p+q = 3$ or $p = 5/4$, $q = 7/4$,

$$I = B\left(\frac{5}{4}, \frac{7}{4}\right) = \frac{\Gamma\left(\frac{5}{4}\right)\Gamma\left(\frac{7}{4}\right)}{\Gamma(3)} = \frac{1}{2}\Gamma\left(\frac{5}{4}\right)\Gamma\left(\frac{7}{4}\right) = 0.4165$$

and with Eqs.(1.20) and (1.35),

$$\int_0^\infty \frac{x^{\frac{1}{4}}\,dx}{(1+x)^3} = \frac{3}{32}\Gamma\left(\frac{1}{4}\right)\Gamma\left(\frac{3}{4}\right) = \frac{3}{32}\frac{\pi}{\sin\frac{\pi}{4}} = \frac{3\pi\sqrt{2}}{32} = 0.4165.$$

2.3.10. b. Evaluate the integral

$$I = \int_0^1 \sqrt{1-x^4}\,dx.$$

Letting $x^4 = y$, $x = y^{\frac{1}{4}}$, $dx = \dfrac{dy}{4x^3} = \dfrac{dy}{4y^{\frac{3}{4}}}$, we have

$$
\begin{aligned}
I &= \int_0^1 \sqrt{1-y}\frac{dy}{4y^{\frac{3}{4}}} = \frac{1}{4}\int_0^1 y^{-\frac{3}{4}}(1-y)^{\frac{1}{2}}\,dy \\
&= \frac{1}{4}\int_0^1 y^{\left(1-\frac{3}{4}\right)-1}(1-y)^{\left(1+\frac{1}{2}\right)-1}\,dy = \frac{1}{4}\int_0^1 y^{\frac{1}{4}-1}(1-y)^{\frac{3}{2}-1}\,dy \\
&= \frac{1}{4}B\left(\frac{1}{4},\frac{3}{2}\right) = \frac{1}{4}B\left(\frac{1}{4},1+\frac{1}{2}\right) \\
&= \frac{1}{4}\frac{\frac{1}{2}}{\frac{1}{4}+\frac{1}{2}}B\left(\frac{1}{4},\frac{1}{2}\right) \\
&= \frac{1}{6}\frac{\Gamma\left(\frac{1}{4}\right)\Gamma\left(\frac{1}{2}\right)}{\Gamma\left(\frac{1}{4}+\frac{1}{2}\right)} = \frac{\sqrt{\pi}}{6}\frac{\Gamma\left(\frac{1}{4}\right)}{\Gamma\left(\frac{3}{4}\right)} \\
&= \frac{\sqrt{\pi}}{6}\frac{\left[\Gamma\left(\frac{1}{4}\right)\right]^2}{\pi\sqrt{2}} = \frac{\left[\Gamma\left(\frac{1}{4}\right)\right]^2}{6\sqrt{2\pi}}.
\end{aligned}
$$

2.3.11. Study the improper integral

$$I = \int_0^\infty x^m e^{-ax}\,dx, \quad a > 0, \quad m \in R. \tag{1}$$

Solution. Using the substitution

$$ax = t, \quad dx = \frac{dt}{a} \tag{2}$$

the Eq.(1.1) yields

$$
\begin{aligned}
I &= \frac{1}{a^{m+1}}\int_0^\infty t^m e^{-t}\,dt = \frac{1}{a^{m+1}}\int_0^\infty t^{(m+1)-1}e^{-t}\,dt \\
&= \frac{1}{a^{m+1}}\Gamma(m+1)
\end{aligned} \tag{3}
$$

and the integral converges for $m > -1$.

Particular cases:

a. Let $m = n \in N \cup \{0\}$ and from (3), we have

$$\int_0^\infty x^n e^{-ax}\, dx = \frac{n!}{a^{n+1}}. \tag{4}$$

b. Let $a = 1$ and $m = n \in N$ and from (3) we have

$$\int_0^\infty x^n e^{-x}\, dx = \Gamma(n+1) = n!. \tag{5}$$

c. Let $n = 0$, $a = 2$ and from (4), we have

$$\int_0^\infty e^{-2x}\, dx = \frac{1}{2}, \quad (0! = 1). \tag{6}$$

d. Let $m = 1/2$, $a = 2$ and from (3), with Eq.(1.13) we have

$$\int_0^\infty x^{\frac{1}{2}} e^{-2x}\, dx = \frac{1}{2^{\frac{3}{2}}}\Gamma\left(\frac{3}{2}\right) = \frac{1}{4}\sqrt{\frac{\pi}{2}}. \tag{7}$$

2.3.12. Study the integral

$$I = \int_0^{\frac{\pi}{2}} \tan^{\frac{2}{5}} \theta\, d\theta. \tag{1}$$

Test for convergence. According to Appendix A1.2, the integral converges at $\theta = \frac{\pi}{2} - 0$, because we can write

$$\tan^{\frac{2}{5}} \theta = \frac{\sin^{\frac{2}{5}} \theta}{\cos^{\frac{2}{5}} \theta} = \frac{\sin^{\frac{2}{5}} \theta}{\sin^{\frac{2}{5}} \left(\frac{\pi}{2} - \theta\right)} = \underset{\theta=\frac{\pi}{2}}{O} \left[\frac{1}{\left(\frac{\pi}{2} - \theta\right)^{\frac{2}{5}}}\right]. \tag{2}$$

Therefore $\alpha = 2/5 < 1$.

Integral evaluation.

Solution 1. With the substitution

$$\sin^2\theta = t, \quad \sin\theta = t^{\frac{1}{2}}, \quad \cos\theta = (1-t)^{\frac{1}{2}}, \quad d\theta = \frac{1}{2}t^{-\frac{1}{2}}(1-t)^{-\frac{1}{2}}\,dt \tag{3}$$

we have

$$\begin{aligned} I &= \frac{1}{2}\int_0^1 t^{-\frac{3}{10}}(1-t)^{-\frac{7}{10}}\,dt = \frac{1}{2}\int_0^1 t^{\left(1-\frac{3}{10}\right)-1}(1-t)^{\left(1-\frac{7}{10}\right)-1}\,dt \\ &= \frac{1}{2}\int_0^1 t^{\frac{7}{10}-1}(1-t)^{\frac{3}{10}-1}\,dt = \frac{1}{2}B\left(\frac{7}{10},\frac{3}{10}\right) = \frac{1}{2}\Gamma\left(\frac{7}{10}\right)\Gamma\left(\frac{3}{10}\right) = 1.942 \end{aligned} \tag{4}$$

and with the complement formula (1.35), we have

$$\int_0^{\frac{\pi}{2}} \tan^{\frac{2}{5}}\theta\,d\theta = \frac{1}{2}\frac{\pi}{\sin\dfrac{3\pi}{10}} = 1.942. \tag{5}$$

Solution 2. Using Eq.(1.124) with $n = 2/5$ yields

$$\int_0^{\frac{\pi}{2}} \tan^{\frac{2}{5}}\theta\,d\theta = \frac{1}{2}B\left(\frac{1+\dfrac{2}{5}}{2},\frac{1-\dfrac{2}{5}}{2}\right) = \frac{1}{2}B\left(\frac{7}{10},\frac{3}{10}\right). \tag{6}$$

2.3.13. Study the improper integral

$$I = \int_0^\infty \frac{x^{m-1}}{1+x}\,dx, \quad m \in R. \tag{1}$$

Solution 1. Let

$$1+x = \frac{1}{t}, \quad x = \frac{1-t}{t}, \quad dx = -\frac{dt}{t^2},$$

and the integral becomes

$$I = \int_0^1 \left(\frac{1-t}{t}\right)^{m-1} \frac{dt}{t} = \int_0^1 \frac{(1-t)^{m-1}}{t^m} dt = \int_0^1 t^{-m}(1-t)^{m-1} dt$$

$$= \int_0^1 t^{(1-m)-1}(1-t)^{m-1} dt = B(1-m, m) = \Gamma(1-m)\Gamma(m).$$

Using Eq.(1.35), we have

$$\int_0^1 \frac{x^{m-1}}{1+x} dx = \frac{\pi}{\sin m\pi} \tag{2}$$

and the integral converges for $1 - m > 0$ and $m > 0$ or $0 < m < 1$. Also, for evaluation of (1), Eq.(1.62) or Eq.(1.63) can be used.

Solution 2. Using the Cauchy's residue theorem.

Test for convergence. According to Appendix A1.2, at the lower limit we can write

$$f(x) = \frac{x^{m-1}}{1+x} = \underset{x=0}{O}\left[\frac{1}{x^{1-m}}\right]$$

and hence $\alpha = 1 - m < 1$ or $m > 0$.

According to Appendix A1.1, at the upper limit we can write

$$f(x) = \frac{x^{m-1}}{1+x} = \underset{x=\infty}{O}\left[\frac{1}{x^{1-m+1}}\right] = \underset{x=\infty}{O}\left[\frac{1}{x^{2-m}}\right]$$

and hence $\alpha = 2 - m > 1$, or $m < 1$ and the integral converges for $0 < m < 1$.

According to the Cauchy's residue theorem,

$$\oint_C \frac{z^{m-1}}{1+x} dz = 2xi \sum_{j=1}^{n} (\mathrm{Res})_j \tag{3}$$

where C is the integral contour shown in Fig.2.10. But, the integrand has only one pole, $z_0 = -1 = e^{i\pi}$, inside the counter, with the residue

$$\mathrm{Res}(-1) = \lim_{z \to z_0} \left[z^{m-1}\right] = z_0^{m-1} = e^{i\pi(m-1)}$$

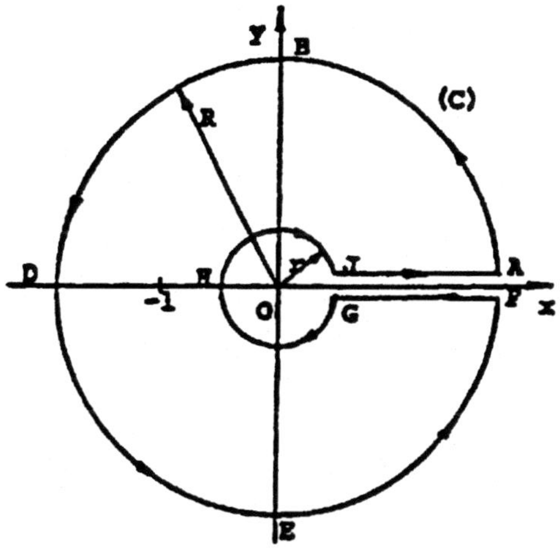

Fig.2.10

and from (3), we have

$$\int_r^R \frac{x^{m-1}}{1+x}\,dx + \int_0^{2\pi} \frac{(Re^{i\theta})^{m-1} iRe^{i\theta}}{1+Re^{i\theta}}\,d\theta + \int_R^r \frac{(xe^{2\pi i})^{m-1}}{1+xe^{2\pi i}}\,dx$$
$$+ \int_{2\pi}^0 \frac{(re^{i\theta})^{m-1} ire^{i\theta}}{1+re^{i\theta}}\,d\theta = 2\pi i e^{\pi i(m-1)}. \tag{4}$$

Taking the limit in (4), for $R \to \infty$ and $r \to 0$, the second and fourth integrals approach zero, and we have

$$\int_0^\infty \frac{x^{m-1}}{1+x}\,dx + \int_\infty^0 \frac{e^{2\pi i(m-1)} x^{m-1}}{1+x}\,dx = 2\pi i e^{\pi i(m-1)} \tag{5}$$

or

$$(1 - e^{2\pi i(m-1)}) \int_0^\infty \frac{x^{m-1}}{1+x}\,dx = 2\pi i e^{\pi i(m-1)}$$

and

$$\int_0^\infty \frac{x^{m-1}}{1+x}\,dx = 2\pi i \frac{e^{i\pi(m-1)}}{1-e^{i2\pi(m-1)}} = -2\pi i \frac{e^{im\pi}}{1-e^{i2\pi m}}$$
$$= \frac{2\pi i}{e^{im\pi} - e^{-im\pi}} = \frac{\pi}{\sin m\pi}, \quad 0 < m < 1. \tag{6}$$

2.3.14. Evaluate the improper integral

$$I = \int_{-\infty}^{+\infty} \frac{dx}{x^4 + 1}. \tag{1}$$

Solution. Since the integrand is an even function, see p.X, the integral can be written

$$I = 2\int_0^{\infty} \frac{dx}{x^4 + 1}. \tag{2}$$

Test for convergence of integral (2). According to Appendix A1.1, the integral converges at the upper limit since we can write

$$f(x) = \frac{1}{x^4+1} = \underset{x=\infty}{O}\left[\frac{1}{x^4}\right]$$

and hence $\alpha = 4 > 1$.

Solution 1. Using the gamma and beta functions. See Fig.2.11.

Let $x^4 = t$, $x = t^{\frac{1}{4}}$, $dx = \dfrac{dt}{4t^{\frac{3}{4}}}$ and we have

$$I = 2\int_0^{\infty} \frac{1}{1+t}\frac{dt}{4t^{\frac{3}{4}}} = \frac{1}{2}\int_0^{\infty} \frac{t^{-\frac{3}{4}}}{1+t}\,dt \tag{3}$$

and with Sec.2.3.13 for $m = 1/4$,

$$I = \frac{1}{2}B\left(\frac{3}{4},\frac{1}{4}\right) = \frac{1}{2}\Gamma\left(\frac{3}{4}\right)\Gamma\left(\frac{1}{4}\right) = \frac{\pi}{2\sin\frac{\pi}{4}} = \frac{\pi}{\sqrt{2}}. \tag{4}$$

Solution 2. Using the gamma and beta functions. Let

$$x^4 + 1 = \frac{1}{t}, \quad x = \left(\frac{1-t}{t}\right)^{\frac{1}{4}}, \quad dx = -\frac{1}{4}t^{-\frac{5}{4}}(1-t)^{-\frac{3}{4}}\,dt \tag{5}$$

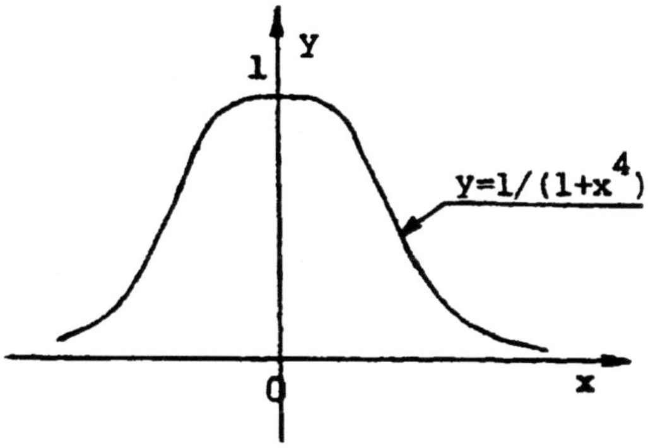

Fig.2.11

and

$$I = \frac{1}{2}\int_0^1 t^{-\frac{1}{4}}(1-t)^{-\frac{3}{4}}\,dt = \frac{1}{2}\int_0^1 t^{\left(1-\frac{1}{4}\right)-1}(1-t)^{\left(1-\frac{3}{4}\right)-1}\,dt$$

$$= \frac{1}{2}\int_0^1 t^{\frac{3}{4}-1}(1-t)^{\frac{1}{4}-1}\,dt = \frac{1}{2}B\left(\frac{3}{4},\frac{1}{4}\right) = \frac{1}{2}\Gamma\left(\frac{3}{4}\right)\Gamma\left(\frac{1}{4}\right) = 2.2214 \qquad (6)$$

or, using the functional equation (1.35),

$$\int_{-\infty}^{+\infty}\frac{dx}{x^4+1} = \frac{1}{2}\frac{\pi}{\sin\frac{\pi}{4}} = \frac{\pi}{\sqrt{2}} = 2.2214.$$

Solution 3. Using the method of partial functions, we have

$$\frac{1}{x^4+1} = \frac{1}{(x^2+x\sqrt{2}+1)(x^2-x\sqrt{2}+1)} = \frac{Ax+B}{x^2+x\sqrt{2}+1} + \frac{Cx+D}{x^2-x\sqrt{2}+1} \qquad (7)$$

from where the identity

$$(Ax+B)(x^2-x\sqrt{2}+1)+(Cx+D)(x^2+x\sqrt{2}+1) \equiv 1 \qquad (8)$$

and we can write the system of linear equations

$$\begin{cases} A+C=0 \\ B+D-\sqrt{2}A+\sqrt{2}C=0 \\ A-\sqrt{2}B+C+\sqrt{2}D=0 \\ B+D=1 \end{cases} \qquad (9)$$

Solving this system of equations, we have

$$A = \frac{1}{2\sqrt{2}}, \quad B = \frac{1}{2}, \quad C = -\frac{1}{2\sqrt{2}}, \quad D = \frac{1}{2} \tag{10}$$

and (7) becomes

$$\frac{1}{x^4+1} = \frac{1}{2\sqrt{2}}\left(\frac{x+\sqrt{2}}{x^2+x\sqrt{2}+1} - \frac{x-\sqrt{2}}{x^2-x\sqrt{2}+1}\right) \tag{11}$$

and

$$I = \frac{1}{\sqrt{2}}\left[\int_0^\infty \frac{x+\sqrt{2}}{x^2+x\sqrt{2}+1}\,dx - \int_0^\infty \frac{x-\sqrt{2}}{x^2-x\sqrt{2}+1}\,dx\right]. \tag{12}$$

Note that the last two integrals diverge, but their difference converges.

$$\begin{aligned}
I &= \frac{1}{2\sqrt{2}}\left[\int_0^\infty \frac{(2x+\sqrt{2})+\sqrt{2}}{x^2+x\sqrt{2}+1}\,dx - \int_0^\infty \frac{(2x-\sqrt{2})-\sqrt{2}}{x^2-x\sqrt{2}+1}\,dx\right] \\
&= \frac{1}{2\sqrt{2}}\left[\ln\left|\frac{x^2+x\sqrt{2}+1}{x^2-x\sqrt{2}+1}\right|_0^\infty + \sqrt{2}\int_0^\infty \frac{dx}{\left(x+\frac{\sqrt{2}}{2}\right)^2+\frac{1}{2}}\right. \\
&\qquad \left. + \sqrt{2}\int_0^\infty \frac{dx}{\left(x-\frac{\sqrt{2}}{2}\right)^2+\frac{1}{2}}\right] \\
&= \frac{1}{2}\left[\sqrt{2}\tan^{-1}\left(x\sqrt{2}+1\right) + \sqrt{2}\tan^{-1}(x\sqrt{2}-1)\right]\Big|_0^\infty \qquad (13) \\
&= \frac{\sqrt{2}}{2}\left(\frac{\pi}{2} - \frac{\pi}{4} + \frac{\pi}{2} + \frac{\pi}{4}\right) \qquad (14)
\end{aligned}$$

and

$$\int_{-\infty}^{+\infty} \frac{dx}{x^4+1} = \frac{\pi}{\sqrt{2}} = 2.221. \tag{15}$$

Solution 4. Using the Cauchy's residue theorem. The contour of integration is shown in Fig.2.12. The poles of the function

$$f(z) = \frac{1}{z^4+1} \tag{16}$$

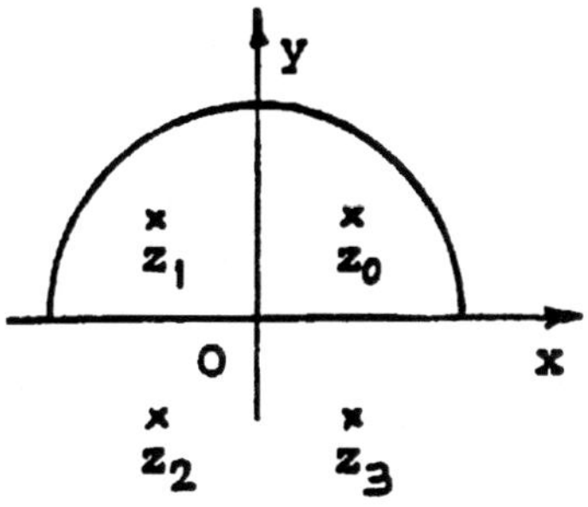

Fig.2.12

above the x axis are

$$z_0 = e^{i\frac{\pi}{4}} = \cos\frac{\pi}{4} + i\sin\frac{\pi}{4} = \frac{1}{\sqrt{2}}(1+i),$$
$$z_1 = e^{i\frac{3\pi}{4}} = \cos\frac{3\pi}{4} + i\sin\frac{3\pi}{4} = \frac{1}{\sqrt{2}}(-1+i)$$

with the residues

$$\mathrm{Res}\,[f(z), z_0] = \lim_{z\to z_0}\frac{z-z_0}{z^4+1} = \lim_{z\to z_0}\frac{1}{4z^3} = \frac{1}{4z_0^3} = \frac{1}{4}e^{-i\frac{3\pi}{4}} = -\frac{1}{4\sqrt{2}}(1+i), \quad (17)$$
$$\mathrm{Res}\,[f(z), z_1] = \lim_{z\to z_1}\frac{z-z_1}{z^4+1} = \lim_{z\to z_1}\frac{1}{4z^3} = \frac{1}{4z_1^3} = \frac{1}{4}e^{-i\frac{9\pi}{4}} = \frac{1}{4\sqrt{2}}(1-i). \quad (18)$$

The integral is

$$I = 2\pi i \cdot \sum_{k=0}^{1} \mathrm{Res}\,[f(z), z_k] \quad (19)$$

or

$$\int_{-\infty}^{+\infty}\frac{dx}{x^4+1} = 2\pi i \cdot \left[\frac{1}{4\sqrt{2}}(1-i-1-i)\right] = \frac{\pi}{\sqrt{2}}. \quad (20)$$

Note. From the above four methods of the evaluation of integral (1), it can be seen that the gamma and beta method is easier to use, and with less possibility of errors. In addition, the generalization can be easily obtained as shown in Sec.2.3.16.

Using all four methods, evaluate the integral substituting x^4 by x^6.

2.3.15. Evaluate the integral

$$I = \int_0^\infty \frac{dx}{1+x^3}. \tag{1}$$

Test for convergence. According to Appendix A1.1, the integral converges for the upper limit since we can write

$$f(x) = \frac{1}{1+x^3} = \underset{x=\infty}{O}\left[\frac{1}{x^3}\right]$$

and hence $\alpha = 3 > 1$.

Integral evaluation

Solution 1. Letting

$$1+x^3 = \frac{1}{t}, \quad 3x^2\,dx = -\frac{dt}{t^2}, \quad dx = -\frac{1}{3}t^{-\frac{4}{3}}(1-t)^{-\frac{2}{3}}\,dt, \tag{2}$$

$$\begin{aligned} I &= \frac{1}{3}\int_0^1 t^{-\frac{1}{3}}(1-t)^{-\frac{2}{3}}\,dt = \frac{1}{3}\int_0^1 t^{\left(1-\frac{1}{3}\right)-1}(1-t)^{\left(1-\frac{2}{3}\right)-1}\,dt \\ &= \frac{1}{3}\int_0^1 t^{\frac{2}{3}-1}(1-t)^{\frac{1}{3}-1}\,dt = \frac{1}{3}B\left(\frac{2}{3},\frac{1}{3}\right) \\ &= \frac{1}{3}\Gamma\left(\frac{1}{3}\right)\Gamma\left(\frac{2}{3}\right) = 1.209. \end{aligned} \tag{3}$$

Using the complement formula (1.35), the integral can be written in another form,

$$\int_0^\infty \frac{dx}{1+x^3} = \frac{1}{3}\frac{\pi}{\sin\frac{\pi}{3}} = \frac{2\pi}{3\sqrt{3}} = 1.209. \tag{4}$$

Solution 2. Let $x^3 = t$, $dx = \frac{1}{3}t^{-\frac{2}{3}}\,dt$, and we have

$$I = \frac{1}{3}\int_0^\infty \frac{t^{-\frac{2}{3}}}{1+t}\,dt$$

and using Sec.2.3.13, $m-1=-2/3$, $m=1/3$,

$$I=\frac{1}{3}B\left(1-\frac{1}{3},\frac{1}{3}\right)=\frac{1}{3}B\left(\frac{2}{3},\frac{1}{3}\right)=\frac{1}{3}\frac{\pi}{\sin\frac{\pi}{3}}.$$

See also Sec.2.3.16, for $n=3$.

Exercise. Evaluate the integral using the method of partial fractions and the Cauchy's residue theorem.

2.3.16. Study the improper integral

$$I=\int_0^\infty \frac{dx}{1+x^n}, \quad n\in R. \tag{1}$$

Solution 1. Let

$$1+x^n=\frac{1}{t}, \quad dx=-\frac{1}{n}t^{-\frac{1}{n}-1}(1-t)^{\frac{1}{n}-1}\,dt \tag{2}$$

and the integral becomes

$$I=\frac{1}{n}\int_0^1 t^{-\frac{1}{n}}(1-t)^{\frac{1}{n}-1}\,dt=\frac{1}{n}\int_0^1 t^{\left(1-\frac{1}{n}\right)-1}(1-t)^{\frac{1}{n}-1}\,dt \tag{3}$$

$$=\frac{1}{n}B\left(1-\frac{1}{n},\frac{1}{n}\right)=\frac{1}{n}\Gamma\left(1-\frac{1}{n}\right)\cdot\Gamma\left(\frac{1}{n}\right)=\frac{\pi}{n\sin\frac{\pi}{n}}, \quad n>1. \tag{4}$$

Solution 2. Let

$$x^n=t, \quad dx=\frac{1}{n}t^{\frac{1}{n}-1}\,dt$$

and the integral (1) becomes

$$I=\frac{1}{n}\int_0^\infty \frac{t^{\frac{1}{n}-1}}{1+t}\,dt$$

and using Sec.2.3.13, we have

$$\int_0^\infty \frac{dx}{1+x^n}=\frac{1}{n}B\left(1-\frac{1}{n},\frac{1}{n}\right), \quad n>1.$$

Particular cases:

a. Let $n = 2$ and we have

$$\int_0^{\infty} \frac{dx}{1+x^2} = \frac{1}{2} B\left(1 - \frac{1}{2}, \frac{1}{2}\right)$$
$$= \frac{1}{2} B\left(\frac{1}{2}, \frac{1}{2}\right) = \frac{1}{2}\left[\Gamma\left(\frac{1}{2}\right)\right]^2 = \frac{\pi}{2}.$$

See also Sec.2.3.1a.

b. Let $n = 4/3$ and from (4) we have

$$\int_0^{\infty} \frac{dx}{1+x^{\frac{4}{3}}} = \frac{3}{4}\Gamma\left(1 - \frac{3}{4}\right)\Gamma\left(\frac{3}{4}\right) = \frac{3\pi}{4\sin\frac{3\pi}{4}} = 3.332.$$

c. Let $n = 3$ and from (4) we have

$$\int_0^{\infty} \frac{dx}{1+x^3} = \frac{1}{3}\Gamma\left(1 - \frac{1}{3}\right)\Gamma\left(\frac{1}{3}\right) = \frac{\pi}{3\sin\frac{\pi}{3}} = \frac{2\pi}{3\sqrt{3}}.$$

d. Let $n = 4$ and from (4) we have

$$\int_0^{\infty} \frac{dx}{1+x^4} = \frac{\pi}{4\sin\frac{\pi}{4}} = \frac{\pi\sqrt{2}}{4}.$$

See also Sec.2.3.14.

2.3.17. Study the improper integral

$$I = \int_{-\infty}^{+\infty} \frac{dx}{(1+x^2)^p}, \quad p \in R. \tag{1}$$

Solution 1. Since x^2 is an even power, the integral can be written

$$I = 2\int_0^{\infty} \frac{dx}{(1+x^2)^p} \tag{2}$$

and using the substitution

$$1+x^2=\frac{1}{t}, \quad dx=-\frac{1}{2}t^{-\frac{3}{2}}(1-t)^{-\frac{1}{2}}\,dt,$$

$$I=\int_0^1 t^{p-\frac{3}{2}}(1-t)^{-\frac{1}{2}}\,dt=\int_0^1 t^{\left(1+p-\frac{3}{2}\right)-1}(1-t)^{\left(1-\frac{1}{2}\right)-1}\,dt$$

$$=\int_0^1 t^{\left(p-\frac{1}{2}\right)-1}(1-t)^{\frac{1}{2}-1}\,dt=B\left(p-\frac{1}{2},\frac{1}{2}\right), \quad p>\frac{1}{2}. \tag{4}$$

Solution 2. Let the substitution $x^2=t$, $x=t^{\frac{1}{2}}$, $dx=\dfrac{dt}{2t^{\frac{1}{2}}}$ and with Sec.2.3.28, we have

$$I=\int_0^\infty \frac{t^{-\frac{1}{2}}}{(1+t)^p}\,dt=\int_0^\infty \frac{t^{\left(1-\frac{1}{2}\right)-1}}{(1+t)^p}\,dt$$

$$=\int_0^\infty \frac{t^{\frac{1}{2}-1}}{(1+t)^p}\,dt=B\left(p-\frac{1}{2},\frac{1}{2}\right), \quad p>\frac{1}{2}.$$

Particular cases:

a. Let $p=n\in N$, and the integral (1) becomes

$$I_n=\int_{-\infty}^{+\infty}\frac{dx}{(1+x^2)^n}=B\left(n-\frac{1}{2},\frac{1}{2}\right) \tag{5}$$

$$=\frac{\sqrt{\pi}}{(n-1)!}\Gamma\left(n-\frac{1}{2}\right), \quad n\geq 1. \tag{6}$$

Using Eq.(1.43), we have

$$\Gamma\left(n-\frac{1}{2}\right)=\Gamma\left[(n-1)+\frac{1}{2}\right]=\frac{\sqrt{\pi}}{2^{n-1}}[2(n-1)-1]!!$$

$$=\frac{\sqrt{\pi}}{2^{n-1}}(2n-3)!! \tag{7}$$

and (6) becomes

$$\int_{-\infty}^{+\infty}\frac{dx}{(1+x^2)^n}=\frac{\pi}{2^{n-1}}\frac{(2n-3)!!}{(n-1)!}, \quad n\geq 2. \tag{8}$$

But $2^{n-1} \cdot (n-1)! = (2n-2)!!$ and (8) can also be written

$$\int_{-\infty}^{+\infty} \frac{dx}{(1+x^2)^n} = \frac{(2n-3)!!}{(2n-2)!!}\pi, \quad n \geq 2. \tag{9}$$

See also Sec.2.3.22 for $p = 2$ and $q = n$.

b. Let $n = 1$, and from (5),

$$\int_{-\infty}^{+\infty} \frac{dx}{1+x^2} = B\left(\frac{1}{2}, \frac{1}{2}\right) = \pi. \tag{10}$$

c. Let $n = 2$, and from (9),

$$\int_{-\infty}^{+\infty} \frac{dx}{(1+x^2)^2} = \frac{\pi}{2}. \tag{11}$$

d. Let $n = 3$, and using (9),

$$\int_{-\infty}^{+\infty} \frac{dx}{(1+x^2)^3} = \frac{3\pi}{8}. \tag{12}$$

2.3.18. Study the integral

$$I = \int_{-\infty}^{+\infty} \frac{dx}{(a^2+x^2)^m}, \quad a > 0, \quad m \in R. \tag{1}$$

Solution. Let

$$x = at, \quad dx = a\,dt \tag{2}$$

and the integral becomes

$$I = \frac{2}{a^{2m-1}} \int_0^{\infty} \frac{dt}{(1+t^2)^m} \tag{3}$$

and using Sec.2.3.17, we have

$$\int_{-\infty}^{+\infty} \frac{dx}{(a^2+x^2)^m} = \frac{1}{a^{2m-1}} B\left(m - \frac{1}{2}, \frac{1}{2}\right), \quad m > \frac{1}{2}, \quad a > 0. \tag{4}$$

2.3.19. Study the improper integral

$$I = \int_{-\infty}^{+\infty} \frac{x^{2p}\,dx}{1+x^{2q}}, \quad q>0, \quad p \in r. \tag{1}$$

Solution 1. Since the integrand is an even function, the integral can be written

$$I = 2\int_{0}^{\infty} \frac{x^{2p}\,dx}{1+x^{2q}}. \tag{2}$$

Using the substitution

$$1+x^{2q} = \frac{1}{t}, \quad dx = -\frac{1}{2q}(1-t)^{\frac{1}{2q}-1}t^{-\frac{1}{2q}-1}\,dt \tag{3}$$

the integral becomes

$$\begin{aligned} I &= \frac{1}{q}\int_0^1 t^{-\frac{2p+1}{2q}}(1-t)^{\frac{2p+1}{2q}-1}\,dt \\ &= \frac{1}{q}\int_0^1 t^{\left(1-\frac{2p+1}{2q}\right)-1}(1-t)^{\frac{2p+1}{2q}-1}\,dt \\ &= \frac{1}{q}B\left(1-\frac{2p+1}{2q}, \frac{2p+1}{2q}\right) && (4) \\ &= \frac{1}{q}\Gamma\left(1-\frac{2p+1}{2q}\right)\cdot\Gamma\left(\frac{2p+1}{2q}\right) && (5) \\ &= \frac{\pi}{q\sin\dfrac{2p+1}{2q}\pi}, \quad 0<2p+1<2q. && (6) \end{aligned}$$

Solution 2. Let

$$x^{2q} = t, \quad x = t^{\frac{1}{2q}}, \quad 2qx^{2q-1}\,dx = dt,$$

$$dx = \frac{1}{2q}\frac{dt}{x^{2q-1}} = \frac{1}{2q}\frac{dt}{(t^{\frac{1}{2q}})^{2q-1}}$$

and the integral (2) becomes

$$I = \frac{1}{q}\int_0^{\infty} \frac{t^{\frac{2p+1}{2q}-1}}{1+t}\,dt$$

and using Sec.2.3.13, we have

$$I = \frac{1}{q} B\left(1 - \frac{2p+1}{2q}, \frac{2p+1}{2q}\right).$$

Particular cases:

a. Let $p = 1/2$, $q = 2$ and the integral becomes

$$\int_{-\infty}^{+\infty} \frac{x\,dx}{1+x^4} = \frac{\pi}{2\sin\frac{\pi}{2}} = \frac{\pi}{2}. \tag{7}$$

b. Let $p = 2$, $q = 3$, and the integral becomes

$$\int_{-\infty}^{+\infty} \frac{x^4\,dx}{1+x^6} = \frac{\pi}{3\sin\frac{5\pi}{6}} = \frac{2\pi}{3}. \tag{8}$$

c. Let $p = 0$ and $q = n \in N$, and the integral becomes

$$\int_{-\infty}^{+\infty} \frac{dx}{1+x^{2n}} = \frac{1}{n} B\left(1 - \frac{1}{2n}, \frac{1}{2n}\right)$$

$$= \frac{1}{n}\Gamma\left(1 - \frac{1}{2n}\right)\Gamma\left(\frac{1}{2n}\right), \quad n \in N, \tag{9}$$

and using the complement formula of the gamma function Eq.(1.35), (9) can be written

$$\int_{-\infty}^{+\infty} \frac{dx}{1+x^{2n}} = \frac{\pi}{n\sin\frac{\pi}{2n}}, \quad n \in N. \tag{10}$$

For $n = 2$, we have

$$\int_{-\infty}^{+\infty} \frac{dx}{1+x^4} = \frac{\pi}{2\sin\frac{\pi}{4}} = \frac{\pi}{\sqrt{2}}, \tag{11}$$

a result found also in Sec.2.3.14.

Exercise. Evaluate the integral at b) using the Cauchy's residual theorem and any other traditional methods.

2.3.20. Study the integral

$$I = \int_0^\infty \frac{x^{m-1}}{1+x^n}\,dx, \quad n > 0, \quad m \in R. \tag{1}$$

Solution 1. Letting

$$1 + x^n = \frac{1}{t}, \quad dx = -\frac{1}{n} t^{-\frac{1}{n}-1}(1-t)^{\frac{1}{n}-1}\,dt, \tag{2}$$

the integral becomes

$$\begin{aligned} I &= \frac{1}{n}\int_0^1 \left(\frac{1-t}{t}\right)^{\frac{m-1}{n}} t \cdot t^{-\frac{1}{n}-1}(1-t)^{\frac{1}{n}-1}\,dt \\ &= \frac{1}{n}\int_0^1 t^{-\frac{m}{n}}(1-t)^{\frac{m}{n}-1}\,dt \\ &= \frac{1}{n}\int_0^1 t^{(1-\frac{m}{n})-1}(1-t)^{\frac{m}{n}-1}\,dt \\ &= \frac{1}{n}B\left(1-\frac{m}{n}, \frac{m}{n}\right) = \frac{1}{n}\Gamma\left(1-\frac{m}{n}\right)\Gamma\left(\frac{m}{n}\right) \end{aligned} \tag{3}$$

and using the complement formula of the gamma function, Eq.(1.35),

$$\int_0^\infty \frac{x^{m-1}}{1+x^n}\,dx = \frac{\pi}{n \sin \dfrac{m\pi}{n}} \tag{4}$$

and from (3), the convergence conditions are $1 - \frac{m}{n} > 0$, $\frac{m}{n} > 0$ or $0 < m < n$.

Solution 2. Let $x^n = t$ and use Sec.2.3.13,

$$I = \frac{1}{n}\int_0^\infty \frac{t^{\frac{m}{n}-1}\,dt}{1+t} = \frac{1}{n}B\left(1-\frac{m}{n}, \frac{m}{n}\right) = \frac{\pi}{n \sin \dfrac{m\pi}{n}}.$$

Particular cases:

a. Let $m = 1$, $n = 4$ and the integral (1) becomes

$$\int_0^\infty \frac{dx}{1+x^4} = \frac{\pi}{4 \sin \dfrac{\pi}{4}} = \frac{\pi}{2\sqrt{2}},$$

a result also found in Sec.2.3.14.

b. Let $m = 5$, $n = 6$ and the integral (1) becomes

$$\int_0^\infty \frac{x^4\,dx}{1+x^6} = \frac{\pi}{6\sin\dfrac{5\pi}{6}} = \frac{\pi}{3}. \tag{5}$$

See also Sec.2.3.19.

c. Let $m = 1$ and the integral (1) becomes

$$\int_0^\infty \frac{dx}{1+x^n} = \frac{1}{n}B\left(1-\frac{1}{n}, \frac{1}{n}\right) = \frac{\pi}{n\sin\dfrac{\pi}{n}}, \qquad n > 1. \tag{6}$$

d. Let $n = 1$, and using the result from (4), we have

$$\int_0^\infty \frac{x^{m-1}}{1+x}\,dx = B(1-m, m) = \frac{\pi}{\sin m\pi}, \qquad 0 < m < 1. \tag{7}$$

2.3.21. Study the improper integral

$$I = \int_0^\infty \frac{x^m\,dx}{a^2+x^{2n}}, \qquad a > 0, \quad n > 0, \quad m \in R. \tag{1}$$

Solution 1. Using the transformation

$$a^2 + x^{2n} = \frac{a^2}{u}, \quad x = a^{\frac{1}{n}} u^{-\frac{1}{2n}}(1-u)^{\frac{1}{2n}}, \quad dx = -\frac{a^{\frac{1}{n}}}{2n} u^{-\frac{1}{2n}-1}(1-u)^{\frac{1}{2n}-1}\,du \tag{2}$$

the integral becomes

$$I = \frac{a^{\frac{m+1}{n}-2}}{2n} \int_0^1 u^{-\frac{m+1}{2n}} (1-u)^{\frac{m+1}{2n}-1}\, du \tag{3}$$

$$= \frac{a^{\frac{m+1}{n}-2}}{2n} \int_0^1 u^{\left(1-\frac{m+1}{2n}\right)-1} (1-u)^{\frac{m+1}{2n}-1}\, du \tag{4}$$

$$= \frac{a^{\frac{m+1}{n}-2}}{2n} B\left(1-\frac{m+1}{2n}, \frac{m+1}{2n}\right) = \frac{a^{\frac{m+1}{n}-2}}{2n} \cdot \frac{\pi}{\sin \frac{m+1}{2n}\pi}, \tag{5}$$

$0 < m+1 < 2n.$

Solution 2. Let

$$x^{2n} = a^2 t, \quad x = (a^2 t)^{\frac{1}{2n}}, \quad 2n x^{2n-1}\, dx = a^2\, dt, \quad dx = \frac{a^2\, dt}{2n(a^2 t)^{\frac{2n-1}{2n}}} \tag{6}$$

and the integral becomes

$$I = \frac{a^{\frac{m-2n+1}{n}}}{2n} \int_0^\infty \frac{t^{\frac{m+1}{2n}-1}}{1+t}\, dt = \frac{a^{\frac{m+1}{n}-2}}{2n} \int_0^\infty \frac{t^{\frac{m+1}{2n}-1}}{1+t}\, dt$$

and with Sec.2.3.13,

$$I = \frac{a^{\frac{m+1}{n}-2}}{2n} B\left(1-\frac{m+1}{2n}, \frac{m+1}{2n}\right), \quad 0 < m+1 < 2n. \tag{7}$$

Particular cases:

a. Let $m = n$,

$$\int_0^\infty \frac{x^n\, dx}{a^2 + x^{2n}} = \frac{a^{\frac{1}{n}-1}}{2n} B\left(1-\frac{n+1}{2n}, \frac{n+1}{2n}\right) = \frac{a^{\frac{1}{n}-1}}{2n} \cdot \frac{\pi}{\sin \frac{n+1}{2n}\pi} \tag{8}$$

and the integral converges for $0 < n+1 < 2n$.

b. Let $m = p - 1$, $n = 1$ and we have

$$\int_0^\infty \frac{x^{p-1}\,dx}{a^2+x^2} = \frac{a^{p-2}}{2}\Gamma\left(1-\frac{p}{2}\right)\Gamma\left(\frac{p}{2}\right) = \frac{a^{p-2}}{2}\cdot\frac{\pi}{\sin\frac{p\pi}{2}}, \quad 0<p<2, \quad a>0. \tag{9}$$

c. Let $a = 1$, $m = 0$, $n = 2$ and from (5) we have

$$\int_0^\infty \frac{dx}{1+x^4} = \frac{1}{4}\frac{\pi}{\sin\frac{\pi}{4}} = \frac{\pi}{2\sqrt{2}}, \tag{10}$$

a result also found in Sec.2.3.14.

2.3.22. Study the integral

$$I = \int_0^\infty \frac{dx}{(1+x^p)^q}, \quad p, q \in R. \tag{1}$$

Solution 1. Letting

$$1+x^p = \frac{1}{t}, \quad x = \left(\frac{1-t}{t}\right)^{\frac{1}{p}}, \quad dx = -\frac{1}{p}t^{-\left(\frac{1}{p}+1\right)}(1-t)^{\frac{1}{p}-1}\,dt \tag{2}$$

the integral becomes

$$\begin{aligned} I &= \frac{1}{p}\int_0^1 t^{q-\frac{p+1}{p}}(1-t)^{\frac{1}{p}-1}\,dt \\ &= \frac{1}{p}\int_0^1 t^{\left(q-\frac{1}{p}\right)-1}(1-t)^{\frac{1}{p}-1}\,dt = \frac{1}{p}B\left(q-\frac{1}{p},\frac{1}{p}\right) \end{aligned} \tag{3}$$

and the integral converges for $p>0$, $pq>1$.

Solution 2. Let

$$x^p = t, \quad px^{p-1}\,dx = dt, \quad dx = \frac{dt}{pt^{\frac{p-1}{p}}}, \quad x = t^{\frac{1}{p}}$$

and the integral (1) becomes

$$I = \frac{1}{p}\int_0^\infty \frac{t^{\frac{1}{p}-1}\,dt}{(1+t)^q}$$

and using Eq.1.62, we have

$$\int_0^\infty \frac{dx}{(1+x^p)^q} = \frac{1}{p} B\left(q - \frac{1}{p}, \frac{1}{p}\right).$$

Particular cases:

a. Let $p = 3$, $q = 1$ and the integral becomes

$$\int_0^\infty \frac{dx}{1+x^3} = \frac{1}{3} B\left(1 - \frac{1}{3}, \frac{1}{3}\right) = \frac{1}{3} B\left(\frac{2}{3}, \frac{1}{3}\right)$$
$$= \frac{1}{3}\Gamma\left(\frac{2}{3}\right)\Gamma\left(\frac{1}{3}\right) = \frac{\pi}{3\sin\frac{\pi}{3}} = \frac{2\pi}{3\sqrt{3}} = 1.209, \tag{4}$$

a result also found in Sec.2.3.15.

b. Let $p = 4$, $q = 1/2$ and the integral becomes

$$I = \frac{1}{4} B\left(\frac{1}{2} - \frac{1}{4}, \frac{1}{4}\right) = \frac{1}{4} B\left(\frac{1}{4}, \frac{1}{4}\right)$$

or

$$\int_0^\infty \frac{dx}{\sqrt{1+x^4}} = \frac{1}{4\sqrt{\pi}}\left[\Gamma\left(\frac{1}{4}\right)\right]^2 = K\left(\frac{1}{\sqrt{2}}\right) = 1.854, \tag{5}$$

a result also found using elliptic integrals, Sec.4.2.19.

c. Let $q = 1/2$ and the integral becomes

$$\int_0^\infty \frac{dx}{\sqrt{1+x^p}} = \frac{1}{p} B\left(\frac{1}{2} - \frac{1}{p}, \frac{1}{p}\right), \quad p > 2, \tag{6}$$

a result also found in Sec.2.3.2.

d. Let $p = 3$, $q = 2$,

$$\int_0^\infty \frac{dx}{(1+x^3)^2} = \frac{1}{3}B\left(2-\frac{1}{3}, \frac{1}{3}\right) = \frac{1}{3}B\left(\frac{5}{3}, \frac{1}{3}\right)$$
$$= \frac{1}{3}\Gamma\left(\frac{5}{3}\right)\Gamma\left(\frac{1}{3}\right) = \frac{1}{3}\Gamma\left(1+\frac{2}{3}\right)\Gamma\left(\frac{1}{3}\right)$$
$$= \frac{2}{9}\Gamma\left(\frac{2}{3}\right)\Gamma\left(\frac{1}{3}\right) = \frac{2\pi}{9\sin\frac{\pi}{3}} = \frac{4\pi}{9\sqrt{3}} = 0.806. \quad (7)$$

Exercise. Evaluate the integral at d., using a traditional method.

2.3.23. Evaluate the integral

$$I = \int_0^\infty \frac{dx}{\sqrt[3]{x}(1+x)^3}.$$

Test for convergence. According to Appendix A1.2, the integral converges at the lower limit, because we can write

$$\frac{1}{\sqrt[3]{x}(1+x)^3} = \underset{x\to 0}{O}\left[\frac{1}{x^{\frac{1}{3}}}\right]$$

and $\alpha = 1/3 < 1$. Also, according to Appendix A1.1, the integral converges at the upper limit, because we can write

$$\frac{1}{\sqrt[3]{x}(1+x)^3} = \underset{x\to 0}{O}\left[\frac{1}{x^{\frac{10}{3}}}\right]$$

and $\alpha = 10/3 > 1$.

Integral evaluation

Solution 1. Let $1+x=\frac{1}{t}$, $x=\frac{1-t}{t}$, $dx=-\frac{dt}{t^2}$,

$$\begin{aligned}I&=\int_0^1 t^{\frac{4}{3}}(1-t)^{-\frac{1}{3}}\,dt=\int_0^1 t^{\left(1+\frac{4}{3}\right)-1}(1-t)^{\left(1-\frac{1}{3}\right)-1}\,dt\\&=\int_0^1 t^{\frac{7}{3}-1}(1-t)^{\frac{2}{3}-1}\,dt=B\left(\frac{7}{3},\frac{2}{3}\right)\\&=\frac{2}{9}B\left(\frac{1}{3},\frac{2}{3}\right)=\frac{2}{9}\Gamma\left(\frac{1}{3}\right)\Gamma\left(\frac{2}{3}\right)\\&=\frac{2}{9}\frac{\pi}{\sin\frac{\pi}{3}}=\frac{4\pi}{9\sqrt{3}}.\end{aligned}$$

Solution 2. We have

$$I=\int_0^\infty\frac{x^{-\frac{1}{3}}dx}{(1+x)^3}$$

and using Eq.(1.63) with $p-1=-1/3$, $p+q=3$ from where $p=2/3$ and $q=7/3$, yields

$$\int_0^\infty\frac{dx}{\sqrt[3]{x}(1+x)^3}=B\left(\frac{7}{3},\frac{2}{3}\right).$$

2.3.24. Evaluate the improper integral

$$I=\int_0^\infty\frac{\sqrt[3]{x}\,dx}{(x^2+a^2)^2},\qquad a\in R^+. \tag{1}$$

Test for convergence. According to Appendix A1.1, the integral converges at the upper limit because we can write

$$f(x)=\frac{\sqrt[3]{x}}{(x^2+a^2)^2}=\underset{x=\infty}{O}\left[\frac{\sqrt[3]{x}}{x^4}\right]=\underset{x=\infty}{O}\left[\frac{1}{x^{\frac{11}{3}}}\right] \tag{2}$$

and $\alpha=11/3>1$.

Integral evaluation.

Solution 1. Let $x = at$, $dx = a\,dt$, and the integral becomes

$$I = \frac{1}{a^{\frac{8}{3}}} \int_0^\infty \frac{t^{\frac{1}{3}}\,dt}{(t^2+1)^2}. \tag{3}$$

Letting

$$t^2 + 1 = \frac{1}{u}, \quad t = u^{-\frac{1}{2}}(1-u)^{\frac{1}{2}}, \quad dt = -\frac{1}{2}u^{-\frac{3}{2}}(1-u)^{-\frac{1}{2}}\,du \tag{4}$$

the integral becomes

$$\begin{aligned} I &= \frac{1}{2a^{\frac{8}{3}}} \int_0^1 u^{\frac{1}{3}}(1-u)^{-\frac{1}{3}}\,du = \frac{1}{2a^{\frac{8}{3}}} \int_0^1 u^{\left(1+\frac{1}{3}\right)-1}(1-u)^{\left(1-\frac{1}{3}\right)-1}\,du \\ &= \frac{1}{2a^{\frac{8}{3}}} \int_0^1 u^{\frac{4}{3}-1}(1-u)^{\frac{2}{3}-1}\,du = \frac{1}{2a^{\frac{8}{3}}} B\left(\frac{4}{3}, \frac{2}{3}\right) = \frac{1}{6a^{\frac{8}{3}}} B\left(\frac{1}{3}, \frac{2}{3}\right) \quad (5) \\ &= \frac{1}{6a^{\frac{8}{3}}} \Gamma\left(\frac{1}{3}\right)\Gamma\left(\frac{2}{3}\right) = \frac{0.6046}{a^{\frac{8}{3}}} \quad (6) \end{aligned}$$

or, using the complement formula of the gamma function Eq.(1.35),

$$\int_0^\infty \frac{\sqrt[3]{x}\,dx}{(x^2+a^2)^2} = \frac{1}{6a^{\frac{8}{3}}} \Gamma\left(\frac{1}{3}\right)\Gamma\left(1-\frac{1}{3}\right) = \frac{1}{6a^{\frac{8}{3}}} \frac{\pi}{\sin\frac{\pi}{3}} = \frac{\pi}{3\sqrt{3}a^{\frac{8}{3}}}. \tag{7}$$

Solution 2. Using Eq.(1.65) with $2q - 1 = 1/3$ and $p + q = 2$ or $q = 2/3$ and $p = 4/3$, the integral (3) becomes

$$I = \frac{1}{a^{\frac{8}{3}}} \int_0^\infty \frac{t^{\frac{1}{3}}\,dt}{(t^2+1)^2} = \frac{1}{2a^{\frac{8}{3}}} B\left(\frac{4}{3}, \frac{2}{3}\right).$$

2.3.25. Study the integral

$$I = \int_0^\infty \frac{x^{p-1}\,dx}{(1+x^q)^2}, \quad p \in R, \quad q > 0. \tag{1}$$

Solution. With the substitution

$$1 + x^q = \frac{1}{t}, \quad x = \left(\frac{1-t}{t}\right)^{\frac{1}{q}}, \quad dx = -\frac{1}{q} t^{-\frac{1}{q}-1}(1-t)^{\frac{1}{q}-1}\,dt \tag{2}$$

we have

$$\begin{aligned}
I &= \frac{1}{q}\int_0^1 \frac{\left(\frac{1-t}{t}\right)^{\frac{p-1}{q}}}{\frac{1}{t^2}} t^{-\frac{1}{q}-1}(1-t)^{\frac{1}{q}-1}\,dt \\
&= \frac{1}{q}\int_0^1 \frac{(1-t)^{\frac{p-1}{q}}}{t^{\frac{p-1}{q}}} t^{-\frac{1}{q}-1+2}(1-t)^{\frac{1}{q}-1}\,dt \\
&= \frac{1}{q}\int_0^1 (1-t)^{\frac{p-1}{q}+\frac{1}{q}-1} t^{-\frac{1}{q}+1} t^{\frac{1-p}{q}}\,dt \\
&= \frac{1}{q}\int_0^1 t^{1-\frac{p}{q}}(1-t)^{\frac{p}{q}-1}\,dt \\
&= \frac{1}{q}\int_0^1 t^{1+\left(1-\frac{p}{q}\right)-1}(1-t)^{\frac{p}{q}-1}\,dt \\
&= \frac{1}{q}\int_0^1 t^{\left(2-\frac{p}{q}\right)-1}(1-t)^{\frac{p}{q}-1}\,dt \\
&= \frac{1}{q}B\left(2-\frac{p}{q}, \frac{p}{q}\right) = \frac{q-p}{q^2}B\left(1-\frac{p}{q}, \frac{p}{q}\right) = \frac{q-p}{q^2}\frac{\pi}{\sin\left(\frac{p}{q}\pi\right)}
\end{aligned} \tag{3}$$

and the integral converges for $0 < p < q$.

Particular cases:

a. Let $p = 1$, $q = 2$, and the integral becomes

$$\int_0^\infty \frac{dx}{(1+x^2)^2} = \frac{1}{4}B\left(\frac{1}{2}, \frac{1}{2}\right) = \frac{\pi}{4}. \tag{4}$$

b. Let $p = 4/3$, $q = 2$, and the integral becomes

$$\int_0^\infty \frac{\sqrt[3]{x}\,dx}{(1+x^2)^2} = \frac{1}{2}B\left(\frac{4}{3}, \frac{2}{3}\right) = \frac{1}{6}B\left(\frac{1}{3}, \frac{2}{3}\right) = \frac{\pi}{3\sqrt{3}}, \tag{5}$$

a result found also in Sec.2.3.24, for $a = 1$.

c. Let $p - 1 = m$, $q = 2$ and (3) becomes

$$\int_0^\infty \frac{x^m\,dx}{(1+x^2)^2} = \frac{1}{2}B\left(2 - \frac{1+m}{2}, \frac{1+m}{2}\right), \quad -1 < m < 3. \tag{6}$$

d. Let $p = 2$, $q = 3$, and from (3),

$$\int_0^\infty \frac{x\,dx}{(1+x^3)^2}\,dx = \frac{1}{9}B\left(\frac{2}{3}, \frac{1}{3}\right) = \frac{\pi}{9\sin\frac{\pi}{3}} = \frac{2\pi}{9\sqrt{3}}. \tag{7}$$

2.3.26. Evaluate the improper integral

$$I = \int_0^\infty \frac{\sqrt[4]{x^3}\,dx}{x^4+1}. \tag{1}$$

Test for convergence. According to Appendix A1.1, the integral converges at the upper limit, since we can write

$$f(x) = \frac{\sqrt[4]{x^3}}{x^4+1} = \underset{x=\infty}{O}\left[\frac{1}{x^{\frac{13}{4}}}\right] \tag{2}$$

and hence, $\alpha = 13/4 > 1$.

Integral evaluation.

Solution 1. See Fig.2.13. Letting

$$x^4 + 1 = \frac{1}{t}, \quad dx = -\frac{1}{4}t^{-\frac{5}{4}}(1-t)^{-\frac{3}{4}}\,dt \tag{3}$$

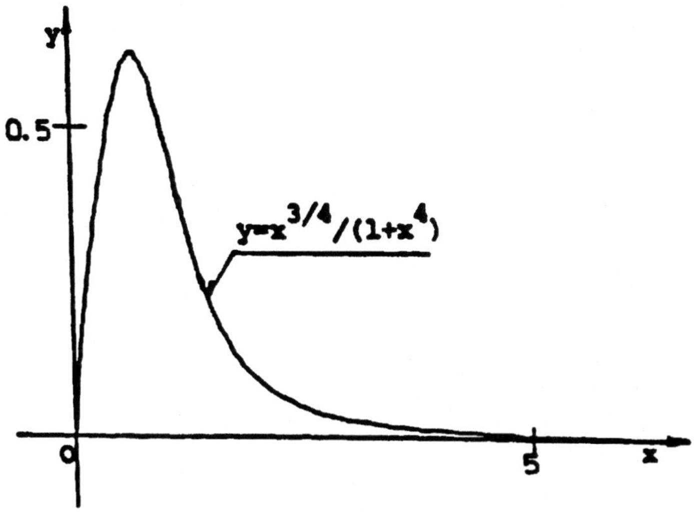

Fig.2.13

we have

$$\begin{aligned} I &= \frac{1}{4}\int_0^1 t^{-\frac{7}{16}}(1-t)^{-\frac{9}{16}}\,dt \\ &= \frac{1}{4}\int_0^1 t^{\left(1-\frac{7}{16}\right)-1}(1-t)^{\left(1-\frac{9}{16}\right)-1}\,dt = \frac{1}{4}\int_0^1 t^{\frac{9}{16}-1}(1-t)^{\frac{7}{16}-1}\,dt \\ &= \frac{1}{4}B\left(\frac{9}{16},\frac{7}{16}\right) = \frac{1}{4}\Gamma\left(\frac{9}{16}\right)\Gamma\left(\frac{7}{16}\right) = 0.801. \end{aligned} \tag{4}$$

Using the complement formula (1.35) the integral can be written in another form,

$$\int_0^\infty \frac{\sqrt[4]{x^3}\,dx}{x^4+1} = \frac{\pi}{4\sin\dfrac{7\pi}{16}} = 0.801. \tag{5}$$

Solution 2. Let

$$x^4 = t, \quad 4x^3\,dx = dt, \quad dx = \frac{dt}{4t^{\frac{3}{4}}}$$

and the integral (1) becomes

$$I = \frac{1}{4}\int_0^\infty \frac{t^{-\frac{9}{16}}\,dt}{t+1} = \frac{1}{4}\int_0^\infty \frac{t^{\left(1-\frac{9}{16}\right)-1}}{t+1}\,dt = \frac{1}{4}\int_0^\infty \frac{t^{\frac{7}{16}-1}}{t+1}\,dt$$

and using Sec.2.3.13, we have

$$I = \frac{1}{4}B\left(1-\frac{7}{16}, \frac{7}{16}\right) = \frac{\pi}{4\sin\dfrac{7\pi}{16}}.$$

Exercise. Evaluate the integral using traditional methods.

2.3.27. Evaluate the integral

$$I = \int_0^\infty \frac{dx}{\sqrt[3]{x^2}(1+\sqrt{x})}. \tag{1}$$

Test for convergence. According to Appendix A1.2, the integral converges at $x = 0+$, since we can write

$$f(x) = \frac{1}{\sqrt[3]{x^2}(1+\sqrt{x})} \underset{x=0+}{=} O\left[\frac{1}{x^{\frac{2}{3}}}\right] \tag{2}$$

and hence, $\alpha = 2/3 < 1$.

Also, according to Appendix A1.1, the integral converges at the upper limit, since we can write

$$f(x) = \frac{1}{\sqrt[3]{x^2}(1+\sqrt{x})} \underset{x=\infty}{=} O\left[\frac{1}{x^{\frac{7}{6}}}\right] \tag{3}$$

and hence, $\alpha = 7/6 > 1$.

Integral evaluation.

Solution 1. With the substitution

$$1+\sqrt{x}=\frac{1}{t}, \quad x=\left(\frac{1-t}{t}\right)^2, \quad dx=-2(1-t)t^{-3}\,dt \tag{4}$$

the integral becomes

$$\begin{aligned} I &= 2\int_0^1 t^{-\frac{2}{3}}(1-t)^{-\frac{1}{3}}\,dt = 2\int_0^1 t^{\left(1-\frac{2}{3}\right)-1}(1-t)^{\left(1-\frac{1}{3}\right)-1}\,dt \\ &= 2\int_0^1 t^{\frac{1}{3}-1}(1-t)^{\frac{2}{3}-1}\,dt = 2B\left(\frac{1}{3},\frac{2}{3}\right) = 2\Gamma\left(\frac{1}{3}\right)\Gamma\left(\frac{2}{3}\right) = 7.255. \end{aligned} \tag{5}$$

Also, using the complement formula (1.35), the integral can be written in another form,

$$\int_0^\infty \frac{dx}{\sqrt[3]{x^2}(1+\sqrt{x})} = 2\frac{\pi}{\sin\frac{\pi}{3}} = \frac{4\pi}{\sqrt{3}} = 7.255. \tag{6}$$

Solution 2. Let $x=t^2$, $dx=2t\,dt$,

$$I = 2\int_0^\infty \frac{t\,dt}{\sqrt[3]{t^4}(1+t)} = 2\int_0^\infty \frac{t^{-\frac{1}{3}}\,dt}{1+t} = 2\int_0^\infty \frac{t^{\left(1-\frac{1}{3}\right)-1}\,dt}{1+t}$$

and using Sec.2.3.13,

$$I = 2\int_0^\infty \frac{t^{\frac{2}{3}-1}}{1+t}\,dt = 2B\left(1-\frac{2}{3},\frac{2}{3}\right) = \frac{2\pi}{\sin\frac{2\pi}{3}} = \frac{4\pi}{\sqrt{3}}. \tag{7}$$

Solution 3. Utilizing Sec.2.3.28 for $p=1/3$, $q=1/2$ and $m=1$.

2.3.28. Study the integral

$$I = \int_0^\infty \frac{x^{p-1}\,dx}{(1+x^q)^m}, \quad q>0. \tag{1}$$

Solution 1. Let $1+x^q=1/t$, $qx^{q-1}\,dx=-dt/t^2$, $x^q=\dfrac{1}{t}-1=\dfrac{1-t}{t}$,

$$x=\left(\frac{1-t}{t}\right)^{\frac{1}{q}},\qquad dx=-\frac{dt}{qx^{q-1}t^2}=-\frac{1}{q}t^{-\frac{1}{q}-1}(1-t)^{\frac{1}{q}-1}\,dt$$

and we have

$$I=\frac{1}{q}\int_0^1 t^{\left(m-\frac{p}{q}\right)-1}(1-t)^{\frac{p}{q}-1}\,dt \tag{2}$$

$$=\frac{1}{q}B\left(m-\frac{p}{q},\frac{p}{q}\right)=\frac{1}{q}\frac{\Gamma\left(m-\frac{p}{q}\right)\Gamma\left(\frac{p}{q}\right)}{\Gamma(m)} \tag{3}$$

with the convergence conditions $m>0$, $p/q>0$, $m>p/q$, or $0<p<mq$.

Solution 2. Let $x^q=t$ and then use Eq.(1.62) or (1.63). We have

$$x=t^{\frac{1}{q}},\quad qx^{q-1}\,dx=dt,\quad dx=\frac{dt}{qt^{1-\frac{1}{q}}},\qquad m=p+q,$$

$$I=\frac{1}{q}\int_0^\infty \frac{t^{\frac{p-1}{q}}}{(1+t)^m}\frac{dt}{t^{1-\frac{1}{q}}}=\frac{1}{q}\int_0^\infty\frac{t^{\frac{p-q}{q}}}{(1+t)^m}\,dt$$

$$=\frac{1}{q}\int_0^\infty\frac{t^{\frac{p}{q}-1}}{(1+t)^m}\,dt=\frac{1}{q}B\left(\frac{p}{q},m-\frac{p}{q}\right). \tag{4}$$

Particular cases:

a. Let $q=2$. We have

$$\int_0^\infty\frac{x^{p-1}\,dx}{(1+x^2)^m}=\frac{1}{2}B\left(\frac{p}{2},m-\frac{p}{2}\right)=\frac{1}{2}\frac{\Gamma\left(\frac{p}{2}\right)\Gamma\left(m-\frac{p}{2}\right)}{\Gamma(m)} \tag{5}$$

with the convergence conditions $0<p<2m$.

b. Let $p=1$, $q=2$. We have

$$\int_0^\infty\frac{dx}{(1+x^2)^m}=\frac{1}{2}B\left(\frac{1}{2},m-\frac{1}{2}\right),\qquad m>\frac{1}{2}. \tag{6}$$

See also Sec.2.3.17.

Exercise. Try to evaluate this integral using a traditional method.

2.3.28. a. Study the integral

$$I = \int_0^\infty \frac{x^m\,dx}{(a+bx^n)^p}, \quad a,b>0, \quad n>0, \quad m\in R. \tag{1}$$

Solution. Let

$$bx^n = at, \quad x^n\frac{a}{b}t, \quad x = \left(\frac{a}{b}\right)^{\frac{1}{n}} t^{\frac{1}{n}}, \quad dx = \frac{1}{n}\left(\frac{a}{b}\right)^{\frac{1}{n}} t^{\frac{1}{n}-1}\,dt$$

and the integral becomes

$$\begin{aligned} I &= \frac{1}{na^p}\int_0^\infty \frac{\left(\frac{a}{b}\right)^{\frac{m}{n}} t^{\frac{m}{n}}}{(1+t)^p}\left(\frac{a}{b}\right)^{\frac{1}{n}} t^{\frac{1}{n}-1}\,dt \\ &= \frac{1}{na^p}\left(\frac{a}{b}\right)^{\frac{m+1}{n}}\int_0^\infty \frac{t^{\frac{m}{n}+\frac{1}{n}-1}}{(1+t)^p}\,dt \\ &= \frac{1}{na^p}\left(\frac{a}{b}\right)^{\frac{m+1}{n}}\int_0^\infty \frac{t^{\frac{m+1}{n}-1}}{(1+t)^p}\,dt \end{aligned} \tag{2}$$

and with Sec.2.3.28 we have

$$\begin{aligned} I &= \frac{1}{na^p}\left(\frac{a}{b}\right)^{\frac{m+1}{n}} B\left(p-\frac{m+1}{n}, \frac{m+1}{n}\right) \\ &= \frac{1}{na^p}\left(\frac{a}{b}\right)^{\frac{m+1}{n}} \frac{\Gamma\left(p-\frac{m+1}{n}\right)\Gamma\left(\frac{m+1}{n}\right)}{\Gamma(p)} \end{aligned} \tag{3}$$

with the convergence conditions $np > m+1 > 0$.

Particular cases:

a. Let $a = b = 1$,

$$\int_0^\infty \frac{x^m\,dx}{(1+x^n)^p} = \frac{1}{n} B\left(p - \frac{m+1}{n}, \frac{m+1}{n}\right)$$

$$= \frac{\Gamma\left(p - \frac{m+1}{n}\right)\Gamma\left(\frac{m+1}{n}\right)}{n\Gamma(p)}, \quad pn > m+1 > 0. \tag{4}$$

b. In (4), let $n = 1$, $p = 2$ and

$$\int_0^\infty \frac{x^m\,dx}{(1+x)^2} = B(1-m, 1+m) = \frac{m\pi}{\sin m\pi}, \tag{5}$$

and for convergence: $1 - m > 0$, $1 + m > 0$ or $-1 < m < 1$.

2.3.29. Study the integral

$$I = \int_0^{\frac{\pi}{2}} \sqrt{\tan x}\,dx. \tag{1}$$

Test for convergence. According to Appendix A1.2, the integral converges at $x = \pi/2$, since we can write

$$\sqrt{\tan x} = \frac{\sqrt{\sin x}}{\sqrt{\sin\left(\frac{\pi}{2} - x\right)}} \underset{x=\frac{\pi}{2}}{=} O\left[\frac{1}{\sqrt{\frac{\pi}{2} - x}}\right] \tag{2}$$

and hence, $\alpha = 1/2 < 1$. The test for convergence can also be obtained from Eq.(1.124), for $n = 1/2 \in (-1, 1)$.

Integral evaluation. The integral can be written

$$I = \int_0^{\frac{\pi}{2}} \tan^{\frac{1}{2}} x\,dx \tag{3}$$

and using Eq.(1.124), for $n = 1/2$, we have

$$I = \frac{1}{2}B\left(\frac{1+\frac{1}{2}}{2}, \frac{1-\frac{1}{2}}{2}\right) = \frac{1}{2}B\left(\frac{3}{4}, \frac{1}{4}\right)$$

$$= \frac{1}{2}\Gamma\left(\frac{3}{4}\right)\Gamma\left(\frac{1}{4}\right) = 2.221. \tag{4}$$

Also, using the complement formula Eq.(1.35),

$$\int_0^{\frac{\pi}{2}} \sqrt{\tan x}\, dx = \frac{1}{2}\frac{\pi}{\sin\frac{\pi}{4}} = \frac{\pi}{\sqrt{2}} = 2.221. \tag{5}$$

Evaluate the integral using Eq.(1.116) for $p = 1/2$, $q = -1/2$.

2.3.30. Evaluate the improper integral

$$I = \int_0^{\frac{\pi}{2}} \frac{dx}{\sqrt{\tan x}}. \tag{1}$$

Test for convergence. According to Appendix A1.2, the integral converges at $x = 0+$, since we can write

$$\frac{1}{\sqrt{\tan x}} = \frac{\sqrt{\cos x}}{\sqrt{\sin x}} = \underset{x=0+}{O}\left[\frac{1}{x^{\frac{1}{2}}}\right] \tag{2}$$

and hence, $\alpha = 1/2 < 1$.

The test for convergence can also be obtained from Eq.(1.124) for $n = -1/2 \in (-1, 1)$.

Integral evaluation. The integral can be written

$$I = \int_0^{\frac{\pi}{2}} \tan^{-\frac{1}{2}} x\, dx \tag{3}$$

and using Eq.(1.124), for $n = -1/2$, we have

$$I = \frac{1}{2}B\left(\frac{1-\frac{1}{2}}{2}, \frac{1+\frac{1}{2}}{2}\right) = \frac{1}{2}B\left(\frac{1}{4}, \frac{3}{4}\right)$$

$$= \frac{1}{2}\Gamma\left(\frac{1}{4}\right)\Gamma\left(\frac{3}{4}\right) = \frac{\pi}{\sqrt{2}} = 2.221. \qquad (4)$$

See also Sec.2.3.29.

Evaluate the integral using Eq.(1.116) for $p = 1/2$, $q = -1/2$.

2.3.31. Study the improper integral

$$I = \int_0^\infty e^{-x^m}\, dx, \quad m \in R. \qquad (1)$$

Solution. With the substitution

$$x^m = t, \quad dx = \frac{1}{m} t^{\frac{1}{m}-1}\, dt \qquad (2)$$

the integral becomes

$$\int_0^\infty e^{-x^m}\, dx = \frac{1}{m}\int_0^\infty e^{-t} t^{\frac{1}{m}-1}\, dt = \frac{1}{m}\Gamma\left(\frac{1}{m}\right), \quad m > 0. \qquad (3)$$

Particular cases:

a. Let $m = 2$ and we have

$$\int_0^\infty e^{-x^2}\, dx = \frac{1}{2}\Gamma\left(\frac{1}{2}\right) = \frac{\sqrt{\pi}}{2}. \qquad (4)$$

b. Let $m = 4$, and the integral becomes

$$\int_0^\infty e^{-x^4}\, dx = \frac{1}{4}\Gamma\left(\frac{1}{4}\right) = 0.906. \qquad (5)$$

Note. See also Eqs.(1.76), (1.77) and Sec.2.7.1 for $m = 2$.

2.3.32. Study the integral

$$I = \int_{-\infty}^{+\infty} \frac{e^{mx}}{1+e^x}\, dx, \quad m \in R. \tag{1}$$

Solution. Let

$$e^x = t, \quad e^x\, dx = dt, \quad dx = \frac{dt}{t} \tag{2}$$

and the integral becomes

$$I = \int_0^{\infty} \frac{t^{m-1}}{1+t}\, dt$$

and with Sec.2.3.13,

$$\int_{-\infty}^{+\infty} \frac{e^{mx}\, dx}{1+e^x} = B(1-m, m) = \frac{\pi}{\sin m\pi}, \quad 0 < m < 1. \tag{3}$$

Particular cases:

a. Let $m = 1/2$ and from (3) we have

$$\int_{-\infty}^{+\infty} \frac{e^{\frac{x}{2}}}{1+e^x}\, dx = \frac{\pi}{\sin\frac{\pi}{2}} = \pi. \tag{4}$$

b. Let $m = 1/3$ and from (3) we have

$$\int_{-\infty}^{+\infty} \frac{e^{\frac{x}{3}}}{1+e^x}\, dx = \frac{\pi}{\sin\frac{\pi}{3}} = \frac{2\pi}{\sqrt{3}}. \tag{5}$$

Exercise. Evaluate the integral using a traditional method.

2.3.32. a. Evaluate the integral

$$\int_0^\infty \frac{x^{m-1}}{e^x-1}\,dx, \quad m>1. \tag{1}$$

Solution. From the integral representation of the Riemann's zeta function, $\zeta(m)$,

$$\zeta(m)=\frac{p^m}{\Gamma(m)}\int_0^\infty \frac{x^{m-1}}{e^{px}-1}\,dx, \quad p>0, \quad m>1, \tag{2}$$

for $p=1$, we have

$$\int_0^\infty \frac{x^{m-1}}{e^x-1}\,dx=\Gamma(m)\zeta(m), \quad m>1, \tag{3}$$

where $\zeta(m)$ can be evaluated with the formula

$$\zeta(2k)=\frac{2^{2k-1}\pi^{2k}|B_{2k}|}{(2k)\,!}, \quad k=1,2,3,\ldots, \tag{4}$$

where B_{2k} are the Bernoulli's numbers,

$$B_0=1,\quad B_1=-\frac{1}{2},\quad B_2=\frac{1}{6},\quad B_3=0,\quad B_4=-\frac{1}{30},$$
$$B_5=0,\quad B_6=\frac{1}{42},\quad B_7=0,\quad B_8=-\frac{1}{30},\ldots. \tag{5}$$

The Bernoulli's numbers of odd index (with the exception of B_1) are equal to zero, $B_{2k+1}=0$, $k=1,2,3,\ldots$.

Particular case. Let $m=4$ and from (3) we have

$$\int_0^\infty \frac{x^3}{e^x-1}\,dx=\Gamma(4)\zeta(4) \tag{6}$$

and with (4), for $k=2$,

$$\zeta(4)=\frac{2^3\pi^4}{4\,!}\frac{1}{30}=\frac{\pi^4}{90}.$$

Therefore from (6),

$$\int_0^\infty \frac{x^3}{e^x-1}\,dx=3\,!\frac{\pi^4}{90}=\frac{\pi^4}{15}. \tag{7}$$

Note that this integral is used in physics regarding the Planck's radiation law and Stefan-Boltzmann's law.

2.3.33. Study the integrals

$$I_1 = \int_0^\infty \frac{\sin x}{x^p}\, dx, \quad I_2 = \int_0^\infty \frac{\cos x}{x^p}\, dx. \tag{1}$$

Solution. In Eq.(1.94), replacing a by x, we have

$$\frac{1}{x^p} = \frac{1}{\Gamma(p)} \int_0^\infty t^{p-1} e^{-xt}\, dt \tag{2}$$

and substituting in (1),

$$I_1 = \frac{1}{\Gamma(p)} \int_0^\infty t^{p-1} \left(\int_0^\infty e^{-xt} \sin x\, dx \right) dt. \tag{3}$$

Using Eq.(1.154), for $a = 1$,

$$\int_0^\infty e^{-xt} \sin x\, dx = \frac{1}{1+t^2}$$

and (3) becomes

$$I_1 = \frac{1}{\Gamma(p)} \int_0^\infty \frac{t^{p-1}}{1+t^2}\, dt \tag{4}$$

and with Sec.2.3.20 for $m = p$ and $n = 2$,

$$I_1 = \frac{1}{2\Gamma(p)} B\left(1 - \frac{p}{2}, \frac{p}{2}\right) = \frac{\pi}{2\Gamma(p) \sin \frac{p\pi}{2}}, \quad 0 < p < 1. \tag{5}$$

Similarly with (2), the integral I_2 becomes

$$I_2 = \frac{1}{\Gamma(p)} \int_0^\infty t^{p-1} \left(\int_0^\infty e^{-xt} \cos x\, dx \right) dt \tag{6}$$

and with Eq.(1.155) for $a = 1$ and then with Sec.2.3.20 for $m = p+1$, $n = 2$,

$$\begin{aligned} I_2 &= \frac{1}{\Gamma(p)} \int_0^\infty \frac{t^p}{1+t^2}\, dt = \frac{1}{2\Gamma(p)} B\left(1 - \frac{p+1}{2}, \frac{p+1}{2}\right) \\ &= \frac{1}{2\Gamma(p)} \Gamma\left(1 - \frac{p+1}{2}\right) \Gamma\left(\frac{p+1}{2}\right) = \frac{\pi}{2\Gamma(p) \sin \frac{p+1}{2}\pi} \\ &= \frac{\pi}{2\Gamma(p) \cos \frac{p\pi}{2}}, \quad 0 < p < 1. \end{aligned} \tag{7}$$

See also Sec.2.7.6.

Note that for $p = 1$, the convergence of integral

$$\int_0^\infty \frac{\sin x}{x}\,dx \tag{8}$$

should be tested separately. The lower limit, $x = 0$, is not a discontinuity because $\lim_{x\to O+}(\sin x)/x = 1$. For convergence of the upper limit the Abel's test can be used. The Laplace transform or functions of complex variables can be used for the integral evaluation of (8).

Particular cases:

a. Let $p = 1/2$, see the integrated graphics Fig.2.14,

$$\int_0^\infty \frac{\sin x}{\sqrt{x}}\,dx = \int_0^\infty \frac{\cos x}{\sqrt{x}}\,dx = \sqrt{\frac{\pi}{2}}.$$

b. Let $p = 1/4$,

$$\int_0^\infty \frac{\sin x}{\sqrt[4]{x}}\,dx = \frac{\pi}{2\Gamma\left(\frac{1}{4}\right)\sin\frac{\pi}{8}} = 1.132$$

and

$$\int_0^\infty \frac{\cos x}{\sqrt[4]{x}}\,dx = \frac{\pi}{2\Gamma\left(\frac{1}{4}\right)\cos\frac{\pi}{8}} = 0.4689.$$

2.3.34. Study the integrals

$$I_1 = \int_0^\infty x^{m-1}\sin(ax)\,dx, \quad I_2 \int_0^\infty x^{m-1}\cos(ax)\,dx, \quad a > 0.$$

Solution. Using the substitution

$$ax = t, \quad a\,dx = dt$$

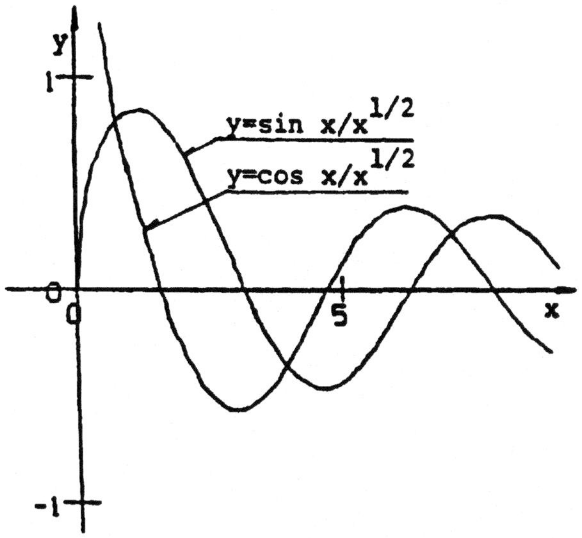

Fig.2.14

yields

$$I_1 = \int_0^\infty \left(\frac{t}{a}\right)^{m-1} \sin t \frac{dt}{a} = \frac{1}{a^m} \int_0^\infty \frac{\sin t}{t^{1-m}}\, dt.$$

Using Sec.2.3.33, we have

$$I_1 = \frac{1}{a^m} \frac{\pi}{2\Gamma(1-m) \sin \dfrac{(1-m)\pi}{2}} = \frac{\pi}{2a^m} \frac{1}{\Gamma(1-m) \cos \dfrac{m\pi}{2}}$$

and with Eq.(1.35),

$$I_1 = \frac{\pi}{2a^m \cos \dfrac{m\pi}{2}} \cdot \frac{\Gamma(m) \sin m\pi}{\pi} = \frac{\Gamma(m)}{2a^m} \frac{\sin m\pi}{\cos \dfrac{m\pi}{2}}$$
$$= \frac{\Gamma(m)}{a^m} \sin \frac{m\pi}{2}, \qquad m \in (0,1).$$

Similarly, we have

$$I_2 = \frac{1}{a^m}\int_0^\infty \frac{\cos t}{t^{1-m}}\,dt = \frac{1}{a^m}\frac{\pi}{2\Gamma(1-m)\sin\dfrac{m\pi}{2}}$$

$$= \frac{\pi}{2a^m \sin\frac{m\pi}{2}}\cdot\frac{\Gamma(m)\sin m\pi}{\pi} = \frac{\Gamma(m)}{2a^m}\cdot\frac{\sin m\pi}{\sin\dfrac{m\pi}{2}}$$

$$= \frac{\Gamma(m)}{2a^m}\cdot\frac{2\sin\dfrac{m\pi}{2}\cos\dfrac{m\pi}{2}}{\sin\dfrac{m\pi}{2}}$$

$$= \frac{\Gamma(m)}{a^m}\cos\frac{m\pi}{2}, \qquad m \in (0,1).$$

Particular cases:

a. Let $a = 1$, $m = 1-p$ and the integral I_1 becomes

$$I_1 = \int_0^\infty \frac{\sin x}{x^p}\,dx = \Gamma(1-p)\sin\frac{\pi(1-p)}{2}$$

and using Eq.(1.35),

$$I_1 = \frac{\pi}{\Gamma(p)\sin p\pi}\cos\frac{p\pi}{2} = \frac{\pi}{\Gamma(p)2\sin\dfrac{p\pi}{2}\cos\dfrac{p\pi}{2}}\cos\frac{p\pi}{2}$$

$$= \frac{\pi}{2\Gamma(p)\sin\dfrac{p\pi}{2}}, \qquad 0 < p < 1.$$

See also Sec.2.3.33.

b. Let $a = 1$, $m = 1-p$ and the integral I_2 becomes

$$I_2 = \int_0^\infty \frac{\cos x}{x^p}\,dx = \Gamma(1-p)\cos\frac{\pi(1-p)}{2}$$

$$= \frac{\pi}{\Gamma(p)\sin p\pi}\sin\frac{p\pi}{2} = \frac{\pi}{\Gamma(p)2\sin\dfrac{p\pi}{2}\cos\dfrac{p\pi}{2}}\sin\frac{p\pi}{2}$$

$$= \frac{\pi}{2\Gamma(p)\cos\dfrac{p\pi}{2}}, \qquad 0 < p < 1.$$

See also Sec.2.3.33.

2.3.35. Study the integrals

$$I_1 = \int_0^\infty x^p \sin(ax+b)\,dx, \quad I_2 = \int_0^\infty x^p \cos(ax+b)\,dx, \quad a > 0. \tag{1}$$

Solution. Using the substitution

$$ax = t, \quad a\,dx = dt$$

yields

$$\begin{aligned} I_1 &= \frac{1}{a^{p+1}} \int_0^\infty t^p \sin(t+b)\,dt \\ &= \frac{1}{a^{p+1}} \int_0^\infty t^p (\sin t \cos b + \cos t \sin b)\,dt \\ &= \frac{1}{a^{p+1}} \left(\cos b \int_0^\infty t^p \sin t\,dt + \sin b \int_0^\infty t^p \cos t\,dt \right) \end{aligned}$$

and with Sec.2.3.34, we have

$$\begin{aligned} I_1 &= \frac{1}{a^{p+1}} \left[\cos b \cdot \Gamma(p+1) \sin \frac{(p+1)\pi}{2} + \sin b \cdot \Gamma(p+1) \cos \frac{(p+1)\pi}{2} \right] \\ &= \frac{\Gamma(p+1)}{a^{p+1}} \left(\cos b \cdot \cos \frac{p\pi}{2} - \sin b \cdot \sin \frac{p\pi}{2} \right) \\ &= \frac{\Gamma(p+1)}{a^{p+1}} \cos\left(b + \frac{p\pi}{2} \right), \quad a > 0, \quad -1 < p < 0. \end{aligned} \tag{2}$$

Similarly, for the second integral, we have

$$\begin{aligned} I_2 &= \frac{1}{a^{p+1}} \int_0^\infty t^p \cos(t+b)\,dt \\ &= \frac{1}{a^{p+1}} \int_0^\infty t^p (cost \cdot \cos b - \sin t \cdot \sin b)\,dt \\ &= \frac{1}{a^{p+1}} \left(\cos b \int_0^\infty t^p \cos t\,dt - \sin b \int_0^\infty t^p \sin t\,dt \right) \end{aligned}$$

and with Sec.2.3.34,

$$I_2 = \frac{1}{a^{p+1}}\left(\cos b \cdot \Gamma(p+1)\cos\frac{(p+1)\pi}{2} - \sin b \cdot \Gamma(p+1)\sin\frac{(p+1)\pi}{2}\right)$$
$$= \frac{\Gamma(p+1)}{a^{p+1}}\left(-\cos b \cdot \sin\frac{p\pi}{2} - \sin b \cdot \cos\frac{p\pi}{2}\right)$$
$$= -\frac{\Gamma(p+1)}{a^{p+1}}\sin\left(b+\frac{p\pi}{2}\right), \quad a > 0, \quad -1 < p < 0. \tag{3}$$

Particular case. Let $a = 1$, $b = 0$, $p = -m$, and (2) becomes

$$I_1 = \int_0^\infty \frac{\sin x}{x^m}\,dx = \Gamma(1-m)\cos\frac{m\pi}{2}$$

and from Eq.(1.35),

$$\Gamma(1-m) = \frac{\pi}{\Gamma(m)\sin m\pi}$$

and

$$\int_0^\infty \frac{\sin x}{x^m}\,dx = \frac{\pi}{\Gamma(m)\sin m\pi}\cdot\cos\frac{m\pi}{2} = \frac{\pi}{2\Gamma(m)\sin\frac{m\pi}{2}},$$

a result also found in Sec.2.3.33.

2.3.36. Evaluate the area bounded by the curve

$$x^4 + y^4 = ax^2y, \quad a > 0. \tag{1}$$

Solution. From (1) comes the result that the curve has symmetry to the y axis and the area bounded by the curve is in the region with $y > 0$. The curve is tangent in origin to the x and y axes. Changing to the polar coordinates, $x = r\cos\theta$, $y = r\sin\theta$, equation (1) becomes

$$r = \frac{a\sin\theta\cos^2\theta}{\sin^4\theta + \cos^4\theta} \tag{2}$$

and the graphic is shown in Fig.2.15. The area bounded by the curve is given by the definite integral

$$A = \frac{2}{2}\int_0^{\frac{\pi}{2}} r^2\,d\theta = \int_0^{\frac{\pi}{2}} \frac{a^2\sin^2\theta\cos^4\theta}{\left(\sin^4\theta + \cos^4\theta\right)^2}\,d\theta. \tag{3}$$

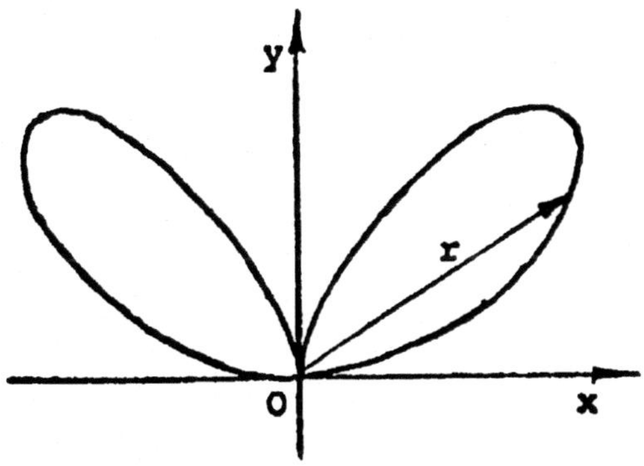

Fig.2.15

Let $\tan\theta = t$, $\sin^2\theta = \cos^2\theta t^2$, $\sin^2\theta = (1-\sin^2\theta)t^2$, $\sin^2\theta = \dfrac{t^2}{1+t^2}$,

$$\cos^2\theta = 1-\sin^2\theta = \frac{1}{1+t^2}, \quad \frac{d\theta}{\cos^2\theta} = dt, \quad d\theta = \cos^2\theta\, d\theta = \frac{dt}{1+t^2}$$

and from (3) we have

$$A = a^2\int_0^\infty \frac{\dfrac{t}{1+t^2}\cdot\dfrac{1}{(1+t^2)^2}\cdot\dfrac{1}{1+t^2}dt}{\left[\dfrac{t^4}{(1+t^2)^2}+\dfrac{1}{(1+t^2)^2}\right]^2} = a^2\int_0^\infty \frac{\dfrac{t^2}{(1+t^2)^4}}{\left[\dfrac{t^4+1}{(1+t^2)^2}\right]^2}dt$$

$$= a^2\int_0^\infty \frac{t^2\,dt}{(t^4+1)^2}. \tag{4}$$

Using Sec.2.3.25, for $p=3$ and $q=4$, we have

$$A = \frac{a^2}{4}B\left(2-\frac{3}{4},\frac{3}{4}\right) = \frac{a^2}{4}B\left(\frac{5}{4},\frac{3}{4}\right)$$
$$= \frac{a^2}{16}\Gamma\left(\frac{1}{4}\right)\Gamma\left(\frac{3}{4}\right) = \frac{a^2}{16}\frac{\pi}{\sin\dfrac{\pi}{4}}$$

and the area is

$$A = \frac{\pi\sqrt{2}}{16}a^2. \tag{5}$$

2.3.37. The χ^2-distribution function, $f(\chi^2)$, with ν degrees of freedom, is defined by

$$f(x) = \begin{cases} 0 \quad \text{for} \quad x \le 0 \\ \dfrac{1}{2^{\frac{\nu}{2}}\Gamma\left(\frac{\nu}{2}\right)} x^{\frac{\nu}{2}-1} e^{-\frac{x}{2}}, \quad \text{where} \quad x = \chi^2, \quad x > 0, \end{cases} \tag{1}$$

and has the graph given in Fig.2.16 for different degrees of freedom (ν). From (1), we have, for $x \ge 0$,

$$f'(x) = \frac{1}{2^{\frac{\nu}{2}+1}\Gamma\left(\frac{\nu}{2}\right)}(\nu - 2 - x)x^{\frac{\nu}{2}-2}e^{-\frac{x}{2}} \tag{2}$$

and hence, $f(x)$ has a maximum at $x = \nu - 2$. In addition, from (2) we have at the origin

$$f'(0) = 0 \quad \text{for} \quad \nu > 4, \tag{3}$$

which means that for $\nu > 4$ the curve is tangent to the x axis at the origin. The graphs in Fig.2.16 have been obtained using Maple V software.

Let $\nu = 1$. From (1),

$$f(x) = \frac{1}{\sqrt{2}\Gamma\left(\frac{1}{2}\right)} x^{-\frac{1}{2}} e^{-\frac{x}{2}}, \quad \text{where} \quad x = \chi^2 \tag{4}$$

$$= 0.39894 x^{-\frac{1}{2}} e^{-\frac{x}{2}}, \quad \text{where} \quad x = \chi^2. \tag{5}$$

Let $\nu = 2$. From (1),

$$f(x) = \frac{1}{2}e^{-\frac{x}{2}}, \quad f'(x) = -\frac{1}{4}e^{-\frac{x}{2}}. \tag{6}$$

Let $\nu = 3$. From (1), we have

$$f(x) = \frac{1}{2^{\frac{3}{2}}\Gamma\left(\frac{3}{2}\right)} x^{\frac{1}{2}} e^{-\frac{x}{2}}$$

$$= 0.39894 x^{\frac{1}{2}} e^{-\frac{x}{2}} \tag{7}$$

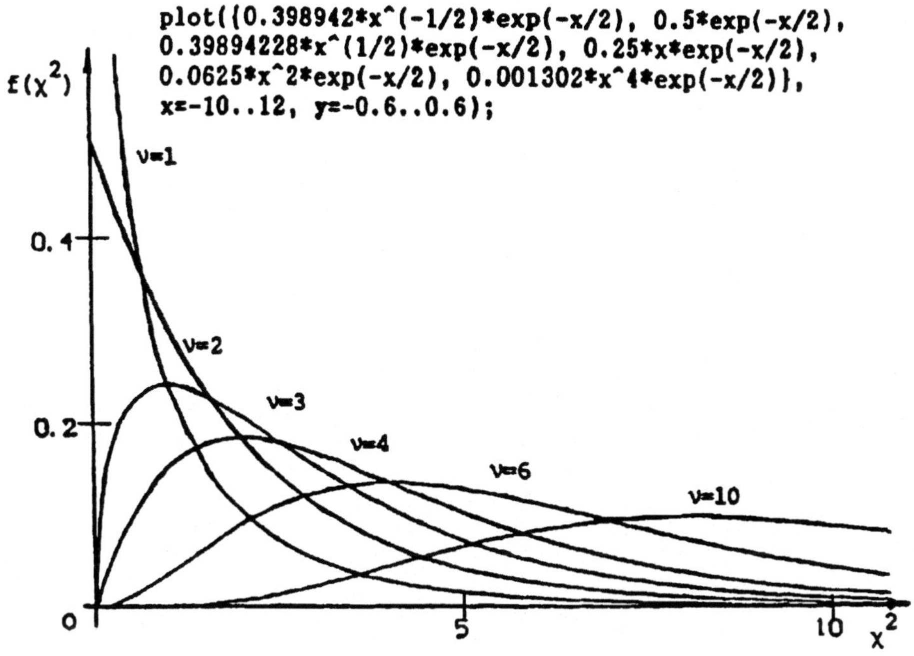

Fig.2.16

and from (2),

$$f'(x) = \frac{1}{2^{\frac{5}{2}}\Gamma\left(\frac{3}{2}\right)}(1-x)x^{-\frac{1}{2}}e^{-\frac{x}{2}}$$

$$= 0.199(1-x)x^{-\frac{1}{2}}e^{-\frac{x}{2}} \tag{8}$$

and hence, the function has a maxim for $x = 1$ and the curve is tangent to the y axis at the origin.

Let $\nu = 4$. From (1) we have (for $x = \chi^2$)

$$f(x) = \frac{1}{4}xe^{-\frac{x}{2}} \tag{9}$$

and from (2),

$$f'(x) = \frac{1}{8}(2 - x)e^{-\frac{x}{2}}. \tag{10}$$

Thus, the function has a maximum for $x = 2$,

$$f(2) = \frac{1}{2e} = 0.1839,$$

$f'(0) = 1/4$ and the curve has the line $f(x) = \frac{1}{4}x$ as tangent at the origin.

For $\nu > 4$, the curve goes through the origin and has the x axis as tangent at the origin. The maximum point is obtained for $x = \nu - 2$ and the location of the maxim is increasing in terms of ν. The graph also shows the curves for $\nu = 6$ and $\nu = 10$.

The probability integral of the χ^2-distribution function is defined by

$$P\left(\frac{\chi^2}{\nu}\right) = \frac{1}{2^{\frac{\nu}{2}}\Gamma\left(\frac{\nu}{2}\right)}\int_0^{\chi^2} x^{\frac{\nu}{2}-1}e^{-\frac{x}{2}}\,dx, \quad \chi^2 \in R^+ \cup \{0\} \tag{11}$$

with the value of the area bounded by the Ox, Oy axes, $f(\chi^2)$, and $x = \chi^2$. Changing the variable

$$\frac{x}{2} = u, \quad x = 2u, \quad dx = 2\,du \tag{12}$$

Eq.(11) becomes

$$P\left(\frac{\chi^2}{\nu}\right) = \frac{1}{\Gamma\left(\frac{\nu}{2}\right)}\int_0^{\frac{\chi^2}{2}} u^{\frac{\nu}{2}-1}e^{-u}\,du. \tag{13}$$

The integral in (13) is an incomplete gamma function $\gamma(\nu/2, \chi^2/2)$. As a particular case, from (13) we have

$$P\left(\frac{\infty}{\nu}\right) = \frac{1}{\Gamma\left(\frac{\nu}{2}\right)}\int_0^{\infty} u^{\frac{\nu}{2}-1}e^{-u}\,du = \frac{\Gamma\left(\frac{\nu}{2}\right)}{\Gamma\left(\frac{\nu}{2}\right)} = 1. \tag{14}$$

Thus, independent of the number of degrees of freedom, the area bounded by the Ox, Oy axes and the curve is always 1.0.

The percentage points of the χ^2-distribution function are defined by the integral

$$Q\left(\frac{\chi^2}{\nu}\right) = \frac{1}{2^{\frac{\nu}{2}}\Gamma\left(\frac{\nu}{2}\right)} \int_{\chi^2}^{\infty} x^{\frac{\nu}{2}-1} e^{-\frac{x}{2}}\, dx, \quad \chi^2 \in R^+ \cup \{0\}. \tag{15}$$

Between $P(\chi^2/\nu)$ and $Q(\chi^2/\nu)$, there is the following relation (from (11) and (15)):

$$P\left(\frac{\chi^2}{\nu}\right) + Q\left(\frac{\chi^2}{\nu}\right) = \frac{1}{2^{\frac{\nu}{2}}\Gamma\left(\frac{\nu}{2}\right)} \int_{0}^{\infty} x^{\frac{\nu}{2}-1} e^{-\frac{x}{2}}\, dx. \tag{16}$$

In the last integral, let $x = 2t$, $dx = 2\,dt$, and

$$\begin{aligned}
\int_0^{\infty} x^{\frac{\nu}{2}-1} e^{-\frac{x}{2}}\, dx &= \int_0^{\infty} (2t)^{\frac{\nu}{2}-1} e^{-t} 2\, dt \\
&= 2^{\frac{\nu}{2}} \int_0^{\infty} t^{\frac{\nu}{2}-1} e^{-t}\, dt \\
&= 2^{\frac{\nu}{2}} \Gamma\left(\frac{\nu}{2}\right)
\end{aligned}$$

and (16) becomes

$$P\left(\frac{\chi^2}{\nu}\right) = Q\left(\frac{\chi^2}{\nu}\right) = 1. \tag{17}$$

2.4. Line and Double Integrals

2.4.1. Evaluate the line integral

$$I = \int_L \frac{2y\, dy - x\, dx}{\sqrt{xy}} \tag{1}$$

where L is the line (see Fig.2.17) defined by

$$L\{x, y \geq 0, \quad x^2 + y^2 = a^2\}. \tag{2}$$

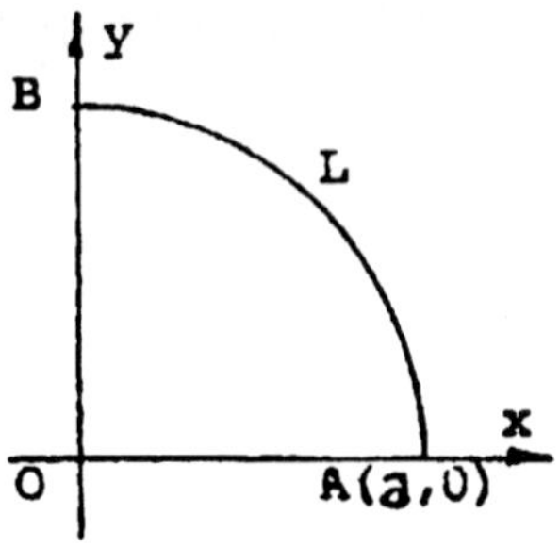

Fig.2.17

Solution. A parametric representation of a circle is

$$x = a\cos\theta, \quad y = a\sin\theta \tag{3}$$

and the integral becomes

$$\begin{aligned} I &= a^2 \int_0^{\frac{\pi}{2}} \frac{2\sin\theta\cos\theta + \cos\theta\sin\theta}{a\sqrt{\sin\theta\cos\theta}}\, d\theta \\ &= 3a \int_0^{\frac{\pi}{2}} \sin^{\frac{1}{2}}\theta \cos^{\frac{1}{2}}\theta\, d\theta, \end{aligned} \tag{4}$$

and using Eq.(1.116), we have

$$\begin{aligned} I &= \frac{3a}{2} B\left(\frac{3}{4}, \frac{3}{4}\right) = \frac{3a}{2} \frac{\Gamma\left(\frac{3}{4}\right)\Gamma\left(\frac{3}{4}\right)}{\Gamma\left(\frac{6}{4}\right)} \\ &= \frac{3a}{2} \frac{\left[\Gamma\left(\frac{3}{4}\right)\right]^2}{\Gamma\left(\frac{3}{2}\right)} = \frac{3a}{2} \frac{\left[\Gamma\left(\frac{3}{4}\right)\right]^2}{\frac{1}{2}\Gamma\left(\frac{1}{2}\right)} = \frac{3a}{\sqrt{\pi}} \left[\Gamma\left(\frac{3}{4}\right)\right]^2 \\ &= 2.542 \times a. \end{aligned} \tag{5}$$

For more line integrals see Sec.4.2.24 to Sec.4.2.28.

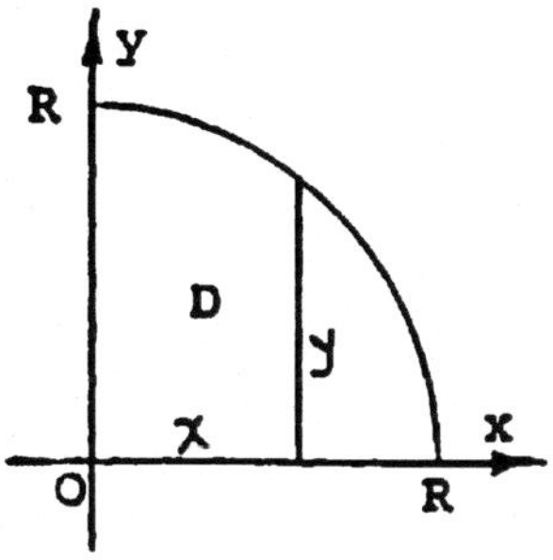

Fig.2.18

2.4.2. Evaluate the double integral

$$I = \iint_D x^{p-1}\, dx\, dy, \quad p \in R,$$
$$D\{x, y \geq 0, \quad x^2 + y^2 \leq R^2\}. \tag{1}$$

Solution 1. We have (see Fig.2.18)

$$I = \int_0^R x^{p-1} \left(\int_0^{\sqrt{R^2 - x^2}} dy \right) dx = \int_0^R x^{p-1}\, dx \cdot y \Big|_0^{\sqrt{R^2 - x^2}}$$
$$= \int_0^R x^{p-1} \sqrt{R^2 - x^2}\, dx. \tag{2}$$

Using the substitution

$$x^2 = R^2 t, \quad x = R\sqrt{t}, \quad dx = \frac{R\, dt}{2\sqrt{t}}, \tag{3}$$

$$\begin{aligned}
I &= \int_0^1 \sqrt{R^2 - R^2 t}\left(Rt^{\frac{1}{2}}\right)^{p-1} \frac{R\,dt}{2t^{\frac{1}{2}}} \\
&= \frac{R^2}{2}\int_0^1 \sqrt{1-t}R^{p-1}t^{\frac{p-1}{2}} \frac{R\,dt}{t^{\frac{1}{2}}} \\
&= \frac{R^{p+1}}{2}\int_0^1 t^{\frac{p}{2}-1}(1-t)^{\frac{1}{2}}\,dt = \frac{R^{p+1}}{2}\int_0^1 t^{\frac{p}{2}-1}(1-t)^{\left(1+\frac{1}{2}\right)-1}\,dt \\
&= \frac{R^{p+1}}{2}\int_0^1 t^{\frac{p}{2}-1}(1-t)^{\frac{3}{2}-1}\,dt \\
&= \frac{R^{p+1}}{2}B\left(\frac{p}{2}, \frac{3}{2}\right) = \frac{R^{p+1}}{2}B\left(\frac{p}{2}, 1+\frac{1}{2}\right)
\end{aligned}$$

and using Eq.(1.58),

$$I = \frac{R^{p+1}}{2}\frac{\frac{1}{2}}{\frac{1}{2}+\frac{p}{2}}B\left(\frac{p}{2}, \frac{1}{2}\right) \tag{4}$$

$$= \frac{R^{p+1}}{2(p+1)} \cdot \frac{\sqrt{\pi}\Gamma\left(\frac{p}{2}\right)}{\Gamma\left(\frac{p+1}{2}\right)}, \quad p > 0. \tag{5}$$

Solution 2. In polar coordinates

$$x = r\cos\theta, \quad y = r\sin\theta, \quad dx\,dy = r\,dr\,d\theta \tag{6}$$

the integral (1) becomes

$$I = \iint_{D'} (r\cos\theta)^{p-1} r\,dr\,d\theta = \iint_{D'} r^p(\cos\theta)^{p-1}\,dr\,d\theta$$

and with Eq.(1.120),

$$\begin{aligned}
I &= \int_0^R r^p\,dr\int_0^{\frac{\pi}{2}} \cos^{p-1}\theta\,d\theta = \frac{R^{p+1}}{p+1}\cdot\frac{1}{2}B\left(\frac{p}{2}, \frac{1}{2}\right) \\
&= \frac{R^{p+1}}{2(p+1)}\frac{\sqrt{\pi}\Gamma\left(\frac{p}{2}\right)}{\Gamma\left(\frac{p+1}{2}\right)}.
\end{aligned} \tag{7}$$

Particular cases:

a. Let $p = 1$ and we have

$$I = \iint_D dx\,dy = \frac{R^2}{2} B\left(\frac{1}{2}, \frac{3}{2}\right) = \frac{\pi}{4} R^2, \tag{8}$$

which is the surface area of a quarter of a circle.

b. Let $p = 2$ and we have

$$I = \iint_D x\,dx\,dy = \frac{R^3}{2 \times 3} \frac{\sqrt{\pi}\Gamma(1)}{\Gamma\left(\dfrac{3}{2}\right)} = \frac{\sqrt{\pi}R^3}{6\dfrac{\sqrt{\pi}}{2}} = \frac{R^3}{3}. \tag{9}$$

2.4.3. Study the double integral

$$I = \iint_D x^{p-1} y^{q-1}\,dx\,dy, \quad p, q \in R \tag{1}$$

$$D\{x, y \geq 0, \quad x + y \leq 1\}.$$

Solution 1. We can write, see Fig.2.19,

$$\begin{aligned} I &= \int_0^1 x^{p-1} \left(\int_0^{1-x} y^{q-1}\,dy \right) dx \\ &= \int_0^1 x^{p-1} \left(\frac{y^q}{q}\bigg|_0^{1-x} \right) dx = \frac{1}{q} \int_0^1 x^{p-1}(1-x)^q\,dx \\ &= \frac{1}{q} \int_0^1 x^{p-1}(1-x)^{(q+1)-1}\,dx = \frac{1}{q} B(p, q+1) \\ &= \frac{1}{q} \cdot \frac{q}{p+q} B(p,q) = \frac{1}{p+q} B(p,q) = \frac{1}{p+q} \frac{\Gamma(p)\Gamma(q)}{\Gamma(p+q)}, \quad p, q > 0. \end{aligned} \tag{2}$$

Solution 2. Let the transformation, see Appendix A4,

$$\begin{cases} x + y = u \\ y = uv \end{cases} \Rightarrow \begin{cases} x = u(1-v) \\ y = uv \end{cases} \begin{cases} F_1 = x + y - u \\ F_2 = y - uv \end{cases}, \tag{3}$$

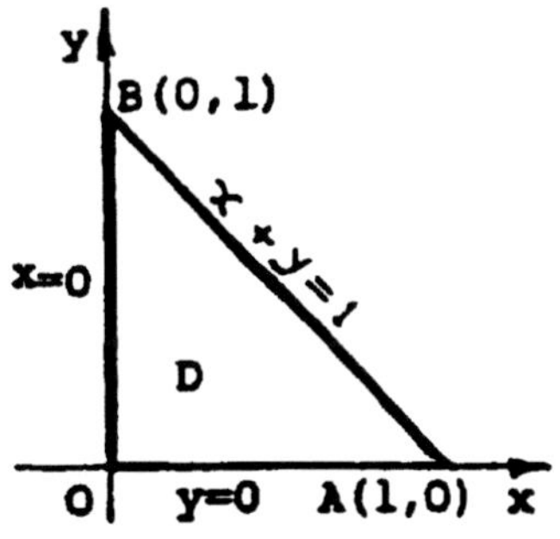

Fig.2.19

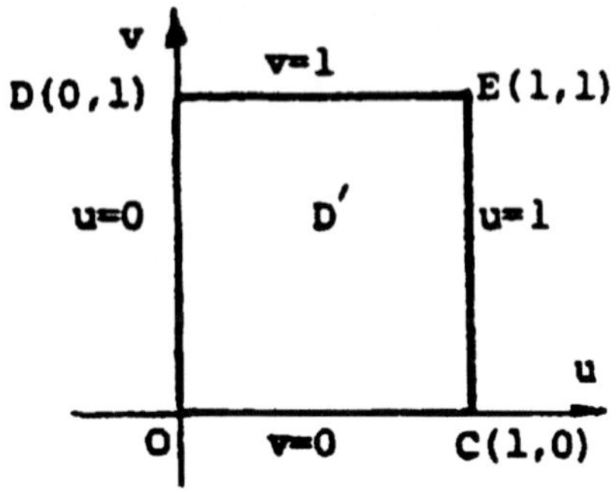

Fig.2.20

where u and v are the new variables and with the Jacobian $J = u$. The triangle OAB, in the xy-plane, has been transformed into a square, see Figs.2.19, 2.20 and Sec.2.4.6,

$$D'[0 \le u \le 1, \quad 0 \le v \le 1\}.$$

The integral becomes

$$\begin{aligned}
I &= \iint_{D'} [u(1-v)]^{p-1} \cdot (uv)^{q-1} \cdot u\, du\, dv \\
&= \iint_{D'} u^{p-1+q-1+1} v^{q-1}(1-v)^{p-1}\, du\, dv \\
&= \int_0^1 u^{p+q-1}\, du \int_0^1 v^{q-1}(1-v)^{p-1}\, dv = \frac{u^{p+q}}{p+q}\bigg|_0^1 B(q,p) \\
&= \frac{1}{p+q} B(q,p), \quad p, q > 0. \qquad (4)
\end{aligned}$$

Particular cases:

a. Let $q = 1$ and from (4),

$$\begin{aligned}
I &= \iint_D x^{p-1}\, dx\, dy = \frac{1}{p+1} B(1,p) = \frac{1}{p+1} \frac{\Gamma(p)}{\Gamma(p+1)} \\
&= \frac{1}{p+1} \frac{\Gamma(p)}{p\Gamma(p)} = \frac{1}{p(p+1)}, \quad p > 0, \qquad (5) \\
& \quad D\{x, y \ge 0, \quad x + y \le 1\}.
\end{aligned}$$

b. Let $p = 1$, $q = 1$ and from (4),

$$\iint_D dx\, dy = \frac{1}{2} B(1,1) = \frac{1}{2}, \tag{6}$$

which is the area of triangle OAB, Fig.2.19.

2.4.4. Study the double integral

$$I = \iint_D x^{p-1} e^{-(x^2+y^2)}\, dx\, dy, \quad p \in R, \tag{1}$$

where

$$D\{x \geq 0, \quad y \geq 0\}.$$

Solution. Changing to polar coordinates, $x = r\cos\theta$, $y = r\sin\theta$, and the Jacobian $J = r$, we have

$$\begin{aligned} I &= \iint_{D'} (r\cos\theta)^{p-1} e^{-r^2} r\, dr\, d\theta \\ &= \int_0^{\frac{\pi}{2}} \cos^{p-1}\theta\, d\theta \int_0^{\infty} r^p e^{-r^2}\, dr \end{aligned} \tag{2}$$

and with $r^2 = t$,

$$\int_0^{\infty} r^p e^{-r^2}\, dr = \frac{1}{2}\int_0^{\infty} t^{\frac{p+1}{2}-1} e^{-t}\, dt = \frac{1}{2}\Gamma\left(\frac{p+1}{2}\right), \quad p + 1 > 0. \tag{3}$$

See also Sec.1.10.2. The second integral can be evaluated using Eq.(1.119) and from (2) we have

$$I = \frac{1}{4} B\left(\frac{p}{2}, \frac{1}{2}\right) \Gamma\left(\frac{p+1}{2}\right) = \frac{\sqrt{\pi}}{4}\Gamma\left(\frac{p}{2}\right), \quad p > 0. \tag{4}$$

Particular cases:

a. Let $p = 1$, and the integral becomes

$$\iint_D e^{-(x^2+y^2)}\, dx\, dy = \frac{\sqrt{\pi}}{4}\Gamma\left(\frac{1}{2}\right) = \frac{\pi}{4}.$$

b. Let $p = 1/2$, and the integral becomes

$$\iint_D x^{-\frac{1}{2}} e^{-(x^2+y^2)}\, dx\, dy = \frac{\sqrt{\pi}}{4}\Gamma\left(\frac{1}{4}\right) = 1.606.$$

c. Let $p = 2$, and the integral becomes

$$\iint_D x e^{-(x^2+y^2)}\, dx\, dy = \frac{\sqrt{\pi}}{4}.$$

2.4.5. Study the improper double integral

$$I = \iint_D e^{-(x^2+y^2)} x^{2p-1} y^{2q-1}\, dx\, dy, \quad p, q \in R, \tag{1}$$

where

$$D\{x, y \geq 0\}.$$

Solution 1. The integral can be written

$$I = \int_0^\infty e^{-x^2} x^{2p-1}\, dx \int_0^\infty e^{-y^2} y^{2q-1}\, dy$$

and using Eq.(1.87), for $m = 2p$ and $n = 2q$, yields

$$I = \frac{1}{4}\Gamma(p)\Gamma(q)$$

and the integral converges for $p, q > 0$.

Solution 2. Changing to polar coordinates,

$$x = r\cos\theta, \quad y = r\sin\theta, \quad J = r,$$

we have

$$\begin{aligned} I &= \iint_{D_1} e^{-r^2} (r\cos\theta)^{2p-1} (r\sin\theta)^{2q-1} r\, dr\, d\theta \\ &= \int_0^{\frac{\pi}{2}} \sin^{2q-1}\theta \cos^{2p-1}\theta\, d\theta \int_0^\infty r^{2(p+q-1)} e^{-r^2} r\, dr. \end{aligned} \tag{2}$$

Using the transformation

$$r^2 = t, \quad r\,dr = \frac{dt}{2} \tag{3}$$

the second integral becomes

$$\int_0^\infty r^{2(p+q-1)} e^{-r^2} r\,dr = \frac{1}{2}\int_0^\infty t^{p+q-1} e^{-t}\,dt = \frac{1}{2}\Gamma(p+q). \tag{4}$$

Using Eq.(1.12), the first integral is

$$\int_0^{\frac{\pi}{2}} \sin^{2q-1}\theta \cos^{2p-1}\theta\,d\theta = \frac{1}{2}B(p,q). \tag{5}$$

Thus,

$$\begin{aligned} I &= \frac{1}{2}B(p,q)\frac{1}{2}\Gamma(p+q) = \frac{1}{4}\frac{\Gamma(p)\Gamma(q)}{\Gamma(p+q)} \cdot \Gamma(p+q) \\ &= \frac{1}{4}\Gamma(p)\Gamma(q) \end{aligned} \tag{6}$$

and the integral converges for $p, q > 0$.

Particular cases:

a. Let $p = 1/2$, $q = 1/2$, and the integral becomes

$$\iint_D e^{-(x^2+y^2)}\,dx\,dy = \frac{1}{4}\Gamma\left(\frac{1}{2}\right)\Gamma\left(\frac{1}{2}\right) = \frac{\pi}{4}. \tag{7}$$

See also Sec.2.4.4.

b. Let $p = 1$, $q = 1$, and the integral becomes

$$\iint_D e^{-(x^2+y^2)} xy\,dx\,dy = \frac{1}{4}, \quad x, y \geq 0. \tag{8}$$

2.4.6. Study the integral

$$I = \iint_D f(x+y)x^{p-1}y^{q-1}\,dx\,dy, \quad p, q \in R, \tag{1}$$

where f is an arbitrary continuous function of its argument and the integration region, see Fig.2.19, is

$$D\{x, y \geq 0, \quad x + y \leq 1\}. \tag{2}$$

Solution. Let the transformation, see Appendix A4,

$$\begin{cases} x + y = u \\ y = uv \end{cases} \qquad \begin{cases} x = u(1-v) \\ y = uv \end{cases}, \tag{3}$$

where u and v are the new variables with the Jacobian $J_2 = u$.

Under the given transformation, if x, y are positive and $x + y \leq 1$, then $0 \leq u \leq 1$ and $0 \leq v \leq 1$. Conversely, if u, v are between 0 and 1, we have $x \geq 0$, $y \geq 0$ and $x + y \leq 1$. Thus, the triangle OAB, on the xy-plane, is replaced by a square, on the uv-plane, as shown in Fig.2.20, and hence,

$$D'\{0 \leq u \leq 1, \quad 0 \leq v \leq 1\}. \tag{4}$$

The integral will become

$$\begin{aligned} I &= \iint_{D'} f(u)\,[u(1-v)]^{p-1}\,(uv)^{q-1}u\,du\,dv \\ &= \iint_{D'} f(u)u^{p+q-1}v^{q-1}(1-v)^{p-1}\,du\,dv \\ &= \int_0^1 f(u)u^{p+q-1}\,du \int_0^1 v^{q-1}(1-v)^{p-1}\,dv \\ &= B(p,q)\int_0^1 f(u)u^{p+q-1}\,du \\ &= \frac{\Gamma(p)\Gamma(q)}{\Gamma(p+q)}\int_0^1 f(u)u^{p+q-1}\,du, \quad p, q \in R^+, \end{aligned} \tag{5}$$

and the double integral has been reduced to a simple integral.

Particular cases:

a. Let $f(x+y)=(x+y)^2$, $p=2$, $q=3$ and the integral becomes

$$I_1=\iint_D (x+y)^2, xy^2\,dx\,dy$$
$$D\{x,y\geq 0,\quad x+y\leq 1\}$$

and from (5) we have

$$I=\frac{\Gamma(2)\Gamma(3)}{\Gamma(5)}\int_0^1 u^2\cdot u^{2+3-1}\,du=\frac{2}{4!}\int_0^1 u^6\,du$$
$$=\frac{1}{12}\frac{u^7}{7}\bigg|_0^1=\frac{1}{84}.$$

b. Let $f(x+y)=e^{x+y}$, $p=q=1$ and the integral becomes

$$I=\iint_D e^{x+y}\,dx\,dy$$
$$D\{x,y\geq 0,\quad x+y\leq 1\}$$

and from (5),

$$I=\frac{\Gamma(1)\Gamma(1)}{\Gamma(2)}\int_0^1 e^u u\,du=\int_0^1 u\,de^u$$
$$=ue^u\bigg|_0^1-\int_0^1 e^u\,du=e-e^u\bigg|_0^1$$

and

$$\iint_D e^{x+y}\,dx\,dy=1.$$

c. Let $f(\,)=1$, $p=q=1$ and the integral (1) becomes

$$I=\iint_D dx\,dy,\quad D\{x,y\geq 0,\quad x+y\leq 1\},$$

which is the area of triangle OAB, Fig.2.19,

$$\iint_D dx\,dy=\frac{1}{\Gamma(2)}\int_0^1 u\,du=\frac{1}{2}u^2\bigg|_0^1=\frac{1}{2}.$$

2.4.7. Evaluate the Dirichlet's integral,

$$I = \iint_D f\left[\left(\frac{x}{a}\right)^m + \left(\frac{y}{b}\right)^n\right] x^{p-1} y^{q-1}\, dx\, dy, \quad a, b > 0, \tag{1}$$

where $m, n, p, q \in R$, and f is an arbitrary continuous function of its argument, and the integration region D is defined by

$$D\left\{x, y \geq 0, \quad \left(\frac{x}{a}\right)^m + \left(\frac{y}{b}\right)^n \leq 1\right\}. \tag{2}$$

Solution. Letting the transformation

$$x = a x_1^{\frac{1}{m}}, \quad y = b y_1^{\frac{1}{n}} \tag{3}$$

with the Jacobian (see Appendix A5),

$$J = \frac{ab}{mn} x_1^{\frac{1}{m}-1} y_1^{\frac{1}{n}-1} \tag{4}$$

the integral becomes

$$\begin{aligned} I &= \iint_{D'} f(x_1 + y_1) \cdot \left(a x_1^{\frac{1}{m}}\right)^{p-1} \left(b y_1^{\frac{1}{n}}\right)^{q-1} \cdot \frac{ab}{mn} x_1^{\frac{1}{m}-1} y_1^{\frac{1}{n}-1}\, dx_1\, dy_1 \\ &= \frac{a^p b^q}{mn} \iint_{D'} f(x_1 + y_1) \cdot x_1^{\frac{p}{m}-1} y_1^{\frac{q}{n}-1}\, dx_1\, dy_1 \end{aligned} \tag{5}$$

$$D'\{x_1, y_1 \geq 0, \quad x_1 + y_1 \leq 1\}.$$

The last integral is similar to the integral given in Sec.2.4.6, and with the transformation

$$\begin{cases} x_1 + y_1 = u \\ y_1 = uv \end{cases} \Rightarrow \begin{cases} x_1 = u(1-v) \\ y_1 = uv \end{cases}, \quad J = u,$$

the new region of integration is D'', shown in Fig.2.20, and we have

$$\begin{aligned}
I &= \frac{a^p b^q}{mn} \iint_{D''} f(u) \cdot [u(1-v)]^{\frac{p}{m}-1} (uv)^{\frac{q}{n}-1} u \, du \, dv \\
&= \frac{a^p b^q}{mn} \iint_{D''} f(u) u^{\frac{p}{m}+\frac{q}{n}-1} (1-v)^{\frac{p}{m}-1} v^{\frac{q}{n}-1} \, du \, dv \\
&= \frac{a^p b^q}{mn} \int_0^1 f(u) u^{\frac{p}{m}+\frac{q}{n}-1} \, du \int_0^1 (1-v)^{\frac{p}{m}-1} v^{\frac{q}{n}-1} \, dv \\
&= \frac{a^p b^q}{mn} \int_0^1 f(u) u^{\frac{p}{m}+\frac{q}{n}-1} \, du \cdot B\left(\frac{p}{m}, \frac{q}{n}\right) \\
&= \frac{a^p b^q}{mn} B\left(\frac{p}{m}, \frac{q}{n}\right) \int_0^1 f(u) \cdot u^{\frac{p}{m}+\frac{q}{n}-1} \, du, \quad m, n, p, q \in R^+,
\end{aligned} \tag{6}$$

under the assumption that the integral on the right converges absolutely. Also, substituting p by p/m and q by q/n, the result in (6) can be obtained using the result of Sec.2.4.6, (5).

Particular case. Let $f[\,] = 1$, $m = n = 2$, $p = q = 1$, and $a, b > 0$ and we have

$$I = \iint_D dx \, dy$$

$$D\left\{x, y \geq 0, \quad \frac{x^2}{a^2} + \frac{y^2}{b^2} \leq 1\right\}.$$

The integral represents the area of a quarter of an ellipse, see Fig.2.21. From (6) we have

$$\frac{A}{4} = \frac{ab}{4} B\left(\frac{1}{2}, \frac{1}{2}\right) \int_0^1 u^{\frac{1}{2}+\frac{1}{2}-1} \, du = \frac{\pi ab}{4} \int_0^1 du = \frac{\pi ab}{4} \tag{7}$$

and for a complete ellipse $A = \pi ab$.

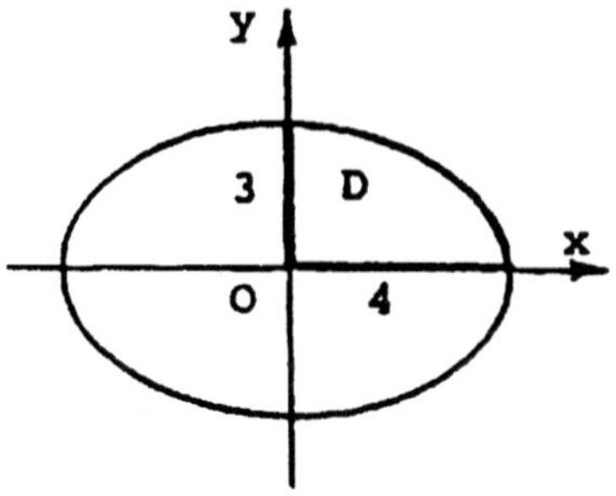

Fig.2.21

2.5. Surface Integrals

2.5.1. Evaluate the surface integral

$$I = \iint_S \frac{dS}{\sqrt{x^2+y^2}} \tag{1}$$

where the integration region S is defined by

$$S\{x, y, z \in R^+ \mid x^2 + y^2 + z^2 = R^2\}.$$

Solution. Because the integration surface is symmetrical with respect to the x, y, z axes, we can write

$$I = 8 \iint_{S_1} \frac{dS}{\sqrt{x^2+y^2}}, \tag{2}$$

$$S_1\{x, y, z \geq 0, \quad x^2 + y^2 + z^2 = R^2\}.$$

The surface element of the sphere is $dS = (R/z)\, dx\, dy$, and (2) becomes

$$I = 8 \iint_D \frac{R\, dx\, dy}{z\sqrt{x^2+y^2}} = 8R \iint_D \frac{dx\, dy}{\sqrt{R^2-x^2-y^2}\sqrt{x^2+y^2}}, \tag{3}$$

$$D\{x, y \geq 0, \quad x^2 + y^2 = R^2\}.$$

Changing to polar coordinates, $x = r\cos\theta$, $y = r\sin\theta$, $dx\, dy = r\, dr\, d\theta$,

$$I = 8R \iint_{D'} \frac{dr\, d\theta}{\sqrt{R^2-r^2}} = 8R \iint_0^{\frac{\pi}{2}} d\theta \int_0^R \frac{dr}{\sqrt{R^2-r^2}} = 4\pi R \int_0^R \frac{dr}{\sqrt{R^2-r^2}},$$

which converges for $r = R$. Using the transformation

$$r^2 = R^2 t, \quad r = R\sqrt{t}, \quad dr = \frac{R\,dt}{2\sqrt{t}}, \tag{4}$$

$$I = 2\pi R \int_0^1 t^{-\frac{1}{2}}(1-t)^{-\frac{1}{2}}\,dt = 2\pi R \int_0^1 t^{\frac{1}{2}-1}(1-t)^{\frac{1}{2}-1}\,dt$$

$$= 2\pi R B\left(\frac{1}{2}, \frac{1}{2}\right) = 2\pi R \frac{\Gamma\left(\frac{1}{2}\right)\Gamma\left(\frac{1}{2}\right)}{\Gamma(1)} = 2\pi^2 R. \tag{5}$$

See also Sec.2.5.2 for $p = 1/2$.

2.5.2. Study the integral

$$I = \iint_S \frac{dS}{(x^2+y^2)^p}, \quad p \in R, \tag{1}$$
$$S\{x, y, z \in R^+ \mid x^2 + y^2 + z^2 = R^2\}.$$

Solution. We have

$$I = 8 \iint_{S_1} \frac{dS}{(x^2+y^2)^p} = 8 \iint_D \frac{R\,dx\,dy}{z(x^2+y^2)^p},$$
$$S_1\{x, y, z \geq 0, \quad x^2 + y^2 + z^2 = R^2\},$$
$$I = 8R \iint_D \frac{dx\,dy}{(R^2 - x^2 - y^2)^{\frac{1}{2}}(x^2+y^2)^p}, \tag{2}$$
$$D\{x, y \geq 0, \quad x^2 + y^2 = R^2\}.$$

Changing to polar coordinates, with the Jacobian $J_2 = r$, see Appendix A3, the integral becomes

$$I = 8R \iint_{D_1} \frac{r\,dr\,d\theta}{(R^2 - r^2)^{\frac{1}{2}} r^{2p}} = 8R \int_0^{\frac{\pi}{2}} d\theta \int_0^R \frac{r^{1-2p}\,dr}{(R^2 - r^2)^{\frac{1}{2}}}$$
$$= 4\pi R \int_0^R r^{1-2p}(R^2 - r^2)^{-\frac{1}{2}}\,dr \tag{3}$$

and with the transformation $r^2 = R^2 t$, we have

$$
\begin{aligned}
I &= 2\pi R^{2(1-p)} \int_0^1 t^{-p}(1-t)^{-\frac{1}{2}}\, dt \\
&= 2\pi R^{2(1-p)} \int_0^1 t^{(1-p)-1}(1-t)^{\left(1-\frac{1}{2}\right)-1}\, dt \\
&= 2\pi R^{2(1-p)} B\left(1-p, \frac{1}{2}\right), \quad 0 < p < 1. \qquad (4)
\end{aligned}
$$

Note. Sec.2.3.9 can also be used for the integral in (3).

2.5.3. Evaluate the surface area, A_n, of a hypersphere in the n-dimensional space,

$$x_1^2 + x_2^2 + \cdots + x_n^2 = a^2, \quad a \in R^+. \qquad (1)$$

Solution. For a hypersphere in the n-dimensional space, the area is given by the integral of the order $n-1$, for which $x_n \geq 0$,

$$A_n = 2\int \cdots \int_S \frac{a\, dx_1\, dx_2 \ldots dx_{n-1}}{x_n}, \quad a \in R^+, \qquad (2)$$

$$S\left\{x_1, x_2, \ldots, x_n \in R^+ \mid x_1^2 + x_2^2 + \cdots + x_n^2 = a^2\right\}. \qquad (3)$$

Let the transformation (see Appendix A3.4)

$$
\begin{cases}
x_1 = a\cos\theta_1 \\
x_2 = a\sin\theta_1\cos\theta_2 \\
x_3 = a\sin\theta_1\sin\theta_2\cos\theta_3 \\
\ldots\ldots\ldots\ldots\ldots\ldots\ldots \\
x_{n-1} = a\sin\theta_1\sin\theta_2\ldots\sin\theta_{n-2}\cos\theta_{n-1} \\
x_n = a\sin\theta_1\sin\theta_2\ldots\sin\theta_{n-2}\sin\theta_{n-1}
\end{cases} \qquad (4)
$$

where $a > 0$, $\theta_1, \theta_2, \ldots, \theta_{n-2} \in [0, \pi]$, $\theta_{n-1} \in [0, 2\pi)$ and with the Jacobian

$$|J_{n-1}| = a^{n-1}\sin^{n-1}\theta_1\sin^{n-2}\theta_2\ldots\sin^2\theta_{n-2}\sin\theta_{n-1}. \qquad (5)$$

Substituting in (2), we have

$$A_n = 2\int \cdots \int_{S_1} \frac{a^n \sin^{n-1}\theta_1 \sin^{n-2}\theta_2 \ldots \sin^2\theta_{n-2}\sin\theta_{n-1}}{a\sin\theta_1\sin\theta_2\ldots\sin\theta_{n-2}\sin\theta_{n-1}}\, d\theta_1\, d\theta_2 \ldots d\theta_{n-1}$$

$$= 2a^{n-1}\int_0^\pi \sin^{n-2}\theta_1\, d\theta_1 \int_0^\pi \sin^{n-3}\theta_2\, d\theta_2 \ldots \int_0^\pi \sin\theta_{n-2}\, d\theta_{n-2}\int_0^\pi d\theta_{n-1}$$

and with Sec.1.12.3,

$$A_n = 2^{n-1}a^{n-1}\pi \int_0^{\frac{\pi}{2}} \sin^{n-2}\theta_1\, d\theta_1 \int_0^{\frac{\pi}{2}} \sin^{n-3}\theta_2\, d\theta_2 \ldots \int_0^{\frac{\pi}{2}} \sin\theta_{n-2}\, d\theta_{n-2}. \tag{6}$$

Using Eq.(1.120), we have

$$A_n = 2^{n-1}a^{n-1}\pi\frac{1}{2}B\left(\frac{n-1}{2},\frac{1}{2}\right)\cdot\frac{1}{2}B\left(\frac{n-2}{2},\frac{1}{2}\right)\cdots\frac{1}{2}B\left(\frac{2}{2},\frac{1}{2},\right)\cdot\frac{\pi}{2}$$

$$= 2a^{n-1}\pi(\sqrt{\pi})^{n-2}\frac{\Gamma\left(\frac{n-1}{2}\right)}{\Gamma\left(\frac{n}{2}\right)}\frac{\Gamma\left(\frac{n-2}{2}\right)}{\Gamma\left(\frac{n-1}{2}\right)}\cdots\frac{\Gamma\left(\frac{2}{2}\right)}{\Gamma\left(\frac{3}{2}\right)}$$

$$= 2a^{n-1}\pi\cdot\pi^{\frac{n-2}{2}}\cdot\frac{\Gamma(1)}{\Gamma\left(\frac{n}{2}\right)}$$

$$= 2a^{n-1}\pi^{1+\frac{n}{2}-1}\cdot\frac{1}{\Gamma\left(\frac{n}{2}\right)}$$

$$= 2a^{n-1}\pi^{\frac{n}{2}}\frac{1}{\Gamma\left(\frac{n}{2}\right)}$$

$$= \frac{2\pi^{\frac{n}{2}}}{\Gamma\left(\frac{n}{2}\right)}\cdot a^{n-1}, \quad a\in R^+, \quad n\in N. \tag{7}$$

Particular cases:

a. Let $n = 3$. From (7), the surface area of a sphere is

$$A_3 = \frac{2\pi^{\frac{3}{2}}}{\Gamma\left(\frac{3}{2}\right)}a^2 = \frac{2\pi^{\frac{3}{2}}}{\frac{1}{2}\Gamma\left(\frac{1}{2}\right)}a^2 = 4\pi a^2. \tag{8}$$

b. Let $n = 2p$, $p \in N$ and from (7) we have

$$A_{2p} = \frac{2\pi^{\frac{2p}{2}}}{\Gamma\left(\frac{2p}{2}\right)} a^{2p-1} = \frac{2\pi^p}{\Gamma(p)} a^{2p-1} = \frac{2\pi^p}{(p-1)\,!} a^{2p-1}. \tag{9}$$

c. Let $n = 2p + 1$ and from (7) we have

$$A_{2p+1} = \frac{2\pi^{\frac{2p+1}{2}}}{\Gamma\left(\frac{2p+1}{2}\right)} a^{2p+1-1} = \frac{2\pi^{p+\frac{1}{2}}}{\Gamma\left(p + \frac{1}{2}\right)} a^{2p} \tag{10}$$

and using Eq.(1.43) we have

$$A_{2p+1} = \frac{2\pi^{p+\frac{1}{2}}}{\frac{\sqrt{\pi}}{2^p}(2p-1)\,!!} a^{2p} = \frac{2^{p+1}\pi^p}{(2p-1)\,!!} a^{2p}, \quad p \in N. \tag{11}$$

See also Sec.2.6.6.

2.6. Multiple Integrals

2.6.1. Evaluate the volume bounded by the surface

$$(x^2 + y^2)^2 + z^4 = a^3(y - x), \quad a > 0. \tag{1}$$

Solution. The surface is only in the region for which we have $y - x \geq 0$, or in spherical coordinates, the volume is bounded by $\theta = \pi/4$ and $\theta = 5\pi/4$. In spherical coordinates (r, θ, ϕ), see Appendix A3.2,

$$x = r \sin\phi \cos\theta, \quad y = r \sin\phi \sin\theta, \quad z = r \cos\phi \tag{2}$$

and we have

$$(r\sin^2\phi\cos^2\phi + r^2\sin^2\phi\sin^2\theta)^2 + r^4\cos^4\phi = a^3(r\sin\phi\sin\theta - r\sin\phi\cos\theta)$$
$$r^4(\sin^2\phi\cos^2\theta + \sin^2\phi\sin^2\theta)^2 + r^4\cos^4\phi = a^3 r\sin\phi(\sin\theta - \cos\theta)$$
$$r^4\left[\sin^2\phi(\cos^2\theta + \sin^2\theta)\right]^2 + r^4\cos^4\phi = a^3 r\sin\phi(\sin\theta - \cos\theta)$$
$$r^3(\sin^4\phi + \cos^4\phi) = a^3\sin\phi(\sin\theta - \cos\theta)$$

and (1) becomes

$$r^3 = a^3\frac{\sin\phi(\sin\theta - \cos\theta)}{\sin^4\phi + \cos^4\phi} \tag{3}$$

with the Jacobian of transformation $J = r^2\sin\phi$. The volume is given by the integral

$$V = \iiint_D dx\,dy\,dz = \iiint_{D'} r^2\sin\phi\,dr\,d\theta\,d\phi$$
$$= \int_0^\pi \sin\phi\,d\phi\int_{\frac{\pi}{4}}^{\frac{5\pi}{4}} d\theta\int_0^r r^2\,dr \tag{4}$$
$$= \int_0^\pi \sin\phi\,d\phi\int_{\frac{\pi}{4}}^{\frac{5\pi}{4}} d\theta\cdot\frac{r^3}{3}$$
$$= \frac{a^3}{3}\int_0^\pi \frac{\sin^2\phi\,d\phi}{\sin^4\phi + \cos^4\phi}\int_{\frac{\pi}{4}}^{\frac{5\pi}{4}}(\sin\theta - \cos\theta)\,d\theta. \tag{5}$$

We have

$$I_1 = \int_{\frac{\pi}{4}}^{\frac{5\pi}{4}}(\sin\theta - \cos\theta)\,d\theta = -\cos\theta - \sin\theta\Big|_{\frac{\pi}{4}}^{\frac{5\pi}{4}} = 2\sqrt{2}. \tag{6}$$

For the evaluation of the second integral, change the variable

$$\tan\phi = \sqrt{t},\quad \frac{d\phi}{\cos^2\phi} = \frac{dt}{2\sqrt{t}},\quad d\phi = \cos^2\phi\frac{dt}{2\sqrt{t}},\quad \frac{\sin^2\phi}{\cos^2\phi} = t,\quad \frac{1-\cos^2\phi}{\cos^2\phi} = t,$$
$$1 - \cos^2\phi = t\cos^2\phi,\quad 1 = (1+t)\cos^2\phi,\quad \cos^2\phi = \frac{1}{1+t},$$
$$\sin^2\phi = t\cos^2\phi = \frac{t}{1+t},\quad d\phi = \frac{1}{1+t}\frac{dt}{2\sqrt{t}} = \frac{1}{2}\frac{dt}{(1+t)\sqrt{t}} \tag{7}$$

and taking into consideration the symmetry,

$$I_2 = 2\int_0^{\frac{\pi}{2}} \frac{\sin^2\phi\, d\phi}{\sin^4\phi + \cos^4\phi} = 2\int_0^{\infty} \frac{\dfrac{t}{1+t}\cdot\dfrac{1}{2}\dfrac{dt}{(1+t)\sqrt{t}}}{\dfrac{t^2}{(1+t)^2} + \dfrac{1}{(1+t)^2}}$$

$$= \int_0^{\infty} \frac{\dfrac{t}{1+t}\cdot\dfrac{1}{(1+t)\sqrt{t}}}{\dfrac{t^2+1}{(1+t)^2}}\, dt = \int_0^{\infty} \frac{\sqrt{t}}{t^2+1}\, dt. \tag{8}$$

Using Sec.2.3.20 with $m = 3/2$, $n = 2$ and Eq.(1.35),

$$I_2 = \frac{1}{2}B\left(\frac{1}{4}, \frac{3}{4}\right) = \frac{1}{2}\Gamma\left(\frac{1}{4}\right)\Gamma\left(\frac{3}{4}\right)$$

$$= \frac{1}{2}\frac{\pi}{\sin\dfrac{\pi}{4}} = \frac{1}{2}\frac{\pi}{\dfrac{1}{\sqrt{2}}} = \frac{\pi}{\sqrt{2}} \tag{9}$$

and the volume is

$$V = \frac{a^3}{3}\cdot\frac{\pi}{\sqrt{2}}\cdot 2\sqrt{2} = \frac{2\pi}{3}a^3, \quad a > 0. \tag{10}$$

2.6.2. Study the integral

$$I = \iiint_D (a^2 - x^2 - y^2 - z^2)^p\, dx\, dy\, dz, \quad a \in r^+, \quad p \in R, \tag{1}$$

$$D\{x, y, z \in R^+ mid x^2 + y^2 + z^2 \le a^2\}.$$

Solution. Taking into consideration the symmetries of the sphere, and using the spherical coordinates (see Appendix A3.2), with the Jacobian of transformation $J = r^2 \sin\phi$, the integral evaluation for all spheres is

$$\begin{aligned} I &= 8 \iiint_{D'} (a^2 - r^2)^p r^2 \sin\phi \, dr \, d\theta \, d\phi \\ &= 8 \int_0^{\frac{\pi}{2}} \sin\phi \, d\phi \int_0^{\frac{\pi}{2}} d\theta \int_0^a r^2 (a^2 - r^2)^p \, dr \\ &= 4\pi \int_0^a r^2 (a^2 - r^2)^p \, dr \end{aligned}$$

and with Sec.2.3.9, we have

$$I = 2\pi a^{2p+3} B\left(\frac{3}{2}, p+1\right), \quad p > -1. \tag{2}$$

Using Eqs.(1.58) and (1.59), the integral becomes

$$\begin{aligned} I &= 2\pi a^{2p+3} B\left(1 + \frac{1}{2}, p+1\right) \\ &= \frac{2\pi a^{2p+3}}{2p+3} B\left(\frac{1}{2}, p+1\right) \\ &= \frac{4p\pi a^{2p+3}}{(2p+3)(2p+1)} B\left(\frac{1}{2}, p\right), \quad p > 0. \end{aligned} \tag{3}$$

Particular cases:

a. Let $p = 1$ and from (1) and (3) we have

$$\begin{aligned} I &= \iiint_D (a^2 - x^2 - y^2 - z^2) \, dx \, dy \, dz \\ &= \frac{4\pi^{\frac{3}{2}} a^5}{15} \frac{\Gamma(1)}{\Gamma\left(\frac{3}{2}\right)} \\ &= \frac{4\pi^{\frac{3}{2}} a^5 2}{15\sqrt{\pi}} = \frac{8\pi}{15} a^5. \end{aligned} \tag{4}$$

b. Let $p = 1/2$, and from (1) and (3) we have

$$I = \iiint_D \sqrt{a^2 - x^2 - y^2 - z^2}\, dx\, dy\, dz$$

$$= \frac{2\pi^{\frac{3}{2}} a^4}{(1+3)(1+1)} \frac{\Gamma\left(\frac{1}{2}\right)}{\Gamma\left(\frac{1}{2} + \frac{1}{2}\right)}$$

$$= \frac{2\pi^{\frac{3}{2}} a^4}{4 \times 2} \sqrt{\pi} = \frac{\pi^2}{4} a^4. \tag{5}$$

2.6.3. Evaluate the volume bounded by the surface

$$\left(\frac{x}{a}\right)^n + \left(\frac{y}{b}\right)^n + \left(\frac{z}{c}\right)^n = 1, \quad a, b, c > 0, \quad n \in R. \tag{1}$$

Solution. The volume bounded by the surface is given by the integral

$$V = 8 \iiint_D dx\, dy\, dz, \tag{2}$$

$$D \left\{ x, y, z \geq 0, \left(\frac{x}{a}\right)^n + \left(\frac{y}{b}\right)^n + \left(\frac{z}{c}\right)^n \leq 1 \right\}. \tag{3}$$

Let

$$x = a u^{\frac{2}{n}}, \quad y = b v^{\frac{2}{n}}, \quad z = c w^{\frac{2}{n}} \tag{4}$$

with the Jacobian (see Appendix A5.2)

$$J_3 = \frac{8abc}{n^3} u^{\frac{2}{n}-1} v^{\frac{2}{n}-1} w^{\frac{2}{n}-1} \tag{5}$$

and (2) becomes

$$V = \frac{64abc}{n^3} \iiint_{D_1} u^{\frac{2}{n}-1} v^{\frac{2}{n}-1} w^{\frac{2}{n}-1}\, du\, dv\, dw, \tag{6}$$

$$D_1\{u, v, w \geq 0, \quad u^2 + v^2 + w^2 \leq 1\}. \tag{7}$$

Changing in (6) to spherical coordinates (see Appendix A3.2),

$$\begin{cases} u = r \sin\phi \cos\theta \\ v = r \sin\phi \sin\theta \\ w = r \cos\phi \end{cases} \tag{8}$$

with $r \in [0,1]$, $\phi \in [0,\pi/2]$, $\theta \in [0,\pi/2]$ and the Jacobian $J = r^2 \sin\phi$, we have

$$\begin{aligned}
V &= \frac{64abc}{n^3} \iiint_{D_2} (r\sin\phi\cos\theta)^{\frac{2}{n}-1} (r\sin\phi\sin\theta)^{\frac{2}{n}-1} (r\cos\phi)^{\frac{2}{n}-1} r^2 \sin\phi\, dr\, d\phi\, d\theta \\
&= \frac{64abc}{n^3} \int_0^1 r^{\frac{6}{n}-1}\, dr \int_0^{\frac{\pi}{2}} \sin^{\frac{4}{n}-1}\phi \cos^{\frac{2}{n}-1}\phi\, d\phi \int_0^{\frac{\pi}{2}} \cos^{\frac{2}{n}-1}\theta \sin^{\frac{2}{n}-1}\theta\, d\theta, \qquad (9)
\end{aligned}$$

but

$$\int_0^1 r^{\frac{6}{n}-1}\, dr = \frac{n}{6} r^{\frac{6}{n}}\Big|_0^1 = \frac{n}{6}$$

$$V = \frac{64abc}{6n^2} \int_0^{\frac{\pi}{2}} \sin^{\frac{4}{n}-1}\phi \cos^{\frac{2}{n}-1}\phi\, d\phi \int_0^{\frac{\pi}{2}} \cos^{\frac{2}{n}-1}\theta \sin^{\frac{2}{n}-1}\theta\, d\theta$$

and using Eq.(1.113),

$$\begin{aligned}
V &= \frac{8abc}{3n^2} B\left(\frac{2}{n}, \frac{1}{n}\right) B\left(\frac{1}{n}, \frac{1}{n}\right) \\
&= \frac{8}{3n^2} \frac{\left[\Gamma\left(\frac{1}{n}\right)\right]^3}{\Gamma\left(\frac{3}{n}\right)} abc, \qquad n > 0. \qquad (10)
\end{aligned}$$

Particular cases:

a. Let $a = b = c = R$, $n = 2$ and the surface is a sphere having the volume from (10),

$$\begin{aligned}
V &= \frac{8R^3}{12} \frac{\left[\Gamma\left(\frac{1}{2}\right)\right]^3}{\Gamma\left(\frac{3}{2}\right)} = \frac{4R^3}{3} \frac{\left[\Gamma\left(\frac{1}{2}\right)\right]^3}{\Gamma\left(\frac{1}{2}\right)} \\
&= \frac{4\pi}{3} R^3. \qquad (11)
\end{aligned}$$

b. Let $n = 2$ and the surface is an ellipsoid of semi-axes a, b, c having the volume

$$V = \frac{8abc}{12} \frac{\left[\Gamma\left(\frac{1}{2}\right)\right]^3}{\Gamma\left(\frac{3}{2}\right)} = \frac{2abc}{3} \frac{\left[\Gamma\left(\frac{1}{2}\right)\right]^3}{\frac{1}{2}\Gamma\left(\frac{1}{2}\right)}$$

$$= \frac{4\pi}{3} abc. \tag{12}$$

Note. For the surface area of an ellipsoid, see Sec.4.2.30.

c. Let $a = b = c$, $n = 2/3$ and the surface is

$$x^{\frac{2}{3}} + y^{\frac{2}{3}} + z^{\frac{2}{3}} = a^{\frac{2}{3}} \tag{13}$$

and the volume bounded by this surface is

$$V = 6a^3 \frac{\left[\Gamma\left(\frac{3}{2}\right)\right]^3}{\Gamma\left(\frac{9}{2}\right)} = \frac{4\pi}{35} a^3. \tag{14}$$

d. Let $a = b = c = R$ and the volume bounded by the surface

$$x^n + y^n + z^n = R^n, \quad n > 0, \tag{15}$$

is given by (10),

$$V = \frac{8}{3n^2} \frac{\left[\Gamma\left(\frac{1}{n}\right)\right]^3}{\Gamma\left(\frac{3}{n}\right)} R^3, \quad n > 0. \tag{16}$$

2.6.4. Evaluate the Dirichlet's integral,

$$I = \iiint_D f\left[\left(\frac{x}{a}\right)^m + \left(\frac{y}{b}\right)^n + \left(\frac{z}{c}\right)^p\right] x^{q-1} y^{r-1} z^{s-1}\, dx\, dy\, dz, \tag{1}$$

where $a, b, c > 0$, $m, n, p > 0$, $q, r, s \in R$, and f is an arbitrary continuous function of its argument and D is the integration region defined by

$$D\left\{x, y, z \geq 0, \quad \left(\frac{x}{a}\right)^m + \left(\frac{y}{b}\right)^n + \left(\frac{z}{c}\right)^p \leq 1\right\}. \tag{2}$$

Solution. Let the transformation (see Appendix A5.1)

$$\begin{cases} x = a x_1^{\frac{1}{m}} \\ y = b y_1^{\frac{1}{n}} \\ z = c z_1^{\frac{1}{p}} \end{cases} \Rightarrow \begin{cases} x_1 = \left(\frac{x}{a}\right)^m \\ y_1 = \left(\frac{y}{b}\right)^n \\ z_1 = \left(\frac{z}{c}\right)^p \end{cases} \qquad J_3 = \frac{abc}{mnp} x_1^{\frac{1}{m}-1} y_1^{\frac{1}{n}-1} z_1^{\frac{1}{p}-1} \tag{3}$$

and we have

$$\begin{aligned}
I &= \iiint_{D'} f(x_1 + y_1 + z_1) \cdot (a x_1^{\frac{1}{m}})^{q-1} (b y_1^{\frac{1}{n}})^{r-1} (c z_1^{\frac{1}{p}})^{s-1} \frac{abc}{mnp} x_1^{\frac{1}{m}-1} y_1^{\frac{1}{n}-1} z_1^{\frac{1}{p}-1} \\
&\quad \times dx_1\, dy_1\, dz_1 \\
&= \iiint_{D'} f(x_1 + y_1 + z_1) \frac{a^{q-1+1} b^{r-1+1} c^{s-1+1}}{mnp} x_1^{\frac{q-1}{m}} y_1^{\frac{r-1}{n}} z_1^{\frac{s-1}{p}} x_1^{\frac{1}{m}-1} y_1^{\frac{1}{n}-1} z_1^{\frac{1}{p}-1} \\
&\quad \times dx_1\, dy_1\, dz_1 \\
&= \iiint_{D'} f(x_1 + y_1 + z_1) \frac{a^q b^r c^s}{mnp} x_1^{\frac{q}{m}-\frac{1}{m}+\frac{1}{m}-1} y_1^{\frac{r-1}{n}+\frac{1}{n}-1} z_1^{\frac{s-1}{p}+\frac{1}{p}-1}\, dx_1\, dy_1\, dz_1 \\
&= \frac{a^q b^r c^s}{mnp} \iiint_{D'} f(x_1 + y_1 + z_1) x_1^{\frac{q}{m}-1} y_1^{\frac{r}{n}-1} z_1^{\frac{s}{p}-1}\, dx_1\, dy_1\, dz_1
\end{aligned} \tag{4}$$

where D' is the region in the first octant bounded by the plane $x_1 + y_1 + z_1 = 1$ and the coordinate planes, as shown in (5) and Fig.2.22,

$$D'\{x_1, y_1, z_1 \geq 0, \quad x_1 + y_1 + z_1 \leq 1\}. \tag{5}$$

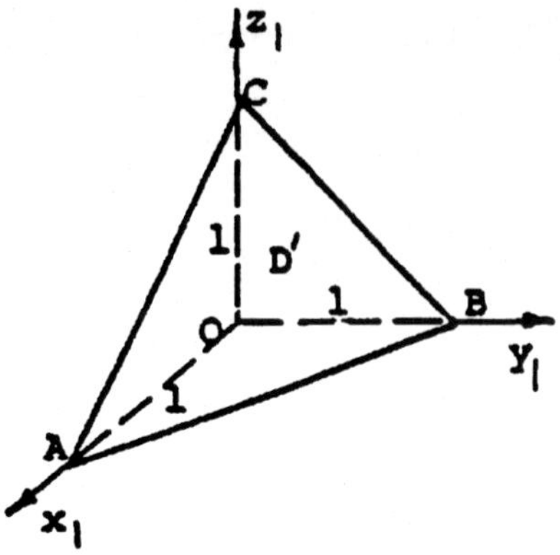

Fig.2.22

Let the second transformation (see Appendix A4) be

$$\begin{cases} x_1 + y_1 + z_1 = u \\ y_1 + z_1 = uv \\ z_1 = uvw \end{cases} \qquad \begin{cases} x_1 = u(1 - v) \\ y_1 = uv(1 - w) \\ z_1 = uvw \end{cases} \tag{6}$$

where u, v, w are the new variables.

Using the notation

$$\begin{cases} F_1 = x_1 + y_1 + z_1 - u \\ F_2 = y_1 + z_1 - uv \\ F_3 = z_1 - uvw \end{cases} \tag{7}$$

the Jacobian of transformation (see Appendix A4) is

$$
\begin{aligned}
J_3 &= \frac{\partial F_1}{\partial u} \cdot \frac{\partial F_2}{\partial v} \cdot \frac{\partial F_3}{\partial w} \\
&= (-1)(-u)(-uv) = -u^2 v, \quad |J_3| = u^2 v, \\
I &= \frac{a^q b^r c^s}{mnp} \iiint_{D''} f(u)\,[u(1-v)]^{\frac{q}{m}-1}\,[uv(1-w)]^{\frac{r}{n}-1}\,(uvw)^{\frac{s}{p}-1} \\
&\quad \times u^2 v\, du\, dv\, dw \qquad (8) \\
&= \frac{a^q b^r c^s}{mnp} \iiint_{D''} f(u) u^{\frac{q}{m}-1+\frac{r}{n}-1+\frac{s}{p}-1} \cdot v^{\frac{r}{n}-1+\frac{s}{p}-1} w^{\frac{s}{p}-1} \\
&\quad \times (1-v)^{\frac{q}{m}-1}(1-w)^{\frac{r}{n}-1} u^2 v\, du\, dv\, dw \qquad (9)
\end{aligned}
$$

where D'' is the region of integration on the uvw space, see Fig.2.23. If x_1, y_1, z_1 are all positive and $x_1 + y_1 + z_1 \leq 1$, then $0 \leq u \leq 1$, $0 \leq v \leq 1$, $0 \leq w \leq 1$ and the tetrahedron D' is replaced by a cube D'' of size 1, in the first octant, as shown in Fig.2.23, and hence

$$D''\{0 \leq u \leq 1, \quad 0 \leq v \leq 1, \quad 0 \leq w \leq 1\}. \qquad (10)$$

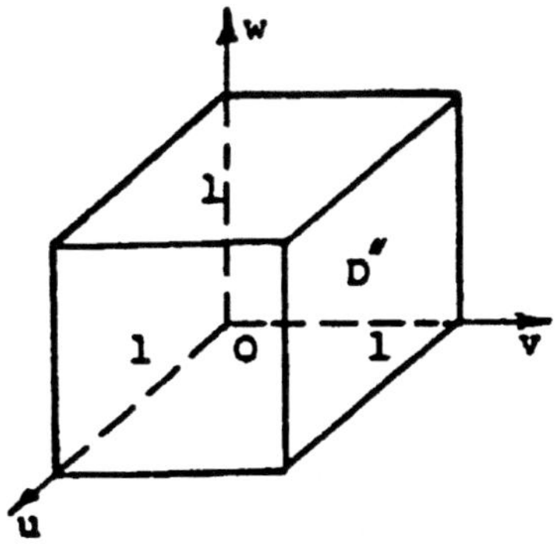

Fig.2.23

Thus, from (9), we have

$$
\begin{aligned}
I &= \frac{a^q b^r c^s}{mnp} \iiint_{D''} f(u) u^{\frac{q}{m}+\frac{r}{n}+\frac{s}{p}-3+2} v^{\frac{r}{n}+\frac{s}{p}-2+1} w^{\frac{s}{p}-1} \\
&\quad \times (1-v)^{\frac{q}{m}-1}(1-w)^{\frac{r}{n}-1}\, du\, dv\, dw \\
&= \frac{a^q b^r c^s}{mnp} \iiint_{D''} f(u) u^{\frac{q}{m}+\frac{r}{n}+\frac{s}{p}-1} v^{\frac{r}{n}+\frac{s}{p}-1} w^{\frac{s}{p}-1} \\
&\quad \times (1-v)^{\frac{q}{m}-1}(1-w)^{\frac{r}{n}-1}\, du\, dv\, dw \\
&= \frac{a^q b^r c^s}{mnp} \int_0^1 f(u) u^{\frac{q}{m}+\frac{r}{n}+\frac{s}{p}-1}\, du \int_0^1 v^{\frac{r}{n}+\frac{s}{p}-1} \\
&\quad \times (1-v)^{\frac{q}{m}-1}\, dv \int_0^1 w^{\frac{s}{p}-1}(1-w)^{\frac{r}{n}-1}\, dw \\
&= \frac{a^q b^r c^s}{mnp} \int_0^1 v^{\frac{r}{n}+\frac{s}{p}-1}(1-v)^{\frac{q}{m}-1}\, dv \int_0^1 w^{\frac{s}{p}-1} \\
&\quad \times (1-w)^{\frac{r}{n}-1}\, dw f(u) u^{\frac{q}{m}+\frac{r}{n}+\frac{s}{p}-1}\, du \\
&= \frac{a^q b^r c^s}{mnp} B\left(\frac{r}{n}+\frac{s}{p}, \frac{q}{m}\right) \cdot B\left(\frac{s}{p}, \frac{r}{n}\right) \int_0^1 f(u) u^{\frac{q}{m}+\frac{r}{n}+\frac{s}{p}-1}\, du \\
&= \frac{a^q b^r c^s}{mnp} \cdot \frac{\Gamma\left(\frac{r}{n}+\frac{s}{p}\right)\Gamma\left(\frac{q}{m}\right)}{\Gamma\left(\frac{r}{n}+\frac{s}{p}+\frac{q}{m}\right)} \cdot \frac{\Gamma\left(\frac{s}{p}\right)\Gamma\left(\frac{r}{n}\right)}{\Gamma\left(\frac{s}{p}+\frac{r}{n}\right)} \int_0^1 f(u) u^{\frac{q}{m}+\frac{r}{n}+\frac{s}{p}-1}\, du \\
&= \frac{a^q b^r c^s}{mnp} \frac{\Gamma\left(\frac{q}{m}\right)\Gamma\left(\frac{r}{n}\right)\Gamma\left(\frac{s}{p}\right)}{\Gamma\left(\frac{q}{m}+\frac{r}{n}+\frac{s}{p}\right)} \int_0^1 f(u) u^{\frac{q}{m}+\frac{r}{n}+\frac{s}{p}-1}\, du \qquad (11)
\end{aligned}
$$

where $q, r, s > 0$. Thus, the integral (1) has been reduced to a simple integral.

Particular cases:

a. Let $a = b = c = R$, $m = n = p = 2$, $q = r = s = 1$, $f(u) = 1$ and we have the volume of $1/8$ of a sphere,

$$
\begin{aligned}
I &= \frac{R^3}{2^3} \frac{\left[\Gamma\left(\frac{1}{2}\right)\right]^3}{\Gamma\left(\frac{3}{2}\right)} \int_0^1 u^{\frac{1}{2}+\frac{1}{2}+\frac{1}{2}-1}\, du \\
&= \frac{R^3}{8} \frac{\pi\sqrt{\pi}}{\frac{1}{2}\Gamma\left(\frac{1}{2}\right)} \int_0^1 u^{\frac{3}{2}-1}\, du \\
&= \frac{\pi R^3}{4} \int_0^1 u^{\frac{1}{2}}\, du = \frac{\pi R^3}{4} \frac{u^{\frac{1}{2}+1}}{\frac{1}{2}+1}\Bigg|_0^1 \\
&= \frac{\pi}{6} R^3
\end{aligned}
$$

and for all volumes of spheres,

$$
V_T = 8\frac{\pi R^3}{6} = \frac{4\pi}{3} R^3.
$$

Evaluate the volume of a sphere using a traditional method.

b. Let $m = n = p = 2$ and the integral (1) becomes

$$
I = \iiint_D f\left(\frac{x^2}{a^2} + \frac{y^2}{b^2} + \frac{z^2}{c^2}\right) x^{q-1} y^{r-1} z^{s-1}\, dx\, dy\, dz,
$$

$$
D\left\{x, y, z \geq 0,\quad \frac{x^2}{a^2} + \frac{y^2}{b^2} + \frac{z^2}{c^2} \leq 1\right\}
$$

and from (11),

$$
I = \frac{a^q b^r c^s}{8} \frac{\Gamma\left(\frac{q}{2}\right)\Gamma\left(\frac{r}{2}\right)\Gamma\left(\frac{s}{2}\right)}{\Gamma\left(\frac{q+r+s}{2}\right)} \int_0^1 f(u) u^{\frac{q+r+s}{2}-1}\, du.
$$

c. Let $f(x,y,z) = 1$, $m = n = p = 2$, $q = r = s = 1$, and the integral (1) becomes

$$I = \iiint_D dx\,dy\,dz,$$

$$D\left\{x, y, z \geq 0, \quad \frac{x^2}{a^2} + \frac{y^2}{b^2} + \frac{z^2}{c^2} \leq 1\right\},$$

which is 1/8 of the volume of an ellipsoid. From (11), we have

$$I = \frac{abc}{8} \frac{\left[\Gamma\left(\frac{1}{2}\right)\right]^3}{\Gamma\left(\frac{3}{2}\right)} \int_0^1 u^{\frac{1}{2}}\,du$$

$$= \frac{abc}{8} \frac{2\pi\sqrt{\pi}}{\sqrt{\pi}} \cdot \frac{2}{3} u^{\frac{3}{2}}\Big|_0^1 = \frac{\pi}{6} abc.$$

The total volume of an ellipsoid is

$$V = 8 \times I = 8\frac{\pi}{6} abc = \frac{4\pi}{3} abc.$$

d. Let $f[] = 1$, $m = n = p$, $q = r = s = 1$ and the integral (1) becomes

$$I = \iiint_D dx\,dy\,dz,$$

$$D\left\{x, y, z \geq 0, \quad \left(\frac{x}{a}\right)^m + \left(\frac{y}{b}\right)^m + \left(\frac{z}{c}\right)^m \leq 1\right\},$$

which is 1/8 the volume bounded by the surface $(x/a)^m + (y/b)^m + (z/c)^m = 1$. Therefore using (11), we have

$$V = \frac{8abc}{m^3} \frac{\left[\Gamma\left(\frac{1}{m}\right)\right]^3}{\Gamma\left(\frac{3}{m}\right)} \int_0^1 u^{\frac{3}{m}-1}\,du = \frac{8abc\left[\Gamma\left(\frac{1}{m}\right)\right]^3}{m^3\Gamma\left(\frac{3}{m}\right)} \frac{u^{\frac{3}{m}}}{\frac{3}{m}}\Bigg|_0^1$$

$$= \frac{8abc\left[\Gamma\left(\frac{1}{m}\right)\right]^3}{m^3\Gamma\left(\frac{3}{m}\right)} \frac{m}{3} = \frac{8}{3m^2} \frac{\left[\Gamma\left(\frac{1}{m}\right)\right]^3}{\Gamma\left(\frac{3}{m}\right)} abc, \quad m > 0,$$

which is also found in Sec.2.6.3.

2.6.5. Study the integral

$$I = \iiint_D (a^2 - x^2 - y^2 - z^2)^m x^p y^q z^s \, dx\, dy\, dz, \tag{1}$$

$$D\{x, y, z \geq 0, \quad x^2 + y^2 + z^2 \leq a^2\} \tag{2}$$

and $a > 0$, $m, p, q, s \in R$.

Solution 1. Using the spherical coordinates (see Appendix A3.2)

$$x = r \sin\phi \cos\theta, \quad y = r \sin\phi \sin\theta, \quad z = r\cos\phi, \tag{3}$$

$$r \geq 0, \quad 0 \leq \phi \leq \pi, \quad 0 \leq \theta \leq 2\pi,$$

with the Jacobian of transformation

$$J_3 = r^2 \sin\phi \tag{4}$$

the integral (1) becomes

$$I = \iiint_{D'} (a^2 - r^2)^m (r \sin\phi\cos\theta)^p (r\sin\phi\sin\theta)^q (r\cos\phi)^s r^2 \sin\phi \, dr\, d\phi\, d\theta$$

$$= \int_0^{\frac{\pi}{2}} (\sin\phi)^{p+q+1} \cos^s\phi \, d\phi \int_0^{\frac{\pi}{2}} \sin^q\theta \cos^p\theta \, d\theta \int_0^a r^{p+q+s+2}(a^2 - r^2)^m \, dr. \tag{5}$$

Using Eq.(1.116), we have

$$I_1 = \int_0^{\frac{\pi}{2}} (\sin\phi)^{p+q+1} \cos^s\phi \, d\phi = \frac{1}{2} B\left(\frac{p+q+2}{2}, \frac{s+1}{2}\right), \tag{6}$$

$$I_2 = \int_0^{\frac{\pi}{2}} \sin^q\theta \cdot \cos^p\theta \, d\theta = \frac{1}{2} B\left(\frac{p+1}{2}, \frac{q+1}{2}\right). \tag{7}$$

For the last integral, using Sec.2.3.9

$$I_3 = \int_0^a r^{p+q+s+2}(a^2 - r^2)^m \, dr$$

$$= \frac{1}{2} a^{p+q+s+2m+3} B\left(\frac{p+q+s+3}{2}, m+1\right). \tag{8}$$

Thus, the evaluation of the integral (1) is

$$
\begin{aligned}
I &= \frac{1}{8}a^{p+q+s+2m+3}B\left(\frac{p+q+s+3}{2}, m+1\right)\cdot B\left(\frac{p+q+2}{2}, \frac{s+1}{2}\right) \\
&\quad \times B\left(\frac{p+1}{2}, \frac{q+1}{2}\right) \\
&= \frac{1}{8}a^{p+q+s+2m+3}\frac{\Gamma(m+1)\Gamma\left(\frac{p+1}{2}\right)\Gamma\left(\frac{q+1}{2}\right)\Gamma\left(\frac{s+1}{2}\right)}{\Gamma\left(\frac{p+q+s+2m+5}{2}\right)} \qquad (9)
\end{aligned}
$$

with the convergence conditions $m, p, q, s > -1$.

Solution 2. Using the transformation

$$
x = ay_1^{\frac{1}{2}}, \quad y = ay_2^{\frac{1}{2}}, \quad z = ay_3^{\frac{1}{2}} \qquad (10)
$$

with the Jacobian (see Appendix A5.1)

$$
J = \frac{a^3}{8}y_1^{-\frac{1}{2}}y_2^{-\frac{1}{2}}y_3^{-\frac{1}{2}} \qquad (11)
$$

the integral (1) becomes

$$
\begin{aligned}
I &= \iiint_{D'} a^{2m}(1-y_1-y_2-y_3)^m \cdot \left(ay_1^{\frac{1}{2}}\right)^p \left(ay_2^{\frac{1}{2}}\right)^q \left(ay_3^{\frac{1}{2}}\right)^s \\
&\quad \times \frac{a^3}{8}y_1^{-\frac{1}{2}}y_2^{-\frac{1}{2}}y_3^{-\frac{1}{2}}\, dy_1\, dy_2\, dy_3 \\
&= \frac{1}{8}a^{p+q+s+2m+3}\iiint_{D'}(1-y_1-y_2-y_3)^m y_1^{\frac{p-1}{2}} y_2^{\frac{q-1}{2}} y_3^{\frac{s-1}{2}}\, dy_1\, dy_2\, dy_3, \qquad (12) \\
&\qquad D'\{y_1, y_2, y_3 \geq 0, \quad y_1+y_2+y_3 \leq 1\}. \qquad (13)
\end{aligned}
$$

Using the transformation

$$
\begin{cases} y_1+y_2+y_3 = u \\ y_2+y_3 = uv \\ y_3 = uvw \end{cases} \qquad \begin{cases} y_1 = u(1-v) \\ y_2 = uv(1-w) \\ y_3 = uvw \end{cases} \qquad (14)
$$

with the Jacobian of transformation (see Appendix A4 or Sec.2.6.4), $|J_3| = u^2 v$, the integral (12) becomes

$$I = \frac{a^{p+q+s+2m+3}}{8} \iiint_{D''} (1-u)^m [u(1-v)]^{\frac{p-1}{2}} [uv(1-w)]^{\frac{q-1}{2}} [uvw]^{\frac{s-1}{2}} \times u^2 v \, du \, dv \, dw \tag{15}$$

where, from (13) and (14), the integration region D'' is a cube with the sizes equal to 1,

$$D''\{u, v, w \geq 0, \quad u \leq 1, \quad v \leq 1, \quad w \leq 1\}. \tag{16}$$

Thus, we have

$$\begin{aligned}
I &= \frac{a^{p+q+s+2m+3}}{8} \int_0^1 (1-u)^m u^{\frac{p+q+s+1}{2}} du \int_0^1 (1-v)^{\frac{p-1}{2}} v^{\frac{q+s}{2}} dv \\
&\quad \times \int_0^1 (1-w)^{\frac{q-1}{2}} w^{\frac{s-1}{2}} dw \\
&= \frac{a^{p+q+s+2m+3}}{8} \int_0^1 (1-u)^{(1+m)-1} u^{\left(1+\frac{p+q+s+1}{2}\right)-1} du \\
&\quad \times \int_0^1 (1-v)^{\left(1+\frac{p-1}{2}\right)-1} v^{\left(1+\frac{q+s}{2}\right)-1} dv \\
&\quad \times \int_0^1 (1-w)^{\left(1+\frac{q-1}{2}\right)-1} w^{\left(1+\frac{s-1}{2}\right)-1} dw \\
&= \frac{a^{p+q+s+2m+3}}{8} B\left(\frac{p+q+s+3}{2}, m+1\right) B\left(\frac{q+s+2}{2}, \frac{p+1}{2}\right) \\
&\quad \times B\left(\frac{s+1}{2}, \frac{q+1}{2}\right) \\
&= \frac{a^{p+q+s+2m+3}}{8} \frac{\Gamma(m+1)\Gamma\left(\frac{p+1}{2}\right)\Gamma\left(\frac{q+1}{2}\right)\Gamma\left(\frac{s+1}{2}\right)}{\Gamma\left(\frac{p+q+s+2m+5}{2}\right)}
\end{aligned} \tag{17}$$

with the convergence conditions $m, p, q, s > -1$.

Particular cases:

a. Let $p = q = s = 0$, and the integral becomes

$$I = \iiint_D (a^2 - x^2 - y^2 - z^2)^m \, dx\, dy\, dz = \frac{a^{2m+3}}{8} \frac{\Gamma(m+1) \cdot \left[\Gamma\left(\frac{1}{2}\right)\right]^3}{\Gamma\left(\frac{2m+5}{2}\right)}, \tag{18}$$

$$D\{x, y, z \geq 0, \quad x^2 + y^2 + z^2 \leq a^2\},$$

$$I = \frac{a^{2m+3}\pi^{\frac{3}{2}}}{8} \frac{\Gamma(m+1)}{\Gamma\left(\frac{2m+5}{2}\right)}, \qquad m > -1. \tag{19}$$

b. Let $p = q = s = m = a = 1$. The integral (1) becomes

$$I = \iiint_D (1 - x^2 - y^2 - z^2) xyz \, dx\, dy\, dz = \frac{1}{8\Gamma(5)} = \frac{1}{192}, \tag{20}$$

$$D\{x, y, z \geq 0, \quad x^2 + y^2 + z^2 \leq 1\}.$$

c. Let $p = q = s = m = 0$, $a = R$, and the integral represents $1/8$ of the volume of a sphere of radius R. From (19) we have

$$V = \iiint_D dx\, dy\, dz = \frac{R^3 \pi^{\frac{3}{2}}}{8} \frac{1}{\Gamma\left(\frac{5}{2}\right)} = \frac{R^3 \pi^{\frac{3}{2}}}{8} \frac{1}{\frac{3}{4}\sqrt{\pi}}$$

$$= \frac{\pi R^3}{6}. \tag{21}$$

Thus, multiplying by 8 in (21), we get the volume of a sphere

$$V_S = 8 \frac{\pi R^3}{6} = \frac{4\pi}{3} R^3. \tag{22}$$

2.6.6. Evaluate the volume and the surface area of a hypersphere of radius R,

$$x_1^2 + x_2^2 + \cdots + x_n^2 = R^2, \quad R > 0. \tag{1}$$

Solution 1. The volume (V_n) of a hypersphere is given by the integral

$$V_n = \int \cdots \int_D dx_1\, dx_2 \ldots dx_n, \tag{2}$$

$$D\{x_1, x_2, \ldots, x_n \in R^+ \mid x_1^2 + x_2^2 + \cdots + x_n^2 \leq R^2\}. \tag{3}$$

Changing to spherical coordinates, we have

$$V_n = \int \cdots \int_{D_1} |J_n|\, dr\, d\theta_1\, d\theta_2 \ldots d\theta_{n-1} \tag{4}$$

where (see Appendix A3.3)

$$J_n = r^{n-1} \sin^{n-2}\theta_1 \sin^{n-3}\theta_2 \ldots \sin\theta_{n-2} \tag{5}$$

and the volume becomes

$$V_n = \int_0^R r^{n-1}\, dr \int_0^\pi \sin^{n-2}\theta_1\, d\theta_1 \int_0^\pi \sin^{n-3}\theta_2\, d\theta_2 \ldots$$
$$\ldots \int_0^\pi \sin\theta_{n-2}\, d\theta_{n-2} \int_0^{2\pi} d\theta_{n-1}. \tag{6}$$

From (6),

$$\int_0^R r^{n-1}\, dr = \frac{r^n}{n}\bigg|_0^R = \frac{R^n}{n}, \tag{7}$$

$$\int_0^{2\pi} d\theta_{n-1} = 2\pi. \tag{8}$$

Using Eq.(1.120), we have from (6)

$$\begin{cases} I_1 = 2\int_0^{\frac{\pi}{2}} \sin^{n-2}\theta_1\, d\theta_1 = B\left(\frac{n-1}{2}, \frac{1}{2}\right) \\ I_2 = 2\int_0^{\frac{\pi}{2}} \sin^{n-3}\theta_2\, d\theta_2 = B\left(\frac{n-2}{2}, \frac{1}{2}\right) \\ \cdots\cdots\cdots\cdots\cdots\cdots \\ I_{n-2} = 2\int_0^{\frac{\pi}{2}} \sin\theta_{n-2}\, d\theta_{n-2} = B\left(\frac{2}{2}, \frac{1}{2}\right) \end{cases} \tag{9}$$

From (6) with (7), (8) and (9), we get the volume of a sphere,

$$V_n = \frac{2\pi R^n}{n} \cdot \frac{\Gamma\left(\frac{n-1}{2}\right)\sqrt{\pi}}{\Gamma\left(\frac{n}{2}\right)} \cdot \frac{\Gamma\left(\frac{n-2}{2}\right)\sqrt{\pi}}{\Gamma\left(\frac{n-1}{2}\right)} \cdots \frac{\Gamma\left(\frac{2}{2}\right)\sqrt{\pi}}{\Gamma\left(\frac{3}{2}\right)}$$

$$= \frac{2\pi R^n}{n} \cdot \frac{(\pi^{\frac{1}{2}})^{n-2}}{\Gamma\left(\frac{n}{2}\right)} = \frac{2\pi R^n}{n} \cdot \frac{\pi^{\frac{n-2}{2}}}{\Gamma\left(\frac{n}{2}\right)} = \frac{2\pi R^n \pi^{\frac{n-2}{2}}}{n\Gamma\left(\frac{n}{2}\right)}$$

$$= \frac{2\pi^{\frac{n-2}{2}+1}}{n\Gamma\left(\frac{n}{2}\right)} R^n = \frac{2\pi^{\frac{n}{2}}}{n\Gamma\left(\frac{n}{2}\right)} R^n, \quad n \in N. \tag{10}$$

Solution 2. The volume V_n of a sphere of radius R, in an n-dimensional space, is proportional to R^n and we have [7]

$$V_n = K \cdot R^n \tag{11}$$

where K is a constant which depends only on the dimensions of the space. Moreover, we can write the volume element dV_n in the following form,

$$dV_n = S_n \cdot dR, \tag{12}$$

from which we have

$$S_n = \frac{dV_n}{dR} = nKR^{n-1} \tag{13}$$

where S_n is the surface area of the hypersphere. To evaluate the constant K, we use Eq.(1.76),

$$\int_{-\infty}^{+\infty} e^{-x_i^2}\, dx_i = \sqrt{\pi}, \quad i = 1, 2, \ldots, n, \tag{14}$$

and multiplying n such integrals, one for each variable, $x_i (i = 1, 2, \ldots, n)$, we get

$$\pi^{\frac{n}{2}} = \int_{x_i=-\infty}^{+\infty} \cdots \int e^{-(x_1^2+x_2^2+\cdots+x_n^2)}\, dx_1\, dx_2 \ldots dx_n$$

$$= \int_0^{\infty} e^{-R^2}\, dV_n \tag{15}$$

and substituting dV_n from (13), we have

$$\pi^{\frac{n}{2}} = nK \int_0^\infty e^{-R^2} R^{n-1}\, dR \tag{16}$$

and with Eq.(1.87) for $m = n$ and $x = R$,

$$\pi^{\frac{n}{2}} = nK \cdot \frac{1}{2} \Gamma\left(\frac{n}{2}\right), \tag{17}$$

and hence

$$K = \frac{2}{n} \frac{\pi^{\frac{n}{2}}}{\Gamma\left(\frac{n}{2}\right)}. \tag{18}$$

Substituting the value of K in (11) and (13), we have the volume and the surface area of the hypersphere in the n-dimensional space,

$$V_n = \frac{2\pi^{\frac{n}{2}}}{n\Gamma\left(\frac{n}{2}\right)} R^n, \quad n \in N, \tag{19}$$

$$S_n = \frac{2\pi^{\frac{n}{2}}}{\Gamma\left(\frac{n}{2}\right)} R^{n-1}, \quad n \in N. \tag{20}$$

See also Sec.2.5.3.

Particular cases:

a. Let $n = 3$, and Eq.(1) becomes

$$x^2 + y^2 + z^2 = R^2, \tag{21}$$

which is a sphere in the 3-dimensional space. Using (19) and (20) we have respectively,

$$V_3 = \frac{2\pi^{\frac{3}{2}}}{3\Gamma\left(\frac{3}{2}\right)} R^3 = \frac{2\pi^{\frac{3}{2}}}{3 \cdot \frac{\sqrt{\pi}}{2}} R^3 = \frac{4\pi R^3}{3}, \tag{22}$$

$$S_3 = \frac{2\pi^{\frac{3}{2}}}{\Gamma\left(\frac{3}{2}\right)} R^2 = \frac{2\pi^{\frac{3}{2}}}{\frac{\sqrt{\pi}}{2}} R^2 = 4\pi R^2. \tag{23}$$

b. Let $n = 2$, and Eq.(1) becomes

$$x^2 + y^2 = R^2, \tag{24}$$

which is a circle. Using (19) and (20), we have

$$V_2 = \frac{2\pi}{2\Gamma(1)} R^2 = \pi R^2, \tag{25}$$

$$S_2 = 2\pi R, \tag{26}$$

which are respectively the area and circumference of a circle of radius R.

c. Let $n = 2p$, $p \in N$, and the volume and surface of this hypersphere become

$$V_{2p} = \frac{\pi^p}{p\Gamma(p)} R^{2p} = \frac{\pi^p}{p\,!} R^{2p}, \quad p \in N, \tag{27}$$

$$S_{2p} = \frac{2\pi^p}{\Gamma(p)} R^{2p-1} = \frac{2\pi^p}{(p-1)\,!} R^{2p-1}, \quad p \in N. \tag{28}$$

d. Let $n = 2p + 1$, $p \in N$, and the volume and surface of a hypersphere are

$$V_{2p+1} = \frac{2\pi^{p+\frac{1}{2}}}{(2p+1)\Gamma\left(p + \frac{1}{2}\right)} R^{2p+1}, \quad p \in N, \tag{29}$$

$$S_{2p+1} = \frac{2\pi^{p+\frac{1}{2}}}{\Gamma\left(p + \frac{1}{2}\right)} R^{2p}, \quad p \in N, \tag{30}$$

or using Eq.(1.43),

$$V_{2p+1} = \frac{2^{p+1}\pi^p}{(2p+1)\,!!} R^{2p+1}, \quad p \in N, \tag{31}$$

$$S_{2p+1} = \frac{2^{p+1}\pi^p}{(2p-1)\,!!} R^{2p}, \quad p \in N. \tag{32}$$

See also Sec.2.5.3.

2.6.7. Evaluate the volume of the hyperellipsoid

$$\left(\frac{x_1}{a_1}\right)^2 + \left(\frac{x_2}{a_2}\right)^2 + \cdots + \left(\frac{x_n}{a_n}\right)^2 = 1, \quad a_1, a_2, \ldots, a_n > 0. \tag{1}$$

Solution. Changing the variables with the transformation,

$$\frac{x_1}{a_1} = \frac{X_1}{R}, \frac{x_2}{a_2} = \frac{X_2}{R}, \ldots, \frac{x_n}{a_n} = \frac{X_n}{R}, \tag{2}$$

the Jacobian of transformation is (see Appendix A5.3)

$$J_n = \frac{a_1 a_2 \ldots a_n}{R^n} \tag{3}$$

and the equation of hyperellipsoid becomes

$$X_1^2 + X_2^2 + \cdots + X_n^2 = R^2, \tag{4}$$

which is the equation of a hypersphere. The volume of the hyperellipsoid can be evaluated by the multiple integral

$$V_n = \int_V J_n \, dV \tag{5}$$

where

$$V\{X_1, X_2, \ldots, X_n \in R^+ \mid X_1^2 + X_2^2 + \cdots + X_n^2 \leq R^2\}$$

and

$$V_n = \frac{a_1 a_2 \ldots a_n}{R^n} \int_V dX_1 \, dX_2 \ldots dX_n. \tag{6}$$

The last integral is the volume of a hypersphere. Using the result from Sec.2.6.6, the volume of the hyperellipsoid is

$$\begin{aligned} V_n &= \frac{a_1 \; a_2 \; \ldots \; a_n}{R^n} \cdot \frac{2\pi^{\frac{n}{2}}}{n\Gamma\left(\frac{n}{2}\right)} R^n \\ &= \frac{2\pi^{\frac{n}{2}}}{n\Gamma\left(\frac{n}{2}\right)} a_1 \; a_2 \; \ldots \; a_n, \quad n \in N. \end{aligned} \tag{7}$$

Particular cases:

a. Let $n = 2$. Eq.(1) becomes

$$\left(\frac{x_1}{a_1}\right)^2 + \left(\frac{x_2}{a_2}\right)^2 = 1, \tag{8}$$

which is an ellipse, with the area

$$V_2 = \frac{2\pi}{2\Gamma(1)} a_1 \cdot a_2 = \pi a_1 a_2. \tag{9}$$

b. Let $n = 3$. Eq.(1) becomes

$$\left(\frac{x_1}{a_1}\right)^2 + \left(\frac{x_2}{a_2}\right)^2 + \left(\frac{x_3}{a_3}\right)^2 = 1, \tag{10}$$

which is an ellipsoid, with the volume

$$V_3 = \frac{2\pi^{\frac{3}{2}}}{3\Gamma\left(\frac{3}{2}\right)} a_1 a_2 a_3 = \frac{4\pi}{3} a_1 a_2 a_3. \tag{11}$$

See also Sec.2.6.3.

c. Let $n = 3$, $a_1 = a_2 = a_3 = R$, and (1) becomes a sphere with the volume, from (11),

$$V_3 = \frac{4\pi}{3} R^3. \tag{12}$$

2.6.8. Evaluate the Dirichlet's integral,

$$I = \int \cdots \int_D f(x_1 + x_2 + \cdots + x_n) x_1^{p_1 - 1} x_2^{p_2 - 1} \ldots x_n^{p_n - 1}\, dx_1\, dx_2 \ldots dx_n, \tag{1}$$

$$D\{x_1, x_2, \ldots, x_n \geq 0, \quad x_1 + x_2 + \cdots + x_n \leq 1\}, \tag{2}$$

$p_1, p_2, \ldots, p_n \in R$, and f is an arbitrary continuous function of its argument.

Solution. Let the transformation

$$\begin{cases} x_1 + x_2 + \cdots + x_n = y_1 \\ x_2 + \cdots + x_n = y_1 y_2 \\ \cdots\cdots\cdots\cdots\cdots\cdots\cdots \\ x_{n-1} + x_n = y_1 y_2 \ldots y_{n-1} \\ x_n = y_1 y_2 \ldots y_{n-1} y_n \end{cases} \Rightarrow \begin{cases} x_1 = y_1(1 - y_2) \\ x_2 = y_1 y_2 (1 - y_3) \\ \cdots\cdots\cdots\cdots\cdots\cdots\cdots \\ x_{n-1} = y_1 y_2 \ldots y_{n-1}(1 - y_n) \\ x_n = y_1 y_2 \ldots y_{n-1} y_n \end{cases} . \tag{3}$$

According to Appendix A4, the Jacobian of transformation is

$$|J_n| = y_1^{n-1} y_2^{n-2} \ldots y_{n-2}^2 y_{n-1} \tag{4}$$

and (1) becomes

$$\begin{aligned} I = \int \cdots \int_{D'} & f(y_1)[y_1(1 - y_2)]^{p_1 - 1}[y_1 y_2 (1 - y_3)]^{p_2 - 1} \ldots \\ & \ldots [y_1 y_2 \ldots y_{n-1}(1 - y_n)]^{p_{n-1} - 1} \\ & \times (y_1 y_2 \ldots y_n)^{p_n - 1} \cdot y_1^{n-1} y_2^{n-2} \ldots y_{n-2}^2 y_{n-1} \cdot dy_1\, dy_2 \ldots dy_n \end{aligned} \tag{5}$$

where the integration region D' is given by (2) and (3),

$$D'\{0 \le y_1 \le 1, 0 \le y_2 \le 1, \ldots, 0 \le y_n \le 1\}, \tag{6}$$

which is a hypercube in the R^n-space. From (5) we have

$$\int_0^1 f(y_1)\cdot y_1^{p_1-1+p_2-1+\cdots+p_n-1+n-1}\,dy_1$$
$$=\int_0^1 f(y_1)\cdot y_1^{p_1+p_2+\cdots+p_{n-1}+p_n-n+n-1}\,dy_1$$
$$=\int_0^1 f(y_1)\cdot y_1^{p_1+p_2+\cdots+p_n-1}\,dy_1. \tag{7}$$

$$\int_0^1 y_2^{p_2-1+\cdots+p_n-1+n-2}(1-y_2)^{p_1-1}\,dy_2$$
$$=\int_0^1 y_2^{p_2+p_3+\cdots+p_n-(n-1)+n-2}(1-y_2)^{p_1-1}\,dy_2$$
$$=\int_0^1 y_2^{p_2+\cdots+p_n-n+1+n-2}(1-y_2)^{p_1-1}\,dy_2$$
$$=\int_0^1 y_2^{p_2+\cdots+p_n-1}(1-y_2)^{p_1-1}\,dy_2. \tag{8}$$

$$\int_0^1 y_3^{p_3+\cdots+p_n-(n-2)+n-3}(1-y_3)^{p_2-1}\,dy_3$$
$$=\int_0^1 y_3^{p_3+\cdots+p_n+2-3}(1-y_3)^{p_2-1}\,dy_3$$
$$=\int_0^1 y_3^{p_3+\cdots+p_n-1}(1-y_3)^{p_2-1}\,dy_3$$
$$\cdots\int_0^1 y_n^{p_n-1}(1-y_n)^{p_{n-1}-1}\,dy_n \tag{9}$$

and

$$I=\int_0^1 y_2^{p_2+\cdots+p_n-1}(1-y_2)^{p_1-1}\,dy_2\cdot\int_0^1 y_3^{p_3+\cdots+p_n-1}(1-y_3)^{p_2-1}\,dy_3\times\cdots$$
$$\times\int_0^1 y_n^{p_n-1}(1-y_n)^{p_n-1}\,dy_n\int_0^1 f(y_1)y_1^{p_1+p_2+\cdots+p_n-1}\,dy_1 \tag{10}$$
$$=B(p_2+\cdots+p_n,p_1)B(p_3+\cdots+p_{n-1},p_2)\ldots B(p_n,p_{n-1})$$
$$\times\int_0^1 f(y_1)y_1^{p_1+p_2+\cdots+p_n-1}\,dy_1$$

$$= \frac{\Gamma(p_2+p_3+\cdots+p_n)\Gamma(p_1)}{\Gamma(p_1+p_2+\cdots+p_n)} \cdot \frac{\Gamma(p_3+\cdots+p_n)\Gamma(p_2)}{\Gamma(p_2+p_3+\cdots+p_n)} \times \ldots$$

$$\times \frac{\Gamma(p_{n-1})\Gamma(p_n)}{\Gamma(p_{n-1}+p_n)} \int_0^1 f(y_1) y_1^{p_1+p_2+\cdots+p_n-1}\, dy_1$$

$$= \frac{\Gamma(p_1)\Gamma(p_2)\cdots\Gamma(p_{n-1})\Gamma(p_n)}{\Gamma(p_1+p_2+\cdots+p_n)} \int_0^1 f(y_1) y_1^{p_1+p_2+\cdots+p_n-1}\, dy_1 \tag{11}$$

where $p_1, p_2, \ldots, p_n > 0$. Thus, the integral (1) has been reduced to a simple integral, under the assumption that the integral in (7) converges absolutely. See also Sec.2.6.9.

Particular cases:

a. Let $n = 2$ and the integral becomes

$$I = \iint_D f(x_1 + x_2) x_1^{p_1-1} x_2^{p_2-1}\, dx_1\, dx_2, \tag{12}$$

$$D\{x_1, x_2 \geq 0, \quad x_1 + x_2 \leq 1\}$$

and using (11), we have

$$I = \frac{\Gamma(p_1)\Gamma(p_2)}{\Gamma(p_1+p_2)} \int_0^1 f(y_1) \cdot y_1^{p_1+p_2-1}\, dy_1, \tag{13}$$

a result also obtained in Sec.2.4.6.

b. Let $n = 3$, $p_1 = p_2 = p_3 = 1$, $f = 1$ and the evaluation of the integral is the volume of the pyramid enclosed by the planes $x_1 = 0$, $x_2 = 0$, $x_3 = 0$ and $x_1 + x_2 + x_3 = 1$,

$$\iiint_D dx_1\, dx_2\, dx_3 = \frac{1}{\Gamma(3)} \int_0^1 y_1^2\, dy_1 = \frac{1}{2} \frac{y_1^3}{3}\Big|_0^1 = \frac{1}{6}, \tag{14}$$

$$D\{x_1, x_2, x_3 \geq 0, \quad x_1 + x_2 + x_3 \leq 1\}.$$

2.6.9. Study the integral

$$I = \int \ldots \int_D x_1^{p_1-1} x_2^{p_2-1} \ldots x_n^{p_n-1} (1 - x_1 - x_2 - \cdots - x_n)^{p-1}\, dx_1\, dx_2 \ldots dx_n, \tag{1}$$

$$D\{x_1, x_2, , \ldots, x_n \geq 0, \quad x_1 + x_2 + \cdots + x_n \leq 1\} \tag{2}$$

and $p, p_i \in R$, $i = 1, 2, \ldots, n$.

Solution. This is a Dirichlet integral, see Sec.2.6.8, where $f(x_1 + x_2 + \cdots + x_n) = (1 - x_1 - x_2 - \cdots - x_n)^{p-1}$ and with the transformation

$$\begin{cases} x_1 + x_2 + \cdots + x_n = y_1 \\ x_2 + \cdots + x_n = y_1 y_2 \\ \cdots\cdots\cdots\cdots\cdots\cdots\cdots \\ x_n = y_1 y_2 \ldots y_n \end{cases} \Rightarrow \begin{cases} x_1 = y_1(1 - y_2) \\ x_2 = y_1 y_2 (1 - y_3) \\ \cdots\cdots\cdots \\ x_{n-1} = y_1 y_2 \ldots y_{n-1}(1 - y_n) \\ x_n = y_1 y_2 \ldots y_{n-1} y_n \end{cases} \tag{3}$$

$$|J_n| = y_1^{n-1} y_2^{n-2} \ldots y_{n-2}^2 y_{n-1}$$

the region D will be changed into a hypercube with vertex 1 on the new coordinate axes $y_1, y_2, \ldots, y_n$ and working similarly to Sec.2.6.8, from Eq.(11), we have

$$I = \frac{\Gamma(p_1)\Gamma(p_2)\ldots\Gamma(p_n)}{\Gamma(p_1 + p_2 + \cdots + p_n)} \int_0^1 (1 - y_1)^{p-1} y_1^{(p_1 + p_2 + \cdots + p_n) - 1} \, dy_1 \tag{4}$$

$$= \frac{\Gamma(p_1)\Gamma(p_2)\cdots\Gamma(p_n)}{\Gamma(p_1 + p_2 + \cdots + p_n)} B(p_1 + p_2 + \cdots + p_n, p)$$

$$= \frac{\Gamma(p_1)\Gamma(p_2)\ldots\Gamma(p_n)}{\Gamma(p_1 + p_2 + \cdots + p_n)} \cdot \frac{\Gamma(p_1 + p_2 + \cdots + p_n) \cdot \Gamma(p)}{\Gamma(p_1 + p_2 + \cdots + p_n + p)}$$

$$= \frac{\Gamma(p_1)\Gamma(p_2)\cdots\Gamma(p_n)\Gamma(p)}{\Gamma(p_1 + p_2 + \cdots + p_n + p)}, \quad p_1, p_2, \ldots, p_n, p > 0. \tag{5}$$

Particular cases:

a. Let $p = 1$, and (1) becomes

$$I = \iint \cdots \int_D x_1^{p_1 - 1} x_2^{p_2 - 1} \ldots x_n^{p_n - 1} \, dx_1 \, dx_2 \ldots dx_n, \tag{6}$$

$$D\{x_1, x_2, \ldots, x_n \geq 0, , \quad x_1 + x_2 + \cdots + x_n \leq 1\} \tag{7}$$

and from (5) we have

$$I = \frac{\Gamma(p_1)\Gamma(p_2)\ldots\Gamma(p_n)}{\Gamma(p_1 + p_2 + \cdots + p_n + 1)}, \quad p_1, p_2, \ldots, p_n > 0. \tag{8}$$

b. In (6), let $n = 3$, $p_1 = p_2 = p_3 = 1$ and the integral (6) will give the volume of pyramid D,

$$I = \iiint_D dx_1\, dx_2\, dx_3 = \frac{1}{\Gamma(40} = \frac{1}{6}. \tag{9}$$

See also Sec.2.6.8.

2.6.10. Evaluate the integral

$$I = \int \cdots \int_D \frac{dx_1\, dx_2 \ldots dx_n}{(a^2 - x_1^2 - x_2^2 - \cdots - x_n^2)^{\frac{1}{2}}}, \quad a > 0, \tag{1}$$

$$D\{x_1, x_2, \ldots, x_n \geq 0, \quad x_1^2 + x_2^2 + \cdots + x_n^2 \leq a^2\}. \tag{2}$$

Solution 1. Using the transformation (see Appendix a5)

$$\begin{cases} x_1 = a y_1^{\frac{1}{2}} \\ x_2 = a y_2^{\frac{1}{2}} \\ \cdots\cdots\cdots \\ x_n = a y_n^{\frac{1}{2}} \end{cases} \Rightarrow \begin{cases} dx_1 = \frac{a}{2} y_1^{-\frac{1}{2}}\, dy_1 \\ dx_2 = \frac{a}{2} y_2^{-\frac{1}{2}}\, dy_2 \\ \cdots\cdots\cdots \\ dx_n = \frac{a}{2} y_n^{-\frac{1}{2}}\, dy_n \end{cases} \begin{cases} x_1^2 = a^2 y_1 \\ x_2^2 = a^2 y_2 \\ \cdots\cdots\cdots \\ x_n^2 = a^2 y_n \end{cases} \tag{3}$$

with the Jacobian

$$J_n = \frac{a^n}{2^n} y_1^{-\frac{1}{2}} y_2^{-\frac{1}{2}} \ldots y_n^{-\frac{1}{2}} \tag{4}$$

the integral (1) becomes

$$I = \frac{a^{n-1}}{2^n} \iint \cdots \int_{D'} \frac{y_1^{-\frac{1}{2}} y_2^{-\frac{1}{2}} \ldots y_n^{-\frac{1}{2}}}{(1 - y_1 - \cdots - y_n)^{\frac{1}{2}}}\, dy_1\, dy_2 \ldots dy_n, \tag{5}$$

$$D'\{y_1, y_2, \ldots, y_n \geq 0, \quad y_1 + y_2 + \cdots + y_n \leq 1\}. \tag{6}$$

Using Sec.2.6.9 with $p_1 = p_2 = \cdots = p_n = p = 1/2$, from (5) we have

$$I = \frac{a^{n-1}}{2^n} \int \cdots \int_{D'} y_1^{-\frac{1}{2}} y_2^{-\frac{1}{2}} \ldots y_n^{-\frac{1}{2}} (1 - y_1 - y_2 - \cdots - y_n)^{-\frac{1}{2}} \, dy_1 \, dy_2 \ldots dy_n \quad (7)$$

$$= \frac{a^{n-1}}{2^n} \frac{\left[\Gamma\left(\frac{1}{2}\right)\right]^n}{\Gamma\left(\frac{n}{2}\right)} B\left(\frac{n}{2}, \frac{1}{2}\right) = \frac{a^{n-1}}{2^n} \frac{\pi^{\frac{n+1}{2}}}{\Gamma\left(\frac{n+1}{2}\right)}, \quad n \in N. \quad (8)$$

If the region of the integration is the full interior of the hypersphere, we have

$$\int \cdots \int_{D_1} \frac{dx_1 \, dx_2 \ldots dx_n}{(a^2 - x_1^2 - x_2^2 - \cdots - x_n^2)^{\frac{1}{2}}} = a^{n-1} \frac{\pi^{\frac{n+1}{2}}}{\Gamma\left(\frac{n+1}{2}\right)}, \quad n \in N, \quad (9)$$

$$D_1\{x_1, x_2, \ldots, x_n \in R^+ \mid x_1^2 + x_2^2 + \cdots + x_n^2 \le a^2\}. \quad (10)$$

Solution 2. Using the hyperspherical coordinates,

$$\begin{cases} x_1 = r \cos \theta_1 \\ x_2 = r \sin \theta_1 \cos \theta_2 \\ \cdots\cdots\cdots \\ x_{n-1} = r \sin \theta_1 \sin \theta_2 \ldots \sin \theta_{n-2} \cos \theta_{n-1} \\ x_n = r \sin \theta_1 \sin \theta_2 \ldots \sin \theta_{n-2} \sin \theta_{n-1} \end{cases} \quad (11)$$

with $r \ge 0$, $\theta_1, \theta_2, \ldots, \theta_{n-2} \in [0, \pi]$, $\theta_{n-1} \in [0, 2\pi)$ and with the Jacobian of transformation (see Appendix A3.3)

$$J_n = r^{n-1} \sin^{n-2} \theta_1 \sin^{n-3} \theta_2 \ldots \sin \theta_{n-2} \quad (12)$$

the integral (1) becomes

$$I = \int \cdots \int_{D_1'} \frac{r^{n-1} \sin^{n-2} \theta_1 \sin^{n-3} \theta_2 \ldots \sin \theta_{n-2}}{(a^2 - r^2)^{\frac{1}{2}}} \, d\theta_1 \, d\theta_2 \ldots d\theta_{n-1} \, dr \quad (13)$$

and knowing that

$$\int_0^{\pi} \sin^p \theta \, d\theta = 2 \int_0^{\frac{\pi}{2}} \sin^p \theta \, d\theta \quad (14)$$ [1]

the integral (5) becomes

$$I = 2^n \int_0^{\frac{\pi}{2}} \sin^{n-2}\theta_1\, d\theta_1 \int_0^{\frac{\pi}{2}} \sin^{n-3}\theta_2\, d\theta_2 \ldots$$
$$\ldots \times \int_0^{\frac{\pi}{2}} \sin\theta_{n-2}\, d\theta_{n-2} \int_0^{\frac{\pi}{2}} d\theta_{n-1} \int_0^a \frac{r^{n-1}\, dr}{(a^2 - r^2)^{\frac{1}{2}}}$$

and using Eq.(1.120),

$$I = 2^n \frac{1}{2^{n-2}} B\left(\frac{n-1}{2}, \frac{1}{2}\right) B\left(\frac{n-2}{2}, \frac{1}{2}\right) \ldots B\left(\frac{3}{2}, \frac{1}{2}\right) B\left(\frac{2}{2}, \frac{1}{2}\right) \cdot \frac{\pi}{2} \int_0^a \frac{r^{n-1} dr}{(a^2 - r^2)^{\frac{1}{2}}}$$
$$= 2\pi \left[\Gamma\left(\frac{1}{2}\right)\right]^{n-2} \frac{\Gamma\left(\frac{n-1}{2}\right)\Gamma\left(\frac{n-2}{2}\right)}{\Gamma\left(\frac{n}{2}\right)\Gamma\left(\frac{n-1}{2}\right)} \cdots \frac{\Gamma\left(\frac{3}{2}\right)\Gamma(1)}{\Gamma(2)\Gamma\left(\frac{3}{2}\right)} \int_0^a r^{n-1}(a^2 - r^2)^{-\frac{1}{2}}\, dr$$
$$= \frac{2\pi^{\frac{n}{2}}}{\Gamma\left(\frac{n}{2}\right)} \int_0^a r^{n-1}(a^2 - r^2)^{-\frac{1}{2}}\, dr. \qquad (15)$$

1

$$\int_0^{\pi} \sin^p\theta\, d\theta = \int_0^{\frac{\pi}{2}} \sin^p\theta\, d\theta + \int_{\frac{\pi}{2}}^{\pi} \sin^p\theta\, d\theta. \qquad (\alpha)$$

Let $\theta = \pi - \phi$, $d\theta = -d\phi$ and

$$\int_{\frac{\pi}{2}}^{\pi} \sin^p\theta\, d\theta = \int_0^{\frac{\pi}{2}} \sin^p(\pi - \phi)\, d\phi = \int_0^{\frac{\pi}{2}} \sin^p\phi\, d\phi$$

and from (α),

$$\int_0^{\pi} \sin^p\theta\, d\theta = 2\int_0^{\frac{\pi}{2}} \sin^p\theta\, d\theta.$$

– The substitution $\theta = \pi/2 + \varphi$ can also be used, see Sec.1.12.3.

Using Sec.2.3.9 with $p = n - 1$, $r = x$, $q = -1/2$,

$$I = \frac{2\pi^{\frac{n}{2}}}{\Gamma\left(\frac{n}{2}\right)} \frac{a^{n-1}}{2} B\left(\frac{n}{2}, 1 - \frac{1}{2}\right) = \frac{\pi^{\frac{n}{2}}}{\Gamma\left(\frac{n}{2}\right)} B\left(\frac{n}{2}, \frac{1}{2}\right) a^{n-1}$$

$$= \frac{\pi^{\frac{n}{2}}}{\Gamma\left(\frac{n}{2}\right)} \frac{\Gamma\left(\frac{n}{2}\right)\Gamma\left(\frac{1}{2}\right)}{\Gamma\left(\frac{n}{2} + \frac{1}{2}\right)} a^{n-1} = \frac{\pi^{\frac{n}{2}}}{\Gamma\left(\frac{n}{2}\right)} \frac{\pi^{\frac{1}{2}}\Gamma\left(\frac{n}{2}\right)}{\Gamma\left(\frac{n+1}{2}\right)} a^{n-1}$$

$$= \frac{\pi^{\frac{n+1}{2}}}{\Gamma\left(\frac{n+1}{2}\right)} a^{n-1}, \quad a > 0. \tag{16}$$

Particular cases:

a. Let $a = 1$ and from (16) we have

$$\int \cdots \int_{D_1} \frac{dx_1\, dx_2 \ldots dx_n}{(1 - x_1^2 - x_2^2 - \cdots - x_n^2)^{\frac{1}{2}}} = \frac{\pi^{\frac{n+1}{2}}}{\Gamma\left(\frac{n+1}{2}\right)}, \tag{17}$$

$$D_1\{x_1, x_2, \ldots, x_n \in R^+ \mid x_1^2 + x_2^2 + \cdots + x_n^2 \leq 1\}.$$

b. Let $n = 2p$, $p \in N$ and from (16) we have

$$I = \frac{\pi^{p+\frac{1}{2}}}{\Gamma\left(p + \frac{1}{2}\right)} a^{2p-1}, \quad p \in N, \quad a > 0, \tag{18}$$

and using Eq.(1.43), we have

$$\int \cdots \int_{D_1} \frac{dx_1\, dx_2 \ldots dx_{2p}}{(a^2 - x_1^2 - x_2^2 - \cdots - x_{2p}^2)^{\frac{1}{2}}} = \frac{2^p \pi^p}{(2p-1)!!} a^{2p-1}, \; p \in N, \; a > 0, \tag{19}$$

$$D_1\{x_1, x_2, \ldots, x_n \in R^+ \mid x_1^2 + x_2^2 + \cdots + x_n^2 \leq a^2\}.$$

c. Let $n = 2p - 1$, $p \in N$, and from (16) we have

$$\int \cdots \int_{D_1} \frac{dx_1\, dx_2 \ldots dx_{2p-1}}{(a^2 - x_1^2 - x_2^2 - \cdots - x_{2p-1}^2)^{\frac{1}{2}}} = \frac{\pi^p a^{2p-2}}{\Gamma(p)} = \frac{\pi^p a^{2(p-1)}}{(p-1)\,!}, \quad p \in N, \quad (20)$$

$$D_1\{x_1, x_2, \ldots, x_{2p-1} \in R \mid x_1^2 + x_2^2 + \cdots + x_{2p-1}^2 \le a^2\}.$$

2.6.11. Study the integral

$$I = \int \cdots \int_D x_1^{p_1} x_2^{p_2} \ldots x_n^{p_n}\, dx_1\, dx_2 \ldots dx_n, \quad (1)$$

$$D\left\{x_1, x_2, \ldots, x_n \ge 0, \quad \left(\frac{x_1}{a_1}\right)^{m_1} + \left(\frac{x_2}{a_2}\right)^{m_2} + \cdots + \left(\frac{x_n}{a_n}\right)^{m_n} \le 1\right\}$$

$$a_1, a_2, \ldots, a_n > 0, \quad m_1, m_2, \ldots, m_n > 0, \quad p_1, p_2, \ldots, p_n \in R.$$

Solution. Using the transformation

$$x_1 = a_1 u_1^{\frac{1}{m_1}}, x_2 = a_2 u_2^{\frac{1}{m_2}}, \ldots, x_n = a_n u_n^{\frac{1}{m_n}} \quad (2)$$

with the Jacobian (see Appendix A5.1)

$$J_n = \frac{a_1 a_2 \ldots a_n}{m_1 m_2 \ldots m_n} u_1^{\frac{1}{m_1} - 1} u_2^{\frac{1}{m_2} - 1} \ldots u_n^{\frac{1}{m_n} - 1} \quad (3)$$

the integral becomes

$$I = \frac{a_1^{p_1+1} a_2^{p_2+1} \ldots a_n^{p_n+1}}{m_1 m_2 \ldots m_n}$$

$$\times \int \cdots \int_{D_1} u_1^{\frac{p_1+1}{m_1} - 1} u_2^{\frac{p_2+1}{m_2} - 1} \ldots u_n^{\frac{p_n+1}{m_n} - 1}\, du_1\, du_2 \ldots du_n, \quad (4)$$

$$D_1\{u_1, u_2, \ldots, u_n \ge 0, \quad u_1 + u_2 + \cdots + u_n \le 1\}.$$

The last multiple integral is similar to Sec.2.6.9(a) and (4) can be written

$$I = \frac{a_1^{p_1+1} a_2^{p_2+1} \ldots a_n^{p_n+1}}{m_1 m_2 \ldots m_n} \cdot \frac{\Gamma\left(\frac{p_1+1}{m_1}\right) \Gamma\left(\frac{p_2+1}{m_2}\right) \cdots \Gamma\left(\frac{p_n+1}{m_n}\right)}{\Gamma\left(\frac{p_1+1}{m_1} + \frac{p_2+1}{m_2} + \cdots + \frac{p_n+1}{m_n} + 1\right)} \quad (5)$$

where $p_1, p_2, \ldots, p_n > -1$.

Particular cases:

a. In R^3-space, let $p_1 = \alpha - 1$, $p_2 = \beta - 1$, $p_3 = \gamma - 1$, $a_1 = a$, $a_2 = b$, $a_3 = c$, $m_1 = p$, $m_2 = q$, $m_3 = r$, and the integral (1) becomes

$$I = \iiint_D x^{\alpha-1} y^{\beta-1} z^{\gamma-1}\, dx\, dy\, dz, \tag{6}$$

$$D\left\{x, y, z \geq 0, \quad \left(\frac{x}{a}\right)^p + \left(\frac{y}{b}\right)^q + \left(\frac{z}{c}\right)^r \leq 1\right\}$$

and from (5),

$$\iiint_D x^{\alpha-1} y^{\beta-1} z^{\gamma-1}\, dx\, dy\, dz = \frac{a^\alpha b^\beta c^\gamma}{pqr} \cdot \frac{\Gamma\left(\frac{\alpha}{p}\right)\Gamma\left(\frac{\beta}{q}\right)\Gamma\left(\frac{\gamma}{r}\right)}{\Gamma\left(\frac{\alpha}{p} + \frac{\beta}{q} + \frac{\gamma}{r} + 1\right)}, \tag{7}$$

$$\alpha, \beta, \gamma > 0.$$

b. In **a.**, let $b = c = a > 0$, $p = q = r = 2$ and from (7) we have

$$I = \iiint_D x^{\alpha-1} y^{\beta-1} z^{\gamma-1}\, dx\, dy\, dz, \tag{8}$$

$$D\{x, y, z \geq 0, \quad x^2 + y^2 + z^2 \leq a^2\},$$

$$I = \frac{a^{\alpha+\beta+\gamma}}{8} \frac{\Gamma\left(\frac{\alpha}{2}\right)\Gamma\left(\frac{\beta}{2}\right)\Gamma\left(\frac{\gamma}{2}\right)}{\Gamma\left(\frac{\alpha+\beta+\gamma}{2} + 1\right)}, \quad \alpha, \beta, \gamma > 0. \tag{9}$$

c. In **b.**, let $\alpha = \beta = \gamma = 1$ and the integral is $1/8$ of the volume of a sphere with the radius a, and from (9),

$$I = \iiint_D dx\, dy\, dz = \frac{a^3}{8} \frac{\left[\Gamma\left(\frac{1}{2}\right)\right]^3}{\Gamma\left(\frac{5}{2}\right)} = \frac{\pi}{6} a^3 \tag{10}$$

and the volume of the sphere is

$$V = 8\frac{\pi}{6} a^3 = \frac{4\pi}{3} a^3, \quad a > 0. \tag{11}$$

2.6.12. Evaluate the Dirichlet's integral

$$I = \int \cdots \int_D f\left(\left(\frac{x_1}{a_1}\right)^{\alpha_1} + \left(\frac{x_2}{a_2}\right)^{\alpha_2} + \cdots + \left(\frac{x_n}{a_n}\right)^{\alpha_n}\right)$$
$$\times\, x_1^{q_1-1} \ldots x_n^{q_n-1}\, dx_1\, dx_2 \ldots dx_n, \tag{1}$$
$$D\left\{x_1, \ldots, x_n \geq 0, \quad \left(\frac{x_1}{a_1}\right)^{\alpha_1} + \cdots + \left(\frac{x_n}{a_n}\right)^{\alpha_n} \leq 1\right\}$$

where f is an arbitrary continuous function of its argument and $a_1, a_2, \ldots, a_n > 0$, $\alpha_1, \alpha_2, \ldots, \alpha_n > 0$, $q_1, q_2, \ldots, q_n \in R$.

Solution. Let the transformation, see Appendix A5.1,

$$x_1 = a_1 u_1^{\frac{1}{\alpha_1}}, x_2 = a_2 u_2^{\frac{1}{\alpha_2}}, \ldots, x_n = a_n u_n^{\frac{1}{\alpha_n}} \tag{2}$$

with the Jacobian

$$J_n = \frac{a_1 a_2 \ldots a_n}{\alpha_1 \alpha_2 \ldots \alpha_n} u_1^{\frac{1}{\alpha_1}-1} u_2^{\frac{1}{\alpha_2}-1} \ldots u_n^{\frac{1}{\alpha_n}-1} \tag{3}$$

and (1) becomes

$$I = \int \cdots \int_{D_1} f(u_1 + u_2 + \cdots + u_n) \cdot (a_1 u_1^{\frac{1}{\alpha_1}})^{q_1-1} \cdot (a_2 u_2^{\frac{1}{\alpha_2}})^{q_2-1} \ldots (a_n u_n^{\frac{1}{\alpha_n}})^{q_n-1}$$
$$\times \frac{a_1 a_2 \ldots a_n}{\alpha_1 \alpha_2 \ldots \alpha_n} u_1^{\frac{1}{\alpha_1}-1} u_2^{\frac{1}{\alpha_2}-1} \ldots u_n^{\frac{1}{\alpha_n}-1} \cdot du_1\, du_2 \ldots du_n$$
$$= \int \cdots \int_{D_1} f(u_1 + \cdots + u_n) \cdot a_1^{q_1-1} \cdot a_2^{q_2-1} \ldots a_n^{q_n-1} u_1^{\frac{q_1-1}{\alpha_1}} \cdot u_2^{\frac{q_2-1}{\alpha_2}} \ldots u_n^{\frac{q_n-1}{\alpha_n}}$$
$$\times \frac{a_1 a_2 \ldots a_n}{\alpha_1 \alpha_2 \ldots \alpha_n} \cdot u_1^{\frac{1}{\alpha_1}-1} u_2^{\frac{1}{\alpha_2}-1} \ldots u_n^{\frac{1}{\alpha_n}-1} \cdot du_1\, du_2 \ldots du_n$$

$$= \frac{a_1^{q_1} a_2^{q_2} \ldots a_n^{q_n}}{\alpha_1 \alpha_2 \ldots \alpha_n} \int \cdots \int_{D_1} f(u_1 + \ldots + u_n) u_1^{\frac{q_1}{\alpha_1}-1} u_2^{\frac{q_2}{\alpha_2}-1} \ldots u_n^{\frac{q_n}{\alpha_n}-1}$$
$$\times\, du_1\, du_2 \ldots du_n, \tag{4}$$
$$D_1\{u_1, u_2, \ldots, u_n \geq 0, \quad u_1 + u_2 + \ldots + u_n \leq 1\}.$$

The last integral is similar to the Dirichlet's integral from Sec.2.6.8, Eq.(11), with $p_1 = q_1/\alpha_1, p_2 = q_2/\alpha_2, \ldots, p_n = q_n/\alpha_n$, and therefore

$$I = \frac{a_1^{q_1} a_2^{q_2} \ldots a_n^{q_n}}{\alpha_1 \alpha_2 \ldots \alpha_n} \frac{\Gamma\left(\frac{q_1}{\alpha_1}\right)\Gamma\left(\frac{q_2}{\alpha_2}\right)\ldots\Gamma\left(\frac{q_n}{\alpha_n}\right)}{\Gamma\left(\frac{q_1}{\alpha_1}+\frac{q_2}{\alpha_2}+\cdots+\frac{q_n}{\alpha_n}\right)} \int_0^1 f(u_1) u_1^{\frac{q_1}{\alpha_1}+\frac{q_2}{\alpha_2}+\cdots+\frac{q_n}{\alpha_n}-1} du_1 \quad (5)$$

where $q_1, q_2, \ldots, q_n > 0$, and under the assumption that the integral in (5) converges absolutely.

Particular case. Letting $n = 3$, $a_1 = a$, $a_2 = b$, $a_3 = c$, $a > b > c > 0$, $\alpha_1 = \alpha_2 = \alpha_3 = 2$, $q_1 = q_2 = q_3 = 1$, $f(u_1) = 1$, from (5) we have the volume of $1/8$ of an ellipsoid,

$$\begin{aligned} I &= \frac{abc}{8} \frac{\left[\Gamma\left(\frac{1}{2}\right)\right]^3}{\Gamma\left(\frac{3}{2}\right)} \int_0^1 u_1^{\frac{3}{2}-1} du_1 \\ &= \frac{abc}{8} \frac{(\sqrt{\pi})^3}{\frac{1}{2}\Gamma\left(\frac{1}{2}\right)} \int_0^1 u_1^{\frac{1}{2}} du_1 \\ &= \frac{abc}{4} \pi \left. \frac{u_1^{\frac{3}{2}}}{\frac{3}{2}} \right|_0^1 \\ &= \frac{abc}{4} \frac{2\pi}{3} = \frac{\pi}{6} abc \end{aligned}$$

and for all volumes of an ellipsoid,

$$V_e = 8\frac{\pi}{6} abc = \frac{4\pi}{3} abc. \quad (6)$$

If $a = b = c = R$, from (6) we have the volume of a sphere,

$$V_s = \frac{4\pi}{3} R^3. \quad (7)$$

2.7. Laplace Transform and the Gamma Function

2.7.1. Evaluate the integral

$$I = \int_0^\infty e^{-x^2}\,dx. \tag{1}$$

Solution. Let

$$F(t) = \int_0^\infty e^{-tx^2}\,dx \tag{2}$$

and taking the Laplace transform, see Sec. 1.15.1,

$$\begin{aligned}\mathcal{L}\{F(t)\} &= \mathcal{L}\left\{\int_0^\infty e^{-tx^2}\,dx\right\} = \int_0^\infty e^{-st}\left(\int_0^\infty e^{-tx^2}\,dx\right)dt \\ &= \int_0^\infty\left(\int_0^\infty e^{-(x^2+s)t}\,dt\right)dx. \end{aligned}\tag{3}$$

But,

$$\int_0^\infty e^{-(x^2+s)t}\,dt = -\frac{e^{-(x^2+s)t}}{x^2+s}\Bigg|_0^\infty = \frac{1}{x^2+s} \tag{4}$$

and

$$\mathcal{L}\{F(t)\} = \int_0^\infty \frac{dx}{x^2+s} = \frac{1}{\sqrt{s}}\tan^{-1}\frac{x}{\sqrt{s}}\Bigg|_0^\infty = \frac{\pi}{2}\frac{1}{\sqrt{s}}. \tag{5}$$

By inverting and using Eq. (1.157) we have

$$\begin{aligned}F(t) &= \frac{\pi}{2}\mathcal{L}^{-1}\left\{\frac{1}{\sqrt{s}}\right\} = \frac{\pi}{2}\frac{t^{-\frac{1}{2}}}{\Gamma\left(\frac{1}{2}\right)} \\ &= \frac{\sqrt{\pi}}{2}t^{-\frac{1}{2}}.\end{aligned}\tag{6}$$

Letting $t = 1$ in (2) and (6) yields

$$\int_0^\infty e^{-x^2}\,dx = \frac{\sqrt{\pi}}{2} = 0.886. \tag{7}$$

Note. For other solutions of integral (1), see Secs. 1.10.1, 1.10.2 and 2.3.31.

2.7.2. Using the convolution theorem, see Sec. 1.15.5, prove the relation

$$B(p,q) = \frac{\Gamma(p)\Gamma(q)}{\Gamma(p+q)}, \quad p,q>0. \tag{1}$$

Solution. By definition, see Secs. 1.2 and 1.15.2,

$$B(p,q) = \int_0^1 x^{p-1}(1-x)^{q-1}\,dx, \quad p,q>0. \tag{2}$$

Let

$$F(x) = x^{p-1} \Rightarrow f(s) = \mathcal{L}\{x^{p-1}\} = \frac{\Gamma(p)}{s^p}$$
$$G(x) = x^{q-1} \Rightarrow g(s) = \mathcal{L}\{x^{q-1}\} = \frac{\Gamma(q)}{s^q}.$$

Then

$$\begin{aligned}\int_0^t F(x)G(t-x)\,dx &= \mathcal{L}^{-1}\{f(s)\cdot g(s)\} = \mathcal{L}^{-1}\left\{\frac{\Gamma(p)\Gamma(q)}{s^{p+q}}\right\}\\ &= \Gamma(p)\Gamma(q)\mathcal{L}^{-1}\left\{\frac{1}{s^{p+q}}\right\} = \Gamma(p)\Gamma(q)\frac{t^{p+q-1}}{\Gamma(p+q)}\\ &= \frac{\Gamma(p)\Gamma(q)}{\Gamma(p+q)}t^{p+q-1}. \end{aligned} \tag{3}$$

Let $t=1$, and from (3),

$$\begin{aligned}\int_0^1 F(x)G(1-x)\,dx &= \int_0^1 x^{p-1}(1-x)^{q-1}\,dx\\ &= B(p,q) = \frac{\Gamma(p)\Gamma(q)}{\Gamma(p+q)}.\end{aligned} \tag{4}$$

Note. For other solutions of (1), see Secs. 1.11.1 and 1.11.2.

2.7.3. Evaluate the Fresnel's integrals

$$I_1 = \int_0^\infty \sin x^2\,dx, \quad I_2 = \int_0^\infty \cos x^2\,dx. \tag{1}$$

Solution 1. Let

$$F(t) = \int_0^\infty \sin tx^2\, dx \tag{2}$$

and with the Laplace transform,

$$\mathcal{L}\{F(t)\} = \int_0^\infty e^{-st}\left(\int_0^\infty \sin tx^2\, dx\right) dt$$
$$= \int_0^\infty dx \int_0^\infty e^{-st} \sin tx^2\, dt = \int_0^\infty \mathcal{L}\{\sin tx^2\}\, dx.$$

Using Eq.(1.154),

$$\mathcal{L}\{\sin tx^2\} = \int_0^\infty e^{-st} \sin x^2 t\, dt = \frac{x^2}{s^2 + x^4}. \tag{3}$$

With Sec.2.3.21 with $m = 2$, $n = 2$ and $a = s$ we have

$$f(s) = \int_0^\infty \frac{x^2}{s^2 + x^4}\, dx = \frac{s^{\frac{1}{2}-1}}{4} \frac{\pi}{\sin \dfrac{\pi}{4}} = \frac{\pi\sqrt{2}}{4} \frac{1}{s^{\frac{1}{2}}}. \tag{4}$$

Taking the inverse Laplace transform of $f(s)$, see Eq.(1.157), with $x + 1 = 1/2$, we have

$$F(t) = \frac{\pi\sqrt{2}}{4} \mathcal{L}^{-1}\left\{\frac{1}{s^{\frac{1}{2}}}\right\} = \frac{\pi\sqrt{2}}{4} \frac{t^{\frac{1}{2}-1}}{\Gamma\left(\dfrac{1}{2}\right)} = \frac{1}{2}\sqrt{\frac{\pi}{2}}\, t^{-\frac{1}{2}}$$

and for $t = 1$,

$$\int_0^\infty \sin x^2\, dx = \frac{1}{2}\sqrt{\frac{\pi}{2}}. \tag{5}$$

Similarly,

$$\int_0^\infty \cos x^2\, dx = \frac{1}{2}\sqrt{\frac{\pi}{2}}. \tag{6}$$

Solution 2. See the integrand graphic in Fig.2.24 and let

$$x^2 = t, \quad 2x\, dx = dt, \quad dx = \frac{dt}{2t^{\frac{1}{2}}}$$

and the integral becomes

$$\int_0^\infty \sin x^2\, dx = \frac{1}{2} \int_0^\infty \frac{\sin t}{t^{\frac{1}{2}}}\, dt. \tag{7}$$

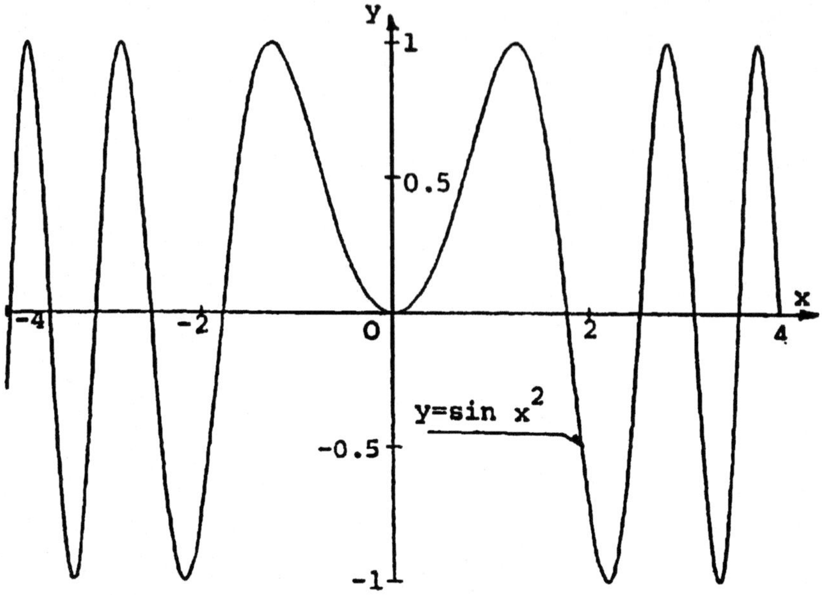

Fig.2.24

Using Sec.2.3.33, for $p = 1/2$, we have

$$\int_0^\infty \frac{\sin t}{t^{\frac{1}{2}}}\,dt = \frac{\pi}{2\Gamma(p)\sin\frac{p\pi}{2}} = \frac{\pi}{2\Gamma\left(\frac{1}{2}\right)\sin\frac{\pi}{4}} \tag{8}$$

and from (7),

$$\int_0^\infty \sin x^2\,dx = \frac{1}{2}\frac{\pi}{2\Gamma\left(\frac{1}{2}\right)\sin\frac{\pi}{4}} = \frac{\pi}{4\Gamma\left(\frac{1}{2}\right)\sin\frac{\pi}{4}} = \frac{1}{2}\sqrt{\frac{\pi}{2}}. \tag{9}$$

Similarly,

$$\int_0^\infty \cos x^2\,dx = \frac{1}{2}\int_0^\infty \frac{\cos t}{t^{\frac{1}{2}}}\,dt = \frac{\pi}{4\Gamma\left(\frac{1}{2}\right)\cos\frac{\pi}{4}} = \frac{1}{2}\sqrt{\frac{\pi}{2}}. \tag{10}$$

See also Secs.2.7.4 and 2.7.5.

2.7.4. Study the integral

$$I = \int_0^\infty \sin(ax^p)\,dx, \quad p \in R, \quad a > 0. \tag{1}$$

Solution 1. Let

$$F(t) = \int_0^\infty \sin(tx^p)\,dx \tag{2}$$

and using the Laplace transform, Eq.(1.150),

$$\begin{aligned} f(s) = \mathcal{L}\{F(t)\} &= \int_0^\infty e^{-st}\left[\int_0^\infty \sin(tx^p)\,dx\right] dt \\ &= \int_0^\infty \left[\int_0^\infty e^{-st}\sin(tx^p)\,dt\right] dx = \int_0^\infty \mathcal{L}\{\sin(tx^p)\}\,dx. \end{aligned} \tag{3}$$

Using Sec.1.15.3, with $a = x^p$,

$$\mathcal{L}\{\sin(x^p t)\} = \frac{x^p}{s^2 + x^{2p}} \tag{4}$$

and

$$f(s) = \int_0^\infty \frac{x^p}{s^2 + x^{2p}}\,dx. \tag{5}$$

Using Sec.2.3.21, for $m = n = p$, $a = s$ and $\sin\frac{p+1}{2p}\pi = \cos\frac{\pi}{2p}$,

$$f(s) = \frac{s^{\frac{1}{p}-1}}{2p}\frac{\pi}{\sin\frac{p+1}{2p}\pi} = \frac{s^{\frac{1}{p}-1}}{2p}\frac{\pi}{\cos\frac{\pi}{2p}} = \frac{\pi s^{\frac{1}{p}-1}}{2p}\sec\frac{\pi}{2p}, \quad p > 1. \tag{6}$$

Taking the inverse Laplace transform of $f(s)$,

$$F(t) = \mathcal{L}^{-1}\{f(s)\} = \frac{\pi}{2p}\sec\frac{\pi}{2p}\mathcal{L}^{-1}\left\{\frac{1}{s^{1-\frac{1}{p}}}\right\} \tag{7}$$

and with Eq.(1.157), we have

$$F(t) = \frac{\pi\sec\frac{\pi}{2p}}{2p\Gamma\left(1 - \frac{1}{p}\right)} t^{-\frac{1}{p}} \tag{8}$$

and using Eq.(1.35),

$$F(t) = \frac{\pi \dfrac{1}{\cos\dfrac{\pi}{2p}} \cdot \sin\dfrac{\pi}{p} \cdot \Gamma\left(\dfrac{1}{p}\right)}{2p\pi} t^{-\frac{1}{p}} = \frac{\Gamma\left(\dfrac{1}{p}\right) \cdot 2\sin\dfrac{\pi}{2p}\cos\dfrac{\pi}{2p}}{2p\cos\dfrac{\pi}{2p}} t^{-\frac{1}{p}} \tag{9}$$

and for $t = a$,

$$\int_0^\infty \sin(ax^p)\,dx = \frac{\Gamma\left(\dfrac{1}{p}\right)\sin\dfrac{\pi}{2p}}{pa^{\frac{1}{p}}}, \quad a > 0, \quad p > 1. \tag{10}$$

Note that the integral (5) can also be evaluated using the substitution

$$x^p = s\tan\theta, \quad dx = \frac{s^{\frac{1}{p}}}{p}\frac{(\tan\theta)^{\frac{1}{p}-1}}{\cos^2\theta}\,d\theta \tag{11}$$

and from (5),

$$f(s) = \frac{1}{p}s^{\frac{1}{p}-1}\int_0^{\frac{\pi}{2}} (\tan\theta)^{\frac{1}{p}}\,d\theta$$

and with Eq.(1.125)

$$f(s) = \frac{s^{\frac{1}{p}-1}}{p}\frac{\pi}{2}\sec\frac{\pi}{2p} = \frac{\pi s^{\frac{1}{p}-1}}{2p}\sec\frac{\pi}{2p}, \quad p > 1. \tag{12}$$

Solution 2. Let

$$ax^p = t, \quad apx^{p-1}\,dx = dt, \quad dx = \frac{1}{pa^{\frac{1}{p}}}\frac{dt}{t^{\frac{p-1}{p}}}$$

and (1) becomes

$$I = \frac{1}{pa^{\frac{1}{p}}}\int_0^\infty \frac{\sin t\,dt}{t^{\frac{p-1}{p}}}.$$

Using Sec.2.3.33, we have

$$I = \frac{1}{pa^{\frac{1}{p}}} \cdot \frac{\pi}{2\Gamma\left(\dfrac{p-1}{p}\right)\sin\left(\dfrac{p-1}{p}\cdot\dfrac{\pi}{2}\right)} = \frac{1}{2pa^{\frac{1}{p}}} \cdot \frac{\pi}{\Gamma\left(1-\dfrac{1}{p}\right)\cos\dfrac{\pi}{2p}}$$

and with the complement formula (1.35),

$$I = \frac{\Gamma\left(\frac{1}{p}\right)\sin\frac{\pi}{p}}{2pa^{\frac{1}{p}}\cos\frac{\pi}{2p}}$$

$$= \frac{2\Gamma\left(\frac{1}{p}\right)\sin\frac{\pi}{2p}\cos\frac{\pi}{2p}}{2pa^{\frac{1}{p}}\cos\frac{\pi}{2p}}$$

$$= \frac{\Gamma\left(\frac{1}{p}\right)\sin\frac{\pi}{2p}}{pa^{\frac{1}{p}}}, \quad a > 0, \quad p > 1.$$

Particular cases:

a. Let $p = 3$ and from (10),

$$\int_0^\infty \sin(ax^3)\,dx = \frac{\Gamma\left(\frac{1}{3}\right)\sin\frac{\pi}{6}}{3a^{\frac{1}{3}}} = \frac{\Gamma\left(\frac{1}{3}\right)}{6a^{\frac{1}{3}}} = \frac{0.446}{a^{\frac{1}{3}}}. \tag{13}$$

b. Letting $a = 1$, $p = 2$, from (10) we have

$$\int_0^\infty \sin x^2\,dx = \frac{1}{2}\Gamma\left(\frac{1}{2}\right)\sin\frac{\pi}{4} = \frac{\sqrt{2\pi}}{4} = 0.627. \tag{14}$$

Note that this integral can also be evaluated with the substitution $x^2 = t$. See Sec.2.7.4.

2.7.5. Evaluate the integral

$$I = \int_0^\infty \cos(ax^p)\,dx, \quad a > 0. \tag{1}$$

Solution. Similarly to Sec.2.7.4, we have

$$F(t) = \int_0^\infty \cos(tx^p)\, dx \tag{2}$$

and

$$\int_0^\infty \cos(ax^p)\, dx = \frac{\Gamma\left(\frac{1}{p}\right)\cos\frac{\pi}{2p}}{pa^{\frac{1}{p}}}, \quad a > 0, \quad p > 1. \tag{3}$$

Particular cases:

a. Let $a = 1$, and from (3),

$$\int_0^\infty \cos x^p\, dx = \frac{1}{p}\Gamma\left(\frac{1}{p}\right)\cos\frac{\pi}{2p}, \quad p > 1. \tag{4}$$

b. Let $p = 3$ in (4),

$$\int_0^\infty \cos x^3\, dx = \frac{1}{3}\Gamma\left(\frac{1}{3}\right)\cos\frac{\pi}{6} = 0.773. \tag{5}$$

c. Let $p = 2$ in (4),

$$\int_0^\infty \cos x^2\, dx = \frac{1}{2}\Gamma\left(\frac{1}{2}\right)\cos\frac{\pi}{4} = \frac{\sqrt{2\pi}}{4} = 0.627. \tag{6}$$

Note that this integral can also be evaluated with the substitution $x^2 = t$.

2.7.6. Study the integrals

$$I_1 = \int_0^\infty \frac{\sin x}{x^p}\, dx, \quad I_2 = \int_0^\infty \frac{\cos x}{x^p}\, dx, \quad p \in R. \tag{1}$$

Solution. For the integral I_1, with Secs.1.15.3 and 1.15.4,

$$G(x) = \sin x \Rightarrow g(s) = \mathcal{L}\{\sin x\} = \frac{1}{s^2+1}$$

$$f(s) = \frac{1}{s^p} \Rightarrow F(t) = \mathcal{L}^{-1}\left\{\frac{1}{s^p}\right\} = \frac{t^{p-1}}{\Gamma(p)}$$

and we have

$$I_1 = \int_0^\infty G(x)f(x)\,dx = \int_0^\infty \frac{\sin x}{x^p}\,dx = \int_0^\infty F(x)g(x)\,dx$$

$$= \frac{1}{\Gamma(p)}\int_0^\infty \frac{x^{p-1}}{x^2+1}\,dx \tag{2}$$

and with Sec.2.3.20,

$$\int_0^\infty \frac{\sin x}{x^p}\,dx = \frac{\pi}{2\Gamma(p)\sin\dfrac{p\pi}{2}}, \quad 0 < p < 1. \tag{3}$$

For the integral I_2, let

$$G(x) = \cos x \Rightarrow g(s) = \mathcal{L}\{\cos x\} = \frac{s}{s^2+1}$$

$$f(s) = \frac{1}{s^p} \Rightarrow F(t) = \mathcal{L}^{-1}\left\{\frac{1}{s^p}\right\} = \frac{t^{p-1}}{\Gamma(p)}$$

and we have

$$I_2 = \int_0^\infty G(x)f(x)\,dx = \int_0^\infty \frac{\cos x}{x^p}\,dx = \int_0^\infty F(x)g(x)\,dx$$

$$= \frac{1}{\Gamma(p)}\int_0^\infty \frac{x^{p}}{x^2+1}\,dx \tag{4}$$

and with Sec.2.3.20,

$$\int_0^\infty \frac{\cos x}{x^p}\,dx = \frac{\pi}{2\Gamma(p)\cos\dfrac{p\pi}{2}}, \quad 0 < p < 1. \tag{5}$$

See also Secs.2.3.33, 2.3.34 and 2.3.35.

Particular cases: Let $p = 1/3$ and from (3),

$$\int_0^\infty \frac{\sin x}{x^{\frac{1}{3}}}\,dx = \frac{\pi}{2\Gamma\left(\dfrac{1}{3}\right)\sin\dfrac{\pi}{6}} = \frac{\pi}{\Gamma\left(\dfrac{1}{3}\right)} = 1.1727$$

and from (5),

$$\int_0^\infty \frac{\cos x}{x^{\frac{1}{3}}}\,dx = \frac{\pi}{2\Gamma\left(\frac{1}{3}\right)\cos\frac{\pi}{6}} = \frac{\pi}{\Gamma\left(\frac{1}{3}\right)\sqrt{3}} = 0.677.$$

2.8. Applications of the Digamma Function and the Function $\beta(x)$

2.8.1. Prove the equality

$$\beta(x) + \beta(1+x) = \frac{1}{x}.$$

Solution. Using Eq.(1.171), we have

$$\begin{aligned} E &= \beta(x) + \beta(1+x) \\ &= \frac{1}{2}\left[\psi\left(\frac{x+1}{2}\right) - \psi\left(\frac{x}{2}\right)\right] + \frac{1}{2}\left[\psi\left(\frac{1+x+1}{2}\right) - \psi\left(\frac{x+1}{2}\right)\right] \\ &= \frac{1}{2}\left[\psi\left(1+\frac{x}{2}\right) - \psi\left(\frac{x}{2}\right)\right] \end{aligned}$$

and with the substitution $x/2 = y$,

$$\begin{aligned} E &= \frac{1}{2}\left[\psi(1+y) - \psi(y)\right] = \frac{1}{2}\left[\psi(y) + \frac{1}{y} - \psi(y)\right] \\ &= \frac{1}{2y} = \frac{1}{x}. \end{aligned}$$

2.8.2. Evaluate the integral

$$I = \int_0^1 \frac{dx}{1+x}.$$

Solution. Using Eq.(1.172) for $p = 1$, and Eq.(1.171) for $x = 1$, we have

$$\int_0^1 \frac{dx}{1+x} = \beta(1) = \frac{1}{2}\left[\psi(1) - \psi\left(\frac{1}{2}\right)\right].$$

But, $\psi(1) = -\gamma = -0.5772$,

$$\psi(0.5) = \psi(0.5+1) - \frac{1}{0.5} = 0.03648 - 2 = -1.96352$$

and

$$\int_0^1 \frac{dx}{1+x} = \frac{1}{2}(-0.5772 + 1.96352) = 0.6931.$$

2.8.3. Evaluate the integral

$$I = \int_0^1 \frac{\sqrt{x}\,dx}{1+x}.$$

Solution. Using Eq.(1.172) for $p = 3/2$ and Eq.(1.171) we have

$$I = \beta\left(\frac{3}{2}\right) = \frac{1}{2}\left[\psi\left(\frac{5}{4}\right) - \psi\left(\frac{3}{4}\right)\right] = \frac{1}{2}\left[\psi(1.25) - \psi(0.75)\right].$$

From [1], $\psi(1.25) = -0.22745$, and with Eq.(1.166),

$$\psi(0.75) = \psi(1.75) - \frac{1}{0.75} = 0.24747 - \frac{1}{0.75} = -1.08586$$

and then,

$$\int_0^1 \frac{\sqrt{x}\,dx}{1+x} = \beta\left(\frac{3}{2}\right) = \frac{1}{2}(-0.22745 + 1.08586) = 0.429.$$

2.8.4. Evaluate the integral

$$I = \int_0^1 \frac{dx}{1+x^3}.$$

Solution 1. Using Eqs.(1.173) and (1.171), with $p = 1,\ q = 3$,

$$I = \frac{1}{3}\beta\left(\frac{1}{3}\right) = \frac{1}{6}\left[\psi\left(\frac{2}{3}\right) - \psi\left(\frac{1}{6}\right)\right]$$

$$= \frac{1}{6}\left[\psi(0.6666) - \psi(0.16666)\right],$$

but

$$\psi(0.6666) = \psi(1.6666) - \frac{1}{0.6666} = 0.1804 - \frac{1}{0.6666} = -1.31975,$$

$$\psi(0.16666) = \psi(1.16666) - \frac{1}{0.16666} = -0.33432 - \frac{1}{0.16666} = -6.33456$$

and the integral is

$$\int_0^1 \frac{dx}{x^3+1} = \frac{1}{6}(-1.31975 + 6.33456) = 0.836.$$

Solution 2. Using the traditional method, see Fig.2.25,

$$\int_0^1 \frac{dx}{x^3+1} = \frac{1}{3}\left(\int_0^1 \frac{dx}{x+1} - \int_0^1 \frac{x-2}{x^2-x+1}\,dx\right).$$

$$I = \frac{1}{6}\ln\frac{(x+1)^2}{x^2-x+1}\Big|_0^1 + \frac{1}{\sqrt{3}}\arctan\frac{2x-1}{\sqrt{3}}\Big|_0^1$$

$$= \frac{1}{6}\ln 4 + \frac{1}{\sqrt{3}}\arctan\frac{1}{\sqrt{3}} - \left(\frac{1}{6}\ln 1 + \frac{1}{\sqrt{3}}\arctan\frac{-1}{\sqrt{3}}\right)$$

$$= \frac{1}{6}\ln 4 + \frac{1}{\sqrt{3}}\arctan\sqrt{3} = \frac{1}{6}\ln 4 + \frac{1}{\sqrt{3}}\cdot\frac{\pi}{3} = 0.836.$$

2.8.5. Evaluate the integral

$$I = \int_0^1 \frac{dx}{1+x^4}.$$

Solution 1. Using Eqs.(1.173) and (1.171) with $p = 1,\ q = 4$, the integral becomes

$$I = \frac{1}{4}\beta\left(\frac{1}{4}\right) = \frac{1}{8}\left[\psi\left(\frac{5}{8}\right) - \psi\left(\frac{1}{8}\right)\right]$$

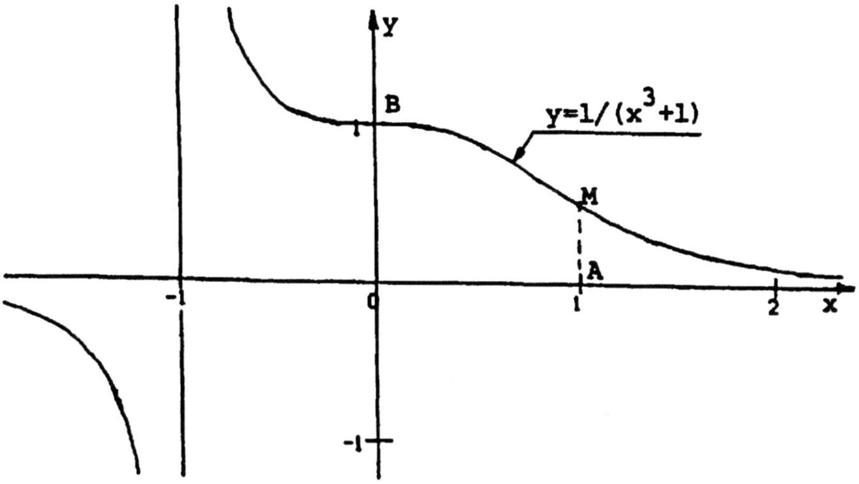

Fig.2.25

but, using Eq.(1.166), we have

$$\psi\left(\frac{5}{8}\right)=\psi\left(\frac{5}{8}+1\right)-\frac{8}{5}=\psi(1.625)-\frac{8}{5}=0.14729-\frac{8}{5}=-1.45271,$$
$$\psi\left(\frac{1}{8}\right)=\psi\left(\frac{1}{8}+1\right)-8=\psi(1.125)-8=-0.38849-8=-8.38849,$$

and

$$I=\frac{1}{8}(-1.45271+8.38849)=0.867.$$

Solution 2. From Sec.2.3.14, we have

$$I = \frac{1}{2\sqrt{2}}\left(\int_0^1 \frac{x+\sqrt{2}}{x^2+x\sqrt{2}+1}\,dx - \int_0^1 \frac{x-\sqrt{2}}{x^2-x\sqrt{2}+1}\right)$$

$$= \frac{1}{2\sqrt{2}}\left[\frac{1}{2}\int_0^1 \frac{(2x+\sqrt{2})+\sqrt{2}}{x^2+x\sqrt{2}+1}\,dx - \frac{1}{2}\int_0^1 \frac{(2x-\sqrt{2})-\sqrt{2}}{x^2-x\sqrt{2}+1}\,dx\right]$$

$$= \frac{1}{4\sqrt{2}}\left[\ln\frac{x^2+x\sqrt{2}+1}{x^2-x\sqrt{2}+1}\Big|_0^1 + \sqrt{2}\left(\int_0^1 \frac{dx}{x^2+x\sqrt{2}+1} + \int_0^1 \frac{dx}{x^2-x\sqrt{2}+1}\right)\right]$$

$$= \frac{1}{4\sqrt{2}}\ln\frac{2+\sqrt{2}}{2-\sqrt{2}} + \frac{1}{4}\left(\int_0^1 \frac{dx}{x^2+x\sqrt{2}+1} + \int_0^1 \frac{dx}{x^2-x\sqrt{2}+1}\right)$$

$$= \frac{1}{4\sqrt{2}}\ln(3+2\sqrt{2}) + \frac{1}{4}\{\int_0^1 \frac{dx}{\left(x+\frac{\sqrt{2}}{2}\right)^2+\frac{1}{2}} + \int_0^1 \frac{dx}{\left(x-\frac{\sqrt{2}}{2}\right)^2+\frac{1}{2}}\}$$

$$= \frac{1}{4\sqrt{2}}\ln(3+2\sqrt{2}) + \frac{1}{4}\left\{\sqrt{2}\arctan\frac{x+\frac{\sqrt{2}}{2}}{\frac{1}{\sqrt{2}}} + \sqrt{2}\arctan\frac{x-\frac{\sqrt{2}}{2}}{\frac{1}{\sqrt{2}}}\right\}\Bigg|_0^1$$

$$= \frac{1}{4\sqrt{2}}\ln(3+2\sqrt{2}) + \frac{\sqrt{2}}{4}\left[\arctan(\sqrt{2}x+1) + \arctan(\sqrt{2}x-1)\right]\Big|_0^1$$

$$= \frac{1}{4\sqrt{2}}\ln(3+2\sqrt{2}) + \frac{\sqrt{2}}{4}\left[\arctan(\sqrt{2}+1) + \arctan(\sqrt{2}-1) - \arctan 1 - \arctan(-1)\right.$$

$$= \frac{1}{4\sqrt{2}}\ln(3+2\sqrt{2}) + \frac{\sqrt{2}}{4}\cdot\frac{\pi}{2} = \frac{1}{4\sqrt{2}}\ln(3+2\sqrt{2}) + \frac{\pi\sqrt{2}}{8} = 0.867.$$

2.8.6. Evaluate the integral

$$\int_0^1 \frac{dx}{1+x^5}.$$

Solution. Using Eq.(1.173) for $p=1$, $q=5$, we have

$$I = \frac{1}{5}\beta\left(\frac{1}{5}\right),$$

but

$$\beta\left(\frac{1}{5}\right) = \frac{1}{2}\left[\psi\left(\frac{6}{10}\right) - \psi\left(\frac{1}{10}\right)\right] = \frac{1}{2}\left[\psi(0.6) - \psi(0.1)\right]$$

$$\psi(0.6) = \psi(1.6) - \frac{1}{0.6} = 0.12604 - \frac{1}{0.6} = -1.54062$$

$$\psi(0.1) = \psi(1.1) - \frac{1}{0.1} = -0.42375 - 10 = -10.42375$$

$$\beta\left(\frac{1}{5}\right) = \frac{1}{2}(-1.54062 + 10.42375) = 4.44156$$

and

$$\int_0^1 \frac{dx}{1+x^5} = \frac{1}{5} \times 4.44156 = 0.888.$$

Exercise. Evaluate the integral using the method of partial fractions.

2.8.7. Evaluate the integral

$$\int_0^1 \frac{x^3\,dx}{1+x^5}.$$

Solution. Using Eq.(1.173), for $p = 4$, $q = 5$, see Fig.2.26, we have

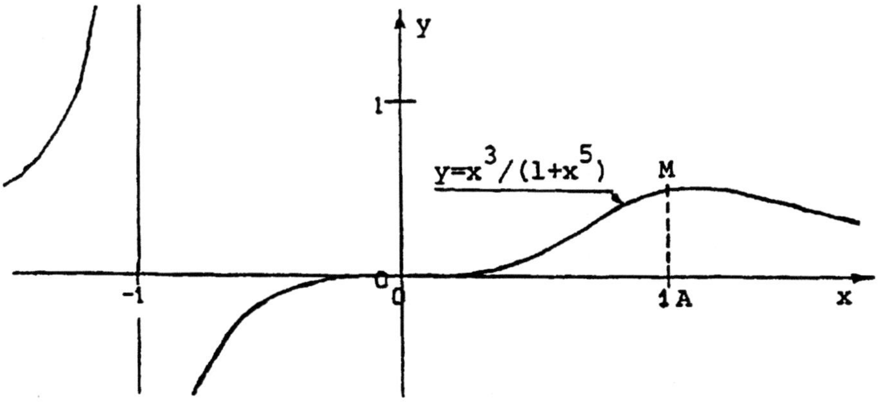

Fig.2.26

$$I = \frac{1}{5}\beta\left(\frac{4}{5}\right).$$

But,

$$\beta\left(\frac{4}{5}\right) = \frac{1}{2}\left[\psi\left(\frac{9}{10}\right) - \psi\left(\frac{4}{10}\right)\right] = \frac{1}{2}\left[\psi(0.9) - \psi(0.4)\right],$$

$$\psi(0.9) = \psi(0.9+1) - \frac{1}{0.9} = \psi(1.9) - \frac{1}{0.9} = 0.35618 - \frac{1}{0.9} = -0.75493,$$

$$\psi(0.4) = \psi(1.4) - \frac{1}{0.4} = -0.06138 - \frac{1}{0.4} = -2.56138,$$

$$\beta\left(\frac{4}{5}\right) = \frac{1}{2}(-0.75493 + 2.56138) = 0.903225$$

and the integral is

$$\int_0^1 \frac{x^3\,dx}{1+x^5} = \frac{1}{5} \times 0.903225 = 0.1806.$$

Evaluate the integral using a traditional method.

2.8.8. Evaluate the integral

$$\int_1^\infty \frac{x^3\,dx}{1+x^5}.$$

Solution. We can write (see Fig.2.26)

$$\int_1^\infty \frac{x^3\,dx}{1+x^5} = \int_0^\infty \frac{x^3\,dx}{1+x^5} - \int_0^1 \frac{x^3\,dx}{1+x^5}.$$

With Sec.2.3.20 for $m = 4$, $n = 5$,

$$\int_0^\infty \frac{x^3\,dx}{1+x^5} = \frac{1}{5}B\left(1 - \frac{4}{5}, \frac{4}{5}\right) = \frac{1}{5}B\left(\frac{1}{5}, \frac{4}{5}\right)$$

$$= \frac{1}{5}\Gamma\left(\frac{1}{5}\right)\Gamma\left(\frac{4}{5}\right) = \frac{\pi}{5\sin\dfrac{\pi}{5}} = 1.0689,$$

$$\int_0^1 \frac{x^3\,dx}{1+x^5} = \frac{1}{5}\beta\left(\frac{4}{5}\right) = 0.1806 \quad \text{(see Sec.2.8.7).}$$

Thus,

$$\int_1^\infty \frac{x^3\,dx}{1+x^5} = \frac{1}{5}\left[B\left(\frac{1}{5},\frac{4}{5}\right) - \beta\left(\frac{4}{5}\right)\right]$$
$$= 1.0689 - 0.1806 = 0.888.$$

More generally,

$$\int_1^\infty \frac{x^{m-1}\,dx}{1+x^n} = \int_0^\infty \frac{x^{m-1}\,dx}{1+x^n} - \int_0^1 \frac{x^m\,dx}{1+x^n}$$
$$= \frac{1}{n}\left[B\left(1-\frac{m}{n},\frac{m}{n}\right) - \beta\left(\frac{m}{n}\right)\right], \quad 0 < m < n.$$

Evaluate the integral using a traditional method.

2.8.9. Evaluate the integral

$$I = \int_0^1 \frac{x\,dx}{1+x^2}.$$

Solution 1. Using Eq.(1.174), for $p = 1$, we have

$$I = \frac{1}{2}\beta(1) = \frac{1}{4}\left[\psi(1) - \psi\left(\frac{1}{2}\right)\right]$$
$$= \frac{1}{4}\left[-0.5772 - \psi(0.5)\right].$$

But,

$$\psi(0.5) = \psi(1.5) - \frac{1}{0.5} = 0.03648 - 2 = -1.96352$$

and therefore

$$\int_0^1 \frac{x\,dx}{1+x^2} = \frac{1}{4}(-0.5772 + 1.96352) = 0.3465.$$

Solution 2. Using a traditional method,

$$\int_0^1 \frac{x\,dx}{1+x^2} = \frac{1}{2}\ln(1+x^2)\Big|_0^1 = \frac{\ln 2}{2} = 0.3465.$$

2.8.10. Evaluate the integral

$$\int_0^1 \frac{x^3\,dx}{1+x^2}.$$

Solution 1. Using Eq.(1.174), for $p=3$,

$$\int_0^1 \frac{x^3\,dx}{1+x^2} = \frac{1}{2}\beta(2),$$

but

$$\beta(2) = \frac{1}{2}\left[\psi\left(\frac{2+1}{2}\right) - \psi\left(\frac{2}{2}\right)\right] = \frac{1}{2}\left[\psi(1.5) - \psi(1)\right]$$
$$= \frac{1}{2}(0.03648 + 0.57721) = 0.306845$$

and

$$\int_0^1 \frac{x^3\,dx}{1+x^2} = \frac{1}{2} \times 0.306845 = 0.1534.$$

Solution 2. Using a traditional method,

$$\frac{x^3}{1+x^2} = x - \frac{x}{1+x^2}$$

and

$$I = \int_0^1 x\,dx - \int_0^1 \frac{x\,dx}{1+x^2} = \left.\frac{x^2}{2}\right|_0^1 - \frac{1}{2}\int_0^1 \frac{2x\,dx}{1+x^2}$$
$$= \frac{1}{2} - \frac{1}{2}\ln(1+x^2)\Big|_0^1 = \frac{1}{2}(1-\ln 2) = 0.1534.$$

2.8.10. a. Evaluate the integral

$$I = \int_0^1 \frac{dx}{\sqrt{2}(1+x)}.$$

a. This integral is a particular case of Eq.(1.64a) for $p=1/2$,

$$I = \int_0^1 \frac{x^{-\frac{1}{2}}}{1+x}\,dx = \frac{1}{2}B\left(\frac{1}{2},\frac{1}{2}\right) = \frac{\pi}{2} = 1.5708.$$

b. The integral can also be evaluated using Eqs.(1.171) and (1.172),

$$I = \int_0^1 \frac{x^{-\frac{1}{2}}}{1+x}\,dx = \beta\left(\frac{1}{2}\right) = \frac{1}{2}\left[\psi\left(\frac{3}{4}\right) - \psi\left(\frac{1}{4}\right)\right],$$

$$\psi\left(\frac{3}{4}\right) = \psi(0.75) = \psi(1.75) - \frac{1}{0.75} = 0.24747 - \frac{1}{0.75} = -1.086,$$

$$\psi\left(\frac{1}{4}\right) = \psi(0.25) = \psi(1.25) - 4 = -0.227 - 4 = -4.227,$$

$$I = \beta\left(\frac{1}{2}\right) = \frac{1}{2}(-1.086 + 4.227) = 1.5705.$$

2.8.11. Evaluate the integral

$$I = \int_0^1 \frac{x^{\frac{3}{2}}\,dx}{1+x^2}.$$

Solution. Using Eq.(1.174), for $p = 3/2$,

$$I = \frac{1}{2}\beta\left(\frac{5}{4}\right) = \frac{1}{2}\beta(1.25),$$

but

$$\beta(1.25) = \frac{1}{2}\left[\psi\left(\frac{1.25+1}{2}\right) - \psi\left(\frac{1.25}{2}\right)\right] = \frac{1}{2}\left[\psi(1.125) - \psi(0.625)\right],$$

$$\psi(1.125) = -0.38849,$$

$$\psi(0.625) = \psi(1.625) - \frac{1}{0.625} = 0.14729 - \frac{1}{0.625} = -1.45271,$$

$$\beta(1.25) = \frac{1}{2}(-0.38849 + 1.45271) = 0.53211.$$

Therefore,

$$\int_0^1 \frac{x^{\frac{3}{2}}\,dx}{1+x^2} = \frac{1}{2} \times 0.53211 = 0.2661.$$

Evaluate the integral using a traditional method.

2.8.12. Evaluate the integral

$$\int_0^1 \frac{x^3\,dx}{(1+x^2)^2}. \tag{1}$$

Solution 1. Using Eq.(1.175), for $p = 3$, we have

$$\int_0^1 \frac{x^3\,dx}{(1+x^2)^2} = -\frac{1}{4} + \frac{1}{2}\beta(1), \tag{2}$$

but

$$\beta(1) = \frac{1}{2}\left[\psi(1) - \psi\left(\frac{1}{2}\right)\right], \tag{3}$$

$$\psi(1) = -\gamma = -0.577215,$$

$$\psi(0.5) = \psi(1.5) - \frac{1}{0.5} = 0.03649 - 2 = -1.96351$$

and

$$\beta(1) = \frac{1}{2}(-0.577215 + 1.96351) = 0.693147. \tag{4}$$

Therefore, from (2),

$$\int_0^1 \frac{x^3\,dx}{(1+x^2)^2} = -\frac{1}{4} + \frac{1}{2}0.693147 = 0.09657. \tag{5}$$

Solution 2. Using the traditional method,

$$\frac{x^3}{(1+x^2)^2} = \frac{Ax+B}{(1+x^2)^2} + \frac{Cx+D}{1+x^2} \tag{6}$$

and multiplying by $(1+x^2)^2$ we get the identity

$$x^3 \equiv Ax + B + (Cx + D)(1+x^2),$$

from where: $A = -1$, $B = 0$, $C = 1$ and $D = 0$, and (6) becomes

$$\frac{x^3}{(1+x^2)^2} = \frac{-x}{(1+x^2)^2} + \frac{x}{1+x^2}.$$

Therefore,

$$\int_0^1 \frac{x^3\,dx}{(1+x^2)^2} = -\int_0^1 \frac{x\,dx}{(1+x^2)^2} + \int_0^1 \frac{x\,dx}{1+x^2} = \frac{1}{2(1+x^2)}\bigg|_0^1 + \frac{1}{2}\ln(1+x^2)\bigg|_0^1$$
$$= \frac{1}{4} - \frac{1}{2} + \frac{1}{2}\ln 2 = \frac{1}{2}\ln 2 - \frac{1}{4} = 0.09657.$$

2.8.13. Evaluate the integral

$$I = \int_0^\infty \frac{e^{-mx}}{1+e^{-x}}\,dx. \tag{1}$$

Solution. Let

$$e^{-x} = t, \quad dx = -\frac{dt}{t} \tag{2}$$

and (1) becomes

$$I = \int_0^1 \frac{t^m}{1+t}\cdot\frac{dt}{t} = \int_0^1 \frac{t^{m-1}}{1+t}\,dt$$

and with Eq.(1.172),

$$\int_0^\infty \frac{e^{-mx}}{1+e^{-x}}\,dx = \beta(m), \quad m > 0. \tag{3}$$

Particular cases:

a. Let $m = 5$ and from (3),

$$\int_0^\infty \frac{e^{-5x}}{1+e^{-x}}\,dx = \beta(5) = \frac{1}{2}\left[\psi(3) - \psi(2.5)\right]. \tag{4}$$

But,

$$\psi(2.5) = \psi(1.5) + \frac{1}{1.5} = 0.03648 + \frac{1}{1.5} = 0.703146$$

and

$$\int_0^\infty \frac{e^{-5x}}{1+e^{-x}}\,dx = \frac{1}{2}(0.92278 - 0.703146) = 0.1098. \tag{5}$$

b. Let $m = 2.4$ and from (3),

$$I = \int_0^\infty \frac{e^{-2.4x}}{1+e^{-x}}\,dx = \beta(2.4) \tag{6}$$

$$= \frac{1}{2}\left[\psi\left(\frac{2.4+1}{2}\right) - \psi\left(\frac{2.4}{2}\right)\right] = \frac{1}{2}[\psi(1.7) - \psi(1.2)]$$

$$= \frac{1}{2}(0.20854 + 0.28903) = 0.2488. \tag{7}$$

CHAPTER 3

Complete Elliptic Integrals $K(k)$ and $E(k)$

3.1. Definition of Elliptic Integrals

The incomplete elliptic integral of the first kind, under Legendre's form, is defined as

$$F(k, \phi) = \int_0^{\phi} \frac{d\theta}{\sqrt{1 - y^2 \sin^2 \theta}}, \quad 0 < k < 1, \quad k = \sin \psi, \tag{3.1}$$

where k is the module of the integral.

The incomplete elliptic integral of the second kind, under Legendre's form, is defined as

$$E(k, \phi) = \int_0^{\phi} \sqrt{1 - k^2 \sin^2 \theta}\, d\theta, \quad 0 < k < 1, \quad k = \sin \psi, \tag{3.2}$$

where k is the module of the integral. Let

$$\int_{\phi_0}^{\phi} \frac{d\theta}{\sqrt{1 - k^2 \sin^2 \theta}} = F_0(k, \phi) = F(k, \phi) - F(k, \phi_0),$$

$$\int_{\phi_0}^{\phi} \sqrt{1 - k^2 \sin^2 \theta}\, d\theta = E_0(k, \phi) = E(k, \phi) - E(k, \phi_0).$$

Using the notation $t = \sin \theta$, with the transformation $x = \sin \phi$, the integrals (3.1) and (3.2) can be reduced to the Jacobi's forms of the elliptic integrals,

$$F_1(k,x) = \int_0^x \frac{dt}{\sqrt{(1-t^2)(1-k^2t^2)}}, \quad 0 < k < 1, \tag{3.3}$$

$$E_1(k,x) = \int_0^x \sqrt{\frac{1-k^2t^2}{1-t^2}}\, dt, \quad 0 < k < 1. \tag{3.4}$$

If $\phi = \pi/2$, the integrals (3.1) and (3.2) are called the *complete elliptic integrals* of the first and second kinds respectively:

$$K(k) = F\left(k, \frac{\pi}{2}\right) = \int_0^{\frac{\pi}{2}} \frac{d\theta}{\sqrt{1-k^2\sin^2\theta}}, \quad 0 < k < 1, \quad K(0) = \frac{\pi}{2}, \tag{3.5}$$

$$E(k) = E\left(k, \frac{\pi}{2}\right) = \int_0^{\frac{\pi}{2}} \sqrt{1-k^2\sin^2\theta}\, d\theta, \quad 0 < k < 1, \quad E(0) = \frac{\pi}{2}. \tag{3.6}$$

In the following, the complete elliptic integrals $K(k)$ and $E(k)$ will be discussed. However, examples with incomplete elliptic integrals have also been given. Note that the elliptic integrals can not be expressed in terms of the elementary functions. The tables of incomplete elliptic integrals can be found in [4].

3.2. Properties of Complete Elliptic Integrals

a. The functions $K(k)$ and $E(k)$ are real functions.

b. The functions $K(k)$ and $E(k)$ are continuous functions for $k \in (0,1)$. The graphs of $K(k)$ and $E(k)$ are given in Figs.3.2 and 3.4.

c. Between K, K', E and E', the Legendre's relation is

$$KE' + EK' - KK' = \frac{\pi}{2} \tag{3.7}$$

where

$$K' = \int_0^{\frac{\pi}{2}} \frac{d\theta}{\sqrt{1 - k'^2 \sin^2\theta}}, \quad E' = \int_0^{\frac{\pi}{2}} \sqrt{1 - k'^2 \sin^2\theta}\, d\theta, \quad k' = \sqrt{1 - k^2}, \quad k^2 < 1.$$

3.3. Complete Elliptic Integrals Outside the Range

To evaluate elliptic integrals for values outside the range contained in Tables 3.1 to 3.4, we have

$$F(k, \pi) = \int_0^{\pi} \frac{d\theta}{\sqrt{1 - k^2 \sin^2\theta}} \tag{3.8}$$

or

$$F(k, \pi) = \int_0^{\frac{\pi}{2}} \frac{d\theta}{\sqrt{1 - k^2 \sin^2\theta}} + \int_{\frac{\pi}{2}}^{\pi} \frac{d\theta}{\sqrt{1 - k^2 \sin^2\theta}}, \tag{3.9}$$

$$F(k, \pi) = K(k) + I \tag{3.10}$$

where

$$I = \int_{\frac{\pi}{2}}^{\pi} \frac{d\theta}{\sqrt{1 - k^2 \sin^2\theta}},$$

and letting $\theta = \pi - \phi$,

$$I = \int_0^{\frac{\pi}{2}} \frac{d\theta}{\sqrt{1 - k^2 \sin^2(\pi - \phi)}} = \int_0^{\frac{\pi}{2}} \frac{d\theta}{\sqrt{1 - k^2 \sin^2\phi}} = K(k). \tag{3.11}$$

Thus,

$$F(k, \pi) = \int_0^{\pi} \frac{d\theta}{\sqrt{1 - k^2 \sin^2\theta}} = 2K(k). \tag{3.12}$$

Similarly, we have

$$E(k, \pi) = \int_0^{\pi} \sqrt{1 - k^2 \sin^2\theta}\, d\theta = 2E(k). \tag{3.13}$$

3.4. Computer Programs to Tabulate $K(k)$

The complete elliptic integral of the first kind is defined by the integral

$$K(k) = \int_0^{\frac{\pi}{2}} \frac{d\theta}{\sqrt{1 - k^2 \sin^2 \theta}}, \qquad k \in (0, 1). \tag{3.14}$$

Using the complementary relation $t = 1 - k^2$, the elliptic integral (3.14) is approximated by the function [1]

$$\begin{aligned} K(k) &= (A_0 t^4 + A_1 t^3 + A_2 t^2 + A_3 t + A_4) \\ &\quad - (B_0 t^4 + B_1 t^3 + B_2 t^2 + B_3 t + B_4) \cdot \ln t + \mathcal{E} \end{aligned} \tag{3.15}$$

where the polynomial coefficients are

$$\begin{aligned} A_0 &= 0.01451196212, & B_0 &= 0.00441787012, \\ A_1 &= 0.03742563713, & B_1 &= 0.03328355346, \\ A_2 &= 0.03590092383, & B_2 &= 0.06880248576, \\ A_3 &= 0.09666344259, & B_3 &= 0.12498593597, \\ A_4 &= 1.38629436112, & B_4 &= 0.5 \end{aligned} \tag{3.16}$$

and $|\mathcal{E}| \leq 2 \times 10^{-8}$. To calculate $K(k)$, a Microsoft QuickBASIC program is given in Fig.3.1. The computer printout is Table 3.1, which gives the values of $K(k)$ for $k \in (0, 1)$.

In addition, a program using Microsoft QuickBASIC programming, to get a table of $K(k)$ in terms of the angle ψ defined by $\sin \psi = k$, has also been given. The program is shown in Fig.3.3, and the computer printout is Table 3.2. The programs are self-explanatory. The error in the tables is $\leq 10^{-5}$, which is sufficiently small for engineering applications. With minor changes in the program, the reader can get a smaller error.

Using Tables 3.1 and 3.2, the graphs of $K(k)$ are shown in Figs.3.2 and 3.4. If a high degree of accuracy is not required, these graphs can be used in engineering work.

With the given algorithms, readers can write their own programs to calculate the values of the complete elliptic integrals. The computer programs given can be easily modified for other PC programming languages. For high-accuracy values of $K(k)$ and $E(k)$ a QuickBASIC program, shown in Fig.3.7, has been given. Tables with elliptic integrals $K(k)$ can also be found in [4].

```
'CMPCK2, Program in QuickBASIC, to tabulate the complete
'elliptic integral of the first kind K(k) in terms of k^2,
'the range 0=<k<1, increment 0.01, approximated by a
'polynomial function. The error is less than 10^-5.
'Refer to Table 3.1.
CLS
DIM I(100) AS INTEGER
DIM P(100), T(100), R1(100), R2(100)
DIM S1(100), S2(100), ELK(100)
   A0 = .01451196212#
   A1 = .03742563713#
   A2 = .03590092383#
   A3 = .09666344259#
   A4 = 1.38629436112#
   B0 = .00441787012#
   B1 = .03328355346#
   B2 = .06880248576#
   B3 = .12498593597#
   B4 = .5
FOR I = 1 TO 99
   P(I) = I / 100
   T(I) = 1 - P(I)
   R1(I) = A0 * T(I) ^ 4 + A1 * T(I) ^ 3 + A2 * T(I) ^ 2
   R2(I) = A3 * T(I) + A4
   S1(I) = B0 * T(I) ^ 4 + B1 * T(I) ^ 3 + B2 * T(I) ^ 2
   S2(I) = B3 * T(I) + B4
   ELK(I) = R1(I) + R2(I) - (S1(I) + S2(I)) * LOG(T(I))
NEXT I
LPRINT CHR$(12)
LPRINT : LPRINT : LPRINT : LPRINT
LPRINT TAB(11); " Table 3.1. Complete elliptic integral of";
LPRINT SPC(1); "the first kind K(k)";
LPRINT TAB(11); "----------------------------------------";
LPRINT SPC(0); "--------------------";
LPRINT : LPRINT
LPRINT TAB(11); "K^2"; SPC(6); "K(k)"; SPC(11); "k^2";
LPRINT SPC(6); "K(k)"; SPC(11); "k^2"; SPC(6); "K(k)";
LPRINT TAB(10); "----"; SPC(4); "--------"; SPC(8); "----";
LPRINT SPC(4); "--------"; SPC(8); "----"; SPC(4); "--------";
LPRINT
LPRINT TAB(10); "0.00"; SPC(4); "1.570796";
LPRINT
   FOR I = 1 TO 33
   LPRINT TAB(10);
   LPRINT USING "#.##"; P(I);
   LPRINT TAB(18);
   LPRINT USING "#.######"; ELK(I);
   LPRINT TAB(34);
   LPRINT USING "#.##"; P(I + 33);
   LPRINT TAB(42);
   LPRINT USING "#.######"; ELK(I + 33);
   LPRINT TAB(58);
   LPRINT USING "#.##"; P(I + 66);
   LPRINT TAB(66);
   LPRINT USING "#.######"; ELK(I + 66);
NEXT I
LPRINT TAB(58); "1.00"; SPC(4); "infinity"
END
```

Fig.3.1

Table 3.1. Complete elliptic integral of the first kind K(k)

K^2	K(k)	k^2	K(k)	k^2	K(k)
0.00	1.570796				
0.01	1.574746	0.34	1.738055	0.67	2.033370
0.02	1.578740	0.35	1.744351	0.68	2.046894
0.03	1.582780	0.36	1.750754	0.69	2.060882
0.04	1.586868	0.37	1.757269	0.70	2.075363
0.05	1.591003	0.38	1.763898	0.71	2.090373
0.06	1.595188	0.39	1.770647	0.72	2.105948
0.07	1.599423	0.40	1.777519	0.73	2.122132
0.08	1.603710	0.41	1.784519	0.74	2.138970
0.09	1.608049	0.42	1.791650	0.75	2.156516
0.10	1.612441	0.43	1.798918	0.76	2.174827
0.11	1.616889	0.44	1.806328	0.77	2.193971
0.12	1.621393	0.45	1.813884	0.78	2.214022
0.13	1.625955	0.46	1.821593	0.79	2.235068
0.14	1.630576	0.47	1.829460	0.80	2.257205
0.15	1.635257	0.48	1.837491	0.81	2.280549
0.16	1.640000	0.49	1.845694	0.82	2.305232
0.17	1.644807	0.50	1.854075	0.83	2.331409
0.18	1.649678	0.51	1.862641	0.84	2.359264
0.19	1.654617	0.52	1.871400	0.85	2.389017
0.20	1.659624	0.53	1.880361	0.86	2.420933
0.21	1.664701	0.54	1.889533	0.87	2.455338
0.22	1.669850	0.55	1.898925	0.88	2.492635
0.23	1.675073	0.56	1.908547	0.89	2.533334
0.24	1.680373	0.57	1.918410	0.90	2.578092
0.25	1.685750	0.58	1.928526	0.91	2.627773
0.26	1.691208	0.59	1.938908	0.92	2.683552
0.27	1.696749	0.60	1.949568	0.93	2.747073
0.28	1.702374	0.61	1.960521	0.94	2.820752
0.29	1.708087	0.62	1.971783	0.95	2.908337
0.30	1.713889	0.63	1.983371	0.96	3.016112
0.31	1.719785	0.64	1.995303	0.97	3.155875
0.32	1.725776	0.65	2.007598	0.98	3.354142
0.33	1.731865	0.66	2.020279	0.99	3.695638
				1.00	infinity

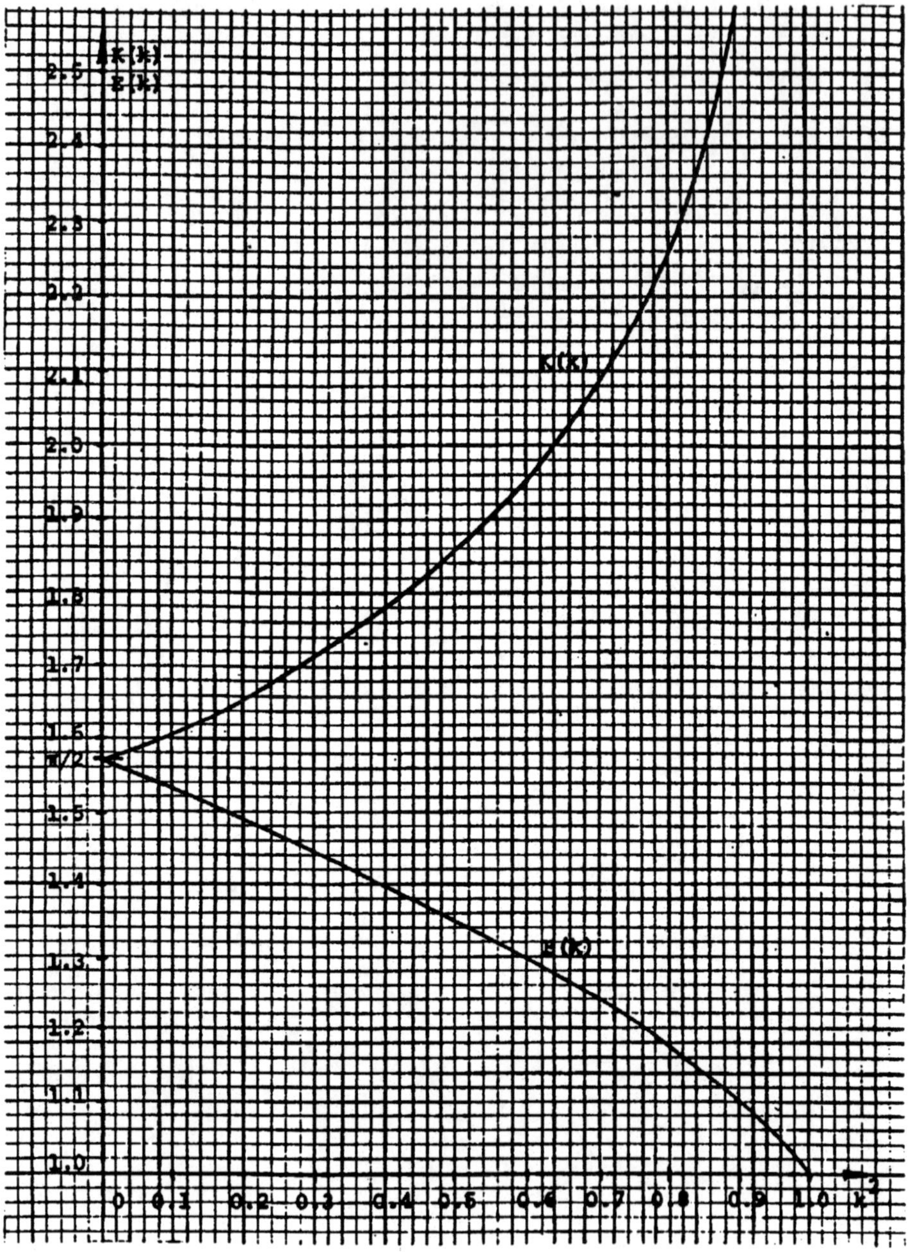
K(k)
E(k)
2.5
2.4
2.3
2.2
2.1
2.0
1.9
1.8
1.7
1.6
π/2
1.5
1.4
1.3
1.2
1.1
1.0
K(k)
E(k)
0
0.1
0.2
0.3
0.4
0.5
0.6
0.7
0.8
0.9
1.0
k²

Fig.3.2

```
'CMPCK3, Program in QuickBASIC, to tabulate the complete
'elliptic integral of the first kind K(k) in terms of angle ψ,
'the range 0 to 90 degrees, increment 1 degree, approximated
'by a polynomial function. The error is less than 10^-5.
'Refer to Table 3.2.
CLS
DIM I(100) AS INTEGER
DIM X(100), K(100), T(100), R1(100)
DIM R2(100), S1(100), S2(100), ELK(100)
   A0 = .01451196212#
   A1 = .03742563713#
   A2 = .03590092383#
   A3 = .09666344259#
   A4 = 1.38629436112#
   B0 = .00441787012#
   B1 = .03328355346#
   B2 = .06880248576#
   B3 = .12498593597#
   B4 = .5
   PI = 3.141592654#
FOR I = 0 TO 89
   X(I) = PI * I / 180
   K(I) = SIN(X(I))
   T(I) = 1 - K(I) ^ 2
   R1(I) = A0 * T(I) ^ 4 + A1 * T(I) ^ 3
   R2(I) = A2 * T(I) ^ 2 + A3 * T(I) + A4
   S1(I) = B0 * T(I) ^ 4 + B1 * T(I) ^ 3
   S2(I) = B2 * T(I) ^ 2 + B3 * T(I) + B4
   ELK(I) = R1(I) + R2(I) - (S1(I) + S2(I)) * LOG(T(I))
NEXT I
LPRINT CHR$(12)
LPRINT : LPRINT : LPRINT : LPRINT
LPRINT TAB(11); "Table 3.2. Complete elliptic integral of the ";
LPRINT SPC(1); "first kind K(k)";
LPRINT TAB(11); "-------------------------------------------";
LPRINT SPC(0); "----------------";
LPRINT : LPRINT
LPRINT TAB(11); "Deg"; SPC(6); "K(k)"; SPC(10); "Deg";
LPRINT SPC(6); "K(k)"; SPC(10); "Deg"; SPC(6); "K(k)";
LPRINT TAB(11); "----"; SPC(3); "---------"; SPC(8); "----";
LPRINT SPC(3); "--------"; SPC(8); "----"; SPC(3); "--------";
LPRINT
FOR I = 0 TO 29
   LPRINT TAB(12);
   LPRINT USING "##"; I;
   LPRINT TAB(18);
   LPRINT USING "#.######"; ELK(I);
   LPRINT TAB(35);
   LPRINT USING "##"; I + 30;
   LPRINT TAB(41);
   LPRINT USING "#.######"; ELK(I + 30);
   LPRINT TAB(58);
   LPRINT USING "##"; I + 60;
   LPRINT TAB(64);
   LPRINT USING "#.######"; ELK(I + 60);
NEXT I
LPRINT TAB(58); "90"; SPC(4); "infinity"
END
```

Fig.3.3

Table 3.2. Complete elliptic integral of the first kind K(k)

Deg	K(k)	Deg	K(k)	Deg	K(k)
0	1.570796	30	1.685750	60	2.156516
1	1.570916	31	1.694114	61	2.184213
2	1.571275	32	1.702836	62	2.213195
3	1.571874	33	1.711925	63	2.243549
4	1.572712	34	1.721391	64	2.275376
5	1.573792	35	1.731245	65	2.308787
6	1.575114	36	1.741499	66	2.343905
7	1.576678	37	1.752165	67	2.380870
8	1.578487	38	1.763256	68	2.419842
9	1.580541	39	1.774786	69	2.461000
10	1.582843	40	1.786769	70	2.504550
11	1.585394	41	1.799222	71	2.550731
12	1.588197	42	1.812160	72	2.599820
13	1.591254	43	1.825602	73	2.652138
14	1.594568	44	1.839567	74	2.708067
15	1.598142	45	1.854075	75	2.768063
16	1.601979	46	1.869148	76	2.832673
17	1.606081	47	1.884809	77	2.902565
18	1.610454	48	1.901083	78	2.978570
19	1.615101	49	1.917997	79	3.061728
20	1.620026	50	1.935581	80	3.153386
21	1.625234	51	1.953865	81	3.255304
22	1.630729	52	1.972882	82	3.369867
23	1.636517	53	1.992670	83	3.500422
24	1.642604	54	2.013267	84	3.651858
25	1.648995	55	2.034715	85	3.831745
26	1.655697	56	2.057062	86	4.052764
27	1.662716	57	2.080358	87	4.338645
28	1.670059	58	2.104658	88	4.742729
29	1.677735	59	2.130021	89	5.434959
				90	infinity

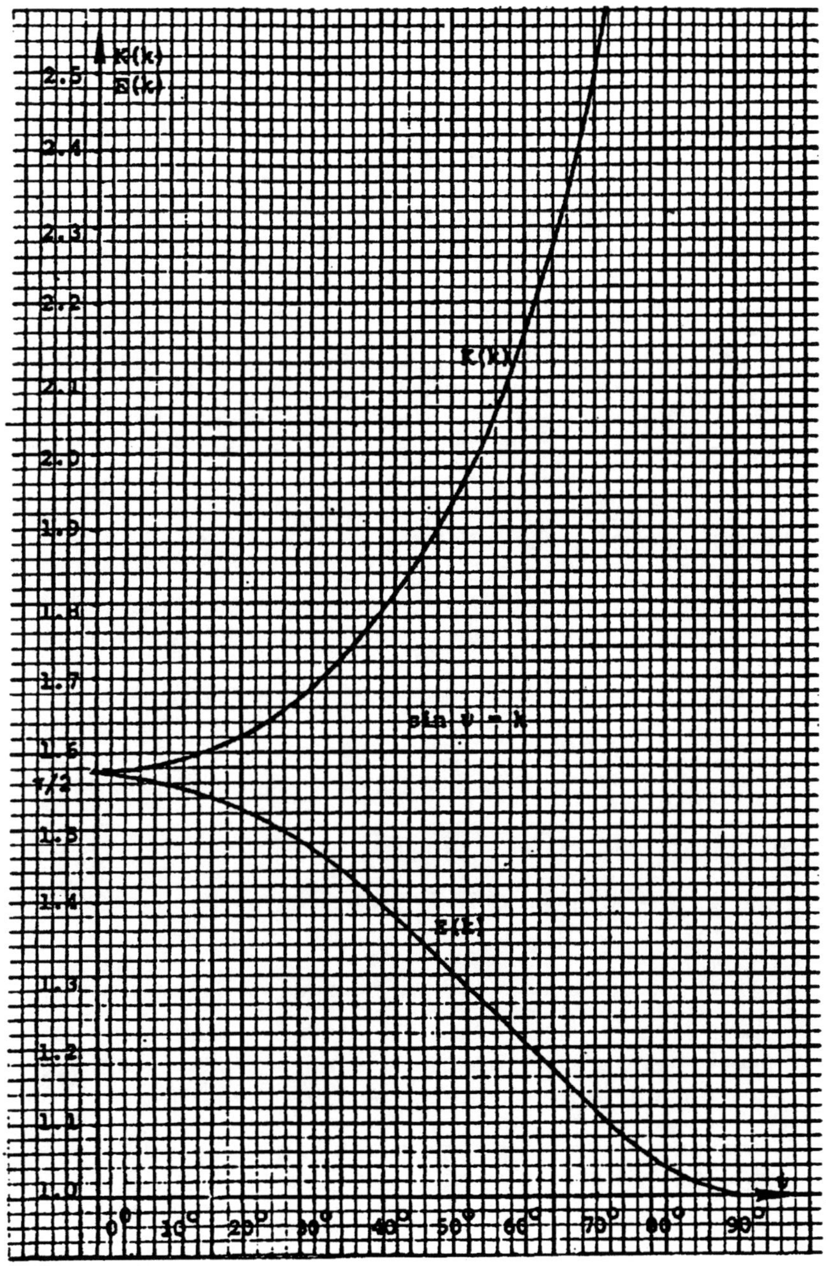
K(k)
E(k)
2.5
2.4
2.3
2.2
2.1
2.0
1.9
1.8
1.7
1.6
π/2
1.5
1.4
1.3
1.2
1.1
1.0
K(k)
sin ψ = k
E(k)
0°
10°
20°
30°
40°
50°
60°
70°
80°
90°
ψ

Fig.3.4

3.5. Computer Programs to Tabulate $E(k)$

The complete elliptic integral of the second kind is defined by the integral

$$E(k) = \int_0^{\frac{\pi}{2}} \sqrt{1 - k^2 \sin^2\theta} \cdot d\theta, \quad k \in (0, 1). \tag{3.17}$$

Using the complementary relation $t = 1 - k^2$, the elliptic integral (3.17) is approximated by the function [1]

$$\begin{aligned} E(k) = {} & (A_0 t^4 + A_1 t^3 + A_2 t^2 + A_3 t + A_4) \\ & - (B_1 t^4 + B_2 t^3 + B_3 t^2 + B_4 t) \cdot \ln t + \mathcal{E} \end{aligned} \tag{3.18}$$

where the polynomial coefficients are

$$\begin{aligned} & A_0 = 0.01736506451, \\ & A_1 = 0.04757383546, \quad B_1 = 0.00526449639, \\ & A_2 = 0.06260601220, \quad B_2 = 0.04069697526, \\ & A_3 = 0.44325141463, \quad B_3 = 0.09200180037, \\ & A_4 = 1.0, \qquad\qquad\quad B_4 = 0.24998368310 \end{aligned} \tag{3.19}$$

and the error $|\mathcal{E}| \leq 2 \times 10^{-8}$. To calculate $E(k)$, a Microsoft QuickBASIC program is given in Fig.3.5. The computer printout is Table 3.3, which gives the values of $E(k)$ for $k \in (0, 1)$.

In addition, to get a table of $E(k)$ in terms of the angle ψ defined by $\sin\psi = k$, a Microsoft QuickBASIC program has also been given. The program is shown in Fig.3.6, and the computer printout is Table 3.4. The programs are self-explanatory. The error in the tables is $\leq 10^{-5}$, which is sufficiently small for most engineering applications. With minor changes in the program, the reader can get a smaller error.

Using Tables 3.3 and 3.4, the graphs of $E(k)$ are shown in Figs.3.2 and 3.4. If a high degree of accuracy is not required, these graphs can be used in engineering work.

With the given algorithms, readers can write their own programs to calculate the values of the complete elliptic integrals. The computer programs given can be easily modified for other PC programming languages. For high-accuracy values of $K(k)$ and $E(k)$ a QuickBASIC program, shown in Fig.3.7, has been given. Tables with elliptic integrals $E(k)$ can also be found in [4].

Example of how to use Tables 3.1 and 3.2. Let

$$K(k) = \int_0^{\frac{\pi}{2}} \frac{d\theta}{\sqrt{1 - \frac{1}{2}\sin^2\theta}}, \qquad 0 < k < 1, \quad k = \sin\psi.$$

Using Table 3.1,

$$k^2 = \frac{1}{2} = 0.5, \qquad K(0.5) = 1.854075.$$

Using Table 3.2,

$$k = \frac{1}{\sqrt{2}}, \qquad \sin\psi = \frac{1}{\sqrt{2}}, \qquad \psi = 45^\circ, \quad K(45^\circ) = 1.854075.$$

The method for Tables 3.3 and 3.4 is similar.

```
'CMPCE2, Program in QuickBASIC, to tabulate the complete
'elliptic integral of the second kind E(k) in terms of k^2,
'the range 0=<k=<1, increment 0.01, approximated by a
'polynomial function. The error is less than 10^-5.
'Refer to Table 3.3.
 CLS
 DIM I(100) AS INTEGER
 DIM P(100), T(100), R1(100), R2(100)
 DIM S1(100), S2(100), ELE(100)
   A0 = .01736506451#
   A1 = .04757383546#
   A2 = 6.260601220000001D-02
   A3 = .44325141463#
   A4 = 1!
   B1 = .00526449639#
   B2 = .04069697526#
   B3 = .09200180037#
   B4 = .24998368310#
FOR I = 1 TO 99
    P(I) = I / 100
    T(I) = 1 - P(I)
    R1(I) = A0 * T(I) ^ 4 + A1 * T(I) ^ 3 + A2 * T(I) ^ 2
    R2(I) = A3 * T(I) + A4
    S1(I) = B1 * T(I) ^ 4 + B2 * T(I) ^ 3 + B3 * T(I) ^ 2
    S2(I) = B4 * T(I)
    ELE(I) = R1(I) + R2(I) - (S1(I) + S2(I)) * LOG(T(I))
NEXT I
LPRINT CHR$(12)
LPRINT : LPRINT : LPRINT : LPRINT
LPRINT TAB(11); "Table 3.3. Complete elliptic integral of";
LPRINT SPC(1); "the second kind E(k)";
LPRINT TAB(11); "----------------------------------------";
LPRINT SPC(0); "-------------------";
LPRINT : LPRINT
LPRINT TAB(11); "K^2"; SPC(6); "E(K)"; SPC(11); "K^2";
LPRINT SPC(6); "E(K)"; SPC(11); "K^2"; SPC(6); "E(K)";
LPRINT TAB(10); "----"; SPC(4); "--------"; SPC(8); "----";
LPRINT SPC(4); "--------"; SPC(8); "----"; SPC(4); "--------";
LPRINT
LPRINT TAB(10); "0.00"; SPC(4); "1.570796";
LPRINT
FOR I = 1 TO 33
   LPRINT TAB(10);
   LPRINT USING "#.##"; P(I);
   LPRINT TAB(18);
   LPRINT USING "#.######"; ELE(I);
   LPRINT TAB(34);
   LPRINT USING "#.##"; P(I + 33);
   LPRINT TAB(42);
   LPRINT USING "#.######"; ELE(I + 33);
   LPRINT TAB(58);
   LPRINT USING "#.##"; P(I + 66);
   LPRINT TAB(66);
   LPRINT USING "#.######"; ELE(I + 66);
NEXT I
LPRINT TAB(58); "1.00"; SPC(4); "1.000000"
END
```

Fig.3.5

Table 3.3. Complete elliptic integral of the second kind E(k)

K^2	E(K)	K^2	E(K)	K^2	E(K)
0.00	1.570796				
0.01	1.566862	0.34	1.427274	0.67	1.259263
0.02	1.562913	0.35	1.422691	0.68	1.253458
0.03	1.558948	0.36	1.418083	0.69	1.247594
0.04	1.554969	0.37	1.413450	0.70	1.241671
0.05	1.550973	0.38	1.408791	0.71	1.235684
0.06	1.546962	0.39	1.404105	0.72	1.229632
0.07	1.542936	0.40	1.399392	0.73	1.223512
0.08	1.538893	0.41	1.394652	0.74	1.217321
0.09	1.534834	0.42	1.389883	0.75	1.211056
0.10	1.530758	0.43	1.385085	0.76	1.204714
0.11	1.526665	0.44	1.380259	0.77	1.198290
0.12	1.522555	0.45	1.375402	0.78	1.191781
0.13	1.518428	0.46	1.370514	0.79	1.185183
0.14	1.514284	0.47	1.365596	0.80	1.178490
0.15	1.510122	0.48	1.360645	0.81	1.171697
0.16	1.505942	0.49	1.355661	0.82	1.164798
0.17	1.501743	0.50	1.350644	0.83	1.157787
0.18	1.497526	0.51	1.345592	0.84	1.150656
0.19	1.493290	0.52	1.340505	0.85	1.143396
0.20	1.489035	0.53	1.335382	0.86	1.135998
0.21	1.484761	0.54	1.330222	0.87	1.128451
0.22	1.480466	0.55	1.325024	0.88	1.120742
0.23	1.476152	0.56	1.319788	0.89	1.112856
0.24	1.471817	0.57	1.314511	0.90	1.104775
0.25	1.467462	0.58	1.309192	0.91	1.096478
0.26	1.463086	0.59	1.303832	0.92	1.087937
0.27	1.458688	0.60	1.298428	0.93	1.079121
0.28	1.454269	0.61	1.292979	0.94	1.069986
0.29	1.449827	0.62	1.287484	0.95	1.060474
0.30	1.445363	0.63	1.281942	0.96	1.050502
0.31	1.440876	0.64	1.276350	0.97	1.039947
0.32	1.436366	0.65	1.270707	0.98	1.028594
0.33	1.431832	0.66	1.265013	0.99	1.015994
				1.00	1.000000

```
'CMPCE3, Program in QuickBASIC, to tabulate the complete
'elliptic integral of the second kind E(k), in terms of angle ψ,
'the range 0 to 90 degrees, increment 1 degree, approximated
'by a polynomial function. The error is less than 10^-5.
'Refer to Table 3.4.
CLS
DIM I(100) AS INTEGER
DIM X(100), K(100), T(100), R1(100)
DIM R2(100), S1(100), S2(100), ELE(100)
   A0 = .01736506451#
   A1 = .04757383546#
   A2 = 6.260601220000001D-02
   A3 = .44325141463#
   A4 = 1!
   B1 = .00526449639#
   B2 = .04069697526#
   B3 = .09200180037#
   B4 = .24998368310#
   PI = 3.141592654#
FOR I = 0 TO 89
    X(I) = PI * I / 180
    K(I) = SIN(X(I))
    T(I) = 1 - K(I) ^ 2
    R1(I) = A0 * T(I) ^ 4 + A1 * T(I) ^ 3
    R2(I) = A2 * T(I) ^ 2 + A3 * T(I) + A4
    S1(I) = B1 * T(I) ^ 4 + B2 * T(I) ^ 3
    S2(I) = B3 * T(I) ^ 2 + B4 * T(I)
    ELE(I) = R1(I) + R2(I) - (S1(I) + S2(I)) * LOG(T(I))
NEXT I
LPRINT CHR$(12)
LPRINT : LPRINT : LPRINT : LPRINT
LPRINT TAB(11); "Table 3.4. Complete elliptic integral of the ";
LPRINT SPC(1); "second kind E(k)";
LPRINT TAB(11); "----------------------------------------";
LPRINT SPC(0); "------------------";
LPRINT : LPRINT
LPRINT TAB(11); "Deg"; SPC(6); "E(k)"; SPC(10); "Deg";
LPRINT SPC(6); "E(k)"; SPC(10); "Deg"; SPC(6); "E(k)";
LPRINT TAB(11); "----"; SPC(3); "--------"; SPC(8); "----";
LPRINT SPC(3); "--------"; SPC(8); "----"; SPC(3); "--------";
LPRINT
FOR I = 0 TO 29
   LPRINT TAB(12);
   LPRINT USING "##"; I;
   LPRINT TAB(18);
   LPRINT USING "#.######"; ELE(I);
   LPRINT TAB(35);
   LPRINT USING "##"; I + 30;
   LPRINT TAB(41);
   LPRINT USING "#.######"; ELE(I + 30);
   LPRINT TAB(58);
   LPRINT USING "##"; I + 60;
   LPRINT TAB(64);
   LPRINT USING "#.######"; ELE(I + 60);
NEXT I
LPRINT TAB(58); "90"; SPC(4); "1.000000"
END
```

Fig.3.6

Table 3.4. Complete elliptic integral of the second kind E(k)

Deg	E(k)	Deg	E(k)	Deg	E(k)
0	1.570796	30	1.467462	60	1.211056
1	1.570677	31	1.460773	61	1.201538
2	1.570318	32	1.453908	62	1.192046
3	1.569720	33	1.446869	63	1.182589
4	1.568884	34	1.439662	64	1.173179
5	1.567809	35	1.432291	65	1.163828
6	1.566497	36	1.424760	66	1.154547
7	1.564947	37	1.417075	67	1.145348
8	1.563162	38	1.409240	68	1.136244
9	1.561142	39	1.401260	69	1.127250
10	1.558887	40	1.393140	70	1.118378
11	1.556400	41	1.384887	71	1.109643
12	1.553681	42	1.376504	72	1.101062
13	1.550732	43	1.367999	73	1.092650
14	1.547554	44	1.359377	74	1.084425
15	1.544151	45	1.350644	75	1.076405
16	1.540522	46	1.341806	76	1.068610
17	1.536670	47	1.332870	77	1.061059
18	1.532597	48	1.323842	78	1.053777
19	1.528306	49	1.314730	79	1.046787
20	1.523799	50	1.305539	80	1.040114
21	1.519078	51	1.296278	81	1.033790
22	1.514147	52	1.286954	82	1.027844
23	1.509007	53	1.277574	83	1.022313
24	1.503662	54	1.268146	84	1.017237
25	1.498115	55	1.258680	85	1.012663
26	1.492369	56	1.249182	86	1.008648
27	1.486427	57	1.239661	87	1.005259
28	1.480293	58	1.230127	88	1.002584
29	1.473970	59	1.220589	89	1.000751
				90	1.000000

3.6. A High-Accuracy Program

Using the algorithms given in Secs.3.4 and 3.5, a program for high-accuracy evaluation for the complete elliptic integrals $K(k)$ and $E(k)$ has been given. The input is the angle ψ defined by $\sin\psi = k$. The program is in Microsoft QuickBASIC for personal computers. The error is $\leq 10^{-5}$, which is sufficiently small for most engineering applications. With this program, $K(k)$ and $E(k)$ will be evaluated for any angle $\psi \in [0, 90°)$, even for values that are not given in the tables. The program can be easily modified for other PC programming languages. See Fig.3.7. Examples of computer printout are given.

```
Enter psi in degrees:
? 30
 30  .5  1.68575  1.467462
? 60
 60  .8660254  2.156516  1.211056
? 38.35
 38.35  .6204636  1.767241  1.406463
?
```

```
'ELKE2, Program in QuickBASIC, to
'calculate K(k) and E(k) in terms
'of ψ=psi in degrees and than transformed
'it in radians, for any psi∈[0,90°).
CLS
PRINT "Enter psi in degrees"
INPUT psi
DO WHILE psi < 90°
'Transform the given psi in radians
LET PI = 3.141592654#
X = PI * psi / 180
k = SIN(X)
T = 1 - k ^ 2
'Calculation of K(k)
A0 = .01451196212#
A1 = .03742563713#
A2 = .03590092383#
A3 = .09666344259#
A4 = 1.38629436112#
B0 = .00441787012#
B1 = .03328355346#
B2 = .06880248576#
B3 = .12498593597#
B4 = .5
P1 = T ^ 2 * (A0 * T ^ 2 + A1 * T + A2)
P2 = A3 * T + A4
P3 = T ^ 2 * (B0 * T ^ 2 + B1 * T + B2)
P4 = B3 * T + B4
k(k) = P1 + P2 - (P3 + P4) * LOG(T)
'Calculation of E(k)
C0 = .01736506451#
C1 = .04757383546#
C2 = 6.260601220000001D-02
C3 = .44325141463#
C4 = 1!
D1 = .0052664496394#
D2 = .04069697526#
D3 = .09200180037#
D4 = .2499836831#
S1 = C0 * T ^ 4 + C1 * T ^ 3
S2 = C2 * T ^ 2 + C3 * T + C4
S3 = D1 * T ^ 4 + D2 * T ^ 3
S4 = D3 * T ^ 2 + D4 * T
E(k) = S1 + S2 - (S3 + S4) * LOG(T)
PRINT psi, k, k(k), E(k)
INPUT psi
LOOP
END
```

Fig.3.7

CHAPTER 4

Applications of Elliptic Integrals

4.1. Introduction

For application of the complete elliptic integrals $K(k)$ and $E(k)$, these integrals will be written as follows:

$$\int_0^{\frac{\pi}{2}} \frac{d\theta}{\sqrt{1-k^2\sin^2\theta}} = K(k), \quad 0 < k < 1, \quad \sin\psi = k, \tag{4.1}$$

$$\int_0^{\frac{\pi}{2}} \sqrt{1-k^2\sin^2\theta}\, d\theta = E(k), \quad 0 < k < 1, \quad \sin\psi = k. \tag{4.2}$$

It is noted that in engineering and physics applications, the modulus k will be $0 < k < 1$. However, for the limiting case $k \to 0$, we take $K(0) = \pi/2$, $E(0) = \pi/2$, and for the limiting case $k \to 1$, we take $K(1) = \infty$ and $E(1) = 1$. For the convergence of elliptic improper integrals, a procedure similar to that of Sec.2.1 can be used.

After an integral has been expressed in terms of the elliptic integrals $K(k)$ and $E(k)$, using Tables 3.1 to 3.4 or the graphs given in Figs.3.2 and 3.4 or the program given in Fig.3.7, the integral can be easily evaluated.

Tables with the numerical values of the incomplete elliptic integrals of the first kind $F(k,\phi)$ and the second kind $E(k,\phi)$ can also be found in [4].

4.2. Applications of Elliptic Integrals

4.2.1. Verify the Legendre's relation Eq.(3.7),

$$KE' + EK' - KK' = \frac{\pi}{2}, \tag{1}$$

for $k = 1/2$.

Solution. We have

$$k' = \sqrt{1-k^2} = \sqrt{1-\frac{1}{4}} = \frac{\sqrt{3}}{2},$$

$$K = K(k) = K\left(\frac{1}{2}\right) = 1.68575, \quad K' = K(k') = K\left(\frac{\sqrt{3}}{2}\right) = 2.156516,$$

$$E = E(k) = E\left(\frac{1}{2}\right) = 1.467462, \quad E' = E(k)' = E\left(\frac{\sqrt{3}}{2}\right) = 1.211056.$$

Thus,

$$KE' + EK' - KK' = 1.68575 \times 1.211056 + 1.467462 \times 2.156516 \\ - 1.68575 \times 2.156516 = 1.570796$$

and $\pi/2 = 1.570796$. Thus, (1) has been verified for $k = 1/2$.

Using a program in QuickBASIC, we have

```
INPUT K, K1, K2, E1, E2
   LET P = 3.141592654#
   M = (1 - K ^ 2) ^ (1 / 2)
   N = (K1 * E2) + (E1 * K2) - (K1 * K2)
   P1 = P / 2
 LPRINT K, K1, K2, E1, E2, M, N, P1

C:\QB45>qb
? .5, 1.68575, 2.156516, 1.467462, 1.211056
 .5            1.68575       2.156516      1.467462      1.211056
 .8660254      1.570796      1.570796E+09
?
```

4.2.2. Evaluate the integrals

$$I_1 = \int_0^{\frac{\pi}{2}} \frac{\sin^2\theta}{\sqrt{1-k^2\sin^2\theta}}\,d\theta, \quad I_2 = \int_0^{\frac{\pi}{2}} \frac{\cos^2\theta}{\sqrt{1-k^2\sin^2\theta}}\,d\theta \tag{1}$$

in terms of elliptic integrals.

Solution. Using the definitions given in Sec.3.1,

$$\begin{aligned} I_1 &= -\frac{1}{k^2}\int_0^{\frac{\pi}{2}} \frac{-1+(1-k^2\sin^2\theta)}{\sqrt{1-k^2\sin^2\theta}}\,d\theta \\ &= \frac{1}{k^2}\left[K(k)-E(k)\right], \quad k \in (0,1), \end{aligned} \tag{2}$$

and

$$\begin{aligned} I_2 &= \int_0^{\frac{\pi}{2}} \frac{1-\sin^2\theta}{\sqrt{1-k^2\sin^2\theta}}\,d\theta = K(k) - \int_0^{\frac{\pi}{2}} \frac{\sin^2\theta}{\sqrt{1-k^2\sin^2\theta}}\,d\theta \\ &= K(k) - \frac{1}{k^2}\left[K(k)-E(k)\right] \\ &= \left(1-\frac{1}{k^2}\right)\cdot K(k) + \frac{1}{k^2}E(k), \quad k \in (0,1). \end{aligned} \tag{3}$$

Particular case. Let $k = 1/2$ and from (2) and (3), we have

$$\begin{aligned} \int_0^{\frac{\pi}{2}} \frac{\sin^2\theta}{\sqrt{1-\frac{1}{4}\sin^2\theta}}\,d\theta &= 4\left[K\left(\frac{1}{2}\right) - E\left(\frac{1}{2}\right)\right] \\ &= 4(1.68575 - 1.467462) = 0.873, \end{aligned} \tag{4}$$

$$\begin{aligned} \int_0^{\frac{\pi}{2}} \frac{\cos^2\theta}{\sqrt{1-\frac{1}{4}\sin^2\theta}}\,d\theta &= -3\times K\left(\frac{1}{2}\right) + 4\times E\left(\frac{1}{2}\right) \\ &= -3\times 1.68575 + 4\times 1.467462 = 0.813. \end{aligned} \tag{5}$$

4.2.3. Evaluate the integral

$$I = \int_0^{\frac{\pi}{2}} \frac{2\sin^2\theta - 1}{\sqrt{1 - k^2\sin^2\theta}}\,d\theta \tag{1}$$

in terms of elliptic integrals.

Solution. The integral can be written

$$\begin{aligned}
I &= \int_0^{\frac{\pi}{2}} \frac{2\sin^2\theta - 1}{\sqrt{1 - k^2\sin^2\theta}}\,d\theta \\
&= \frac{1}{k^2}\int_0^{\frac{\pi}{2}} \frac{(2 - k^2) - 2(1 - k^2\sin^2\theta)}{\sqrt{1 - k^2\sin^2\theta}}\,d\theta \\
&= \frac{1}{k^2}\left[(2 - k^2)\int_0^{\frac{\pi}{2}} \frac{d\theta}{\sqrt{1 - k^2\sin^2\theta}} - 2\int_0^{\frac{\pi}{2}} \sqrt{1 - k^2\sin^2\theta}\,d\theta\right] \\
&= \frac{1}{k^2}\left[(2 - k^2)K(k) - 2E(k)\right].
\end{aligned} \tag{2}$$

Let $k = 1/2$ and we have

$$\begin{aligned}
I &= 4\left[\left(2 - \frac{1}{4}\right)K\left(\frac{1}{2}\right) - 2E\left(\frac{1}{2}\right)\right] \\
&= 4\left[\frac{7}{4}K\left(\frac{1}{2}\right) - 2E\left(\frac{1}{2}\right)\right] \\
&= 7K\left(\frac{1}{2}\right) - 8E\left(\frac{1}{2}\right).
\end{aligned} \tag{3}$$

Using Tables 3.1 and 3.4, we have

$$I = 7 \times 1.68575 - 8 \times 1.467462 = 0.0606. \tag{4}$$

4.2.4. Express the integral

$$I = \int_0^{\sqrt{3}} \frac{dx}{\sqrt{(1 + x^2)(1 + 2x^2)}} \tag{1}$$

in terms of elliptic integrals.

Solution. Let $x = \tan\theta$, $dx = d\theta/\cos^2\theta$ and (1) becomes

$$I = \int_0^{\frac{\pi}{3}} \frac{\dfrac{d\theta}{\cos^2\theta}}{\sqrt{(1+\tan^2\theta)(1+2\tan^2\theta)}} = \int_0^{\frac{\pi}{3}} \frac{\dfrac{d\theta}{\cos^2\theta}}{\sqrt{\left(1+\dfrac{\sin^2\theta}{\cos^2\theta}\right)\left(1+2\dfrac{\sin^2\theta}{\cos^2\theta}\right)}}$$

$$= \int_0^{\frac{\pi}{3}} \frac{\dfrac{d\theta}{\cos^2\theta}}{\sqrt{\dfrac{\sin^2\theta+\cos^2\theta}{\cos^2\theta}\cdot\dfrac{\cos^2\theta+2\sin^2\theta}{\cos^2\theta}}} = \int_0^{\frac{\pi}{3}} \frac{d\theta}{\sqrt{\cos^2\theta+2\sin^2\theta}}$$

$$= \int_0^{\frac{\pi}{3}} \frac{d\theta}{\sqrt{\cos^2\theta+2(1-\cos^2\theta)}}$$

$$= \int_0^{\frac{\pi}{3}} \frac{d\theta}{\sqrt{2-\cos^2\theta}} = \frac{1}{\sqrt{2}}\int_0^{\frac{\pi}{3}} \frac{d\theta}{\sqrt{1-\dfrac{1}{2}\cos^2\theta}}. \tag{2}$$

Let $\theta = \pi/2 - \phi$, $d\theta = -d\phi$, for $\theta = 0$, $\phi = \pi/2$, $\theta = \pi/3$, $\phi = \pi/6$, and from (2) we have

$$I = \frac{1}{\sqrt{2}}\int_{\frac{\pi}{2}}^{\frac{\pi}{6}} \frac{-d\theta}{\sqrt{1-\dfrac{1}{2}\sin^2\phi}} = \frac{1}{\sqrt{2}}\int_{\frac{\pi}{6}}^{\frac{\pi}{2}} \frac{d\phi}{\sqrt{1-\dfrac{1}{2}\sin^2\phi}}$$

$$= \frac{1}{\sqrt{2}}\left[\int_0^{\frac{\pi}{2}} \frac{d\theta}{\sqrt{1-\dfrac{1}{2}\sin^2\phi}} - \int_0^{\frac{\pi}{6}} \frac{d\theta}{\sqrt{1-\dfrac{1}{2}\sin^2\phi}}\right]$$

$$= \frac{1}{\sqrt{2}}\left[K\left(\frac{1}{\sqrt{2}}\right) - F\left(\frac{1}{\sqrt{2}}, \frac{\pi}{6}\right)\right] \tag{3}$$

$$= \frac{1}{\sqrt{2}}(1.854074 - 0.5294) = 0.937. \tag{4}$$

4.2.5. Evalute the integral

$$I = \int_1^{\infty} \frac{dx}{\sqrt{(x^2-1)(x^2+2)}} \tag{1}$$

in terms of the elliptic integrals.

Solution. According to Appendix A1.2, the integral converges at $x = 1$ since we have

$$f(x) = \frac{1}{\sqrt{(x^2-1)(x^2+2)}} = \mathop{O}_{x=1}\left[\frac{1}{(x-1)^{\frac{1}{2}}}\right] \tag{2}$$

and hence $\alpha = 1/2 < 1$.

With the change of variable $x = 1/cos\theta$, $dx = (\sin\theta/\cos^2\theta)\,d\theta$, the integral (1) becomes

$$\begin{aligned}
I &= \int_0^{\frac{\pi}{2}} \frac{\dfrac{\sin\theta}{\cos^2\theta}\,d\theta}{\sqrt{\left(\dfrac{1}{\cos^2\theta}-1\right)\left(\dfrac{1}{\cos^2\theta}+2\right)}} \\
&= \int_0^{\frac{\pi}{2}} \frac{\dfrac{\sin\theta}{\cos^2\theta}\,d\theta}{\sqrt{\dfrac{1-\cos^2\theta}{\cos^2\theta}\cdot\dfrac{1+2\cos^2\theta}{\cos^2\theta}}} \\
&= \int_0^{\frac{\pi}{2}} \frac{\sin\theta\,d\theta}{\sqrt{(1-\cos^2\theta)(1+2\cos^2\theta)}} \\
&= \int_0^{\frac{\pi}{2}} \frac{d\theta}{\sqrt{1+2\cos^2\theta}} = \int_0^{\frac{\pi}{2}} \frac{d\theta}{\sqrt{1+2(1-\sin^2\theta)}} \\
&= \int_0^{\frac{\pi}{2}} \frac{d\theta}{\sqrt{3-2\sin^2\theta}} = \frac{1}{\sqrt{3}}\int_0^{\frac{\pi}{2}} \frac{d\theta}{\sqrt{1-\dfrac{2}{3}\sin^2\theta}} \\
&= \frac{1}{\sqrt{3}}K\left(\sqrt{\frac{2}{3}}\right) = 1.1714.
\end{aligned} \tag{3}$$

4.2.6. Evaluate the integral

$$I = \int_0^{\frac{\pi}{4}} \frac{d\theta}{\sqrt{\cos 2\theta}} \tag{1}$$

in terms of elliptic integrals.

Test for convergence. According to Appendix A1.2, the integral converges at $\theta = \pi/4$, since we can write

$$\frac{1}{\sqrt{\cos 2\theta}} = \frac{1}{\sqrt{\sin\left(\frac{\pi}{2} - 2\theta\right)}} \underset{\theta=\frac{\pi}{4}}{=} O\left[\frac{1}{\sqrt{\frac{\pi}{2} - 2\theta}}\right] \tag{2}$$

and hence $\alpha = 1/2 < 1$.

Integral evaluation. The integral can be written

$$I = \int_0^{\frac{\pi}{4}} \frac{d\theta}{\sqrt{\cos 2\theta}} = \int_0^{\frac{\pi}{4}} \frac{d\theta}{\sqrt{1 - 2\sin^2\theta}}$$

and by substituting $\sqrt{2}\sin\theta = \sin\phi$, we have

$$I = \frac{1}{\sqrt{2}} \int_0^{\frac{\pi}{2}} \frac{d\theta}{\sqrt{1 - \frac{1}{2}\sin^2\phi}} \tag{3}$$

and using Table 3.2, the integral evaluation is

$$\int_0^{\frac{\pi}{4}} \frac{d\theta}{\sqrt{\cos 2\theta}} = \frac{1}{\sqrt{2}} K\left(\frac{1}{\sqrt{2}}\right) = \frac{1}{\sqrt{2}} \cdot 1.854074 = 1.311. \tag{4}$$

Also, from (4) and using Table 3.3,

$$I = \frac{1}{\sqrt{2}} K(\sin 45^\circ) = \frac{1}{\sqrt{2}} \cdot 1.854075 = 1.311. \tag{5}$$

See also Sec.4.3.4.

4.2.7. Express the integral

$$I = \int_0^{\frac{\pi}{2}} \frac{\sin^2\theta\cos^2\theta}{\sqrt{1 - k^2\sin^2\theta}}\, d\theta, \quad k \in (0, 1), \tag{1}$$

in terms of elliptic integrals.

Solution. An integration by parts, see p. xi, with the notations

$$\begin{cases} u = \sin\theta\cos\theta \\ v' = \dfrac{\sin\theta\cos\theta}{\sqrt{1-k^2\sin^2\theta}} \end{cases} \Rightarrow \begin{cases} u' = \cos^2\theta - \sin^2\theta \\ v = -\dfrac{1}{k^2}\sqrt{1-k^2\sin^2\theta} \end{cases} \tag{4.1}$$

we have

$$\begin{aligned}
I &= -\frac{1}{k^2}\left[\sin\theta\cos\theta\sqrt{1-k^2\sin^2\theta}\Big|_0^{\frac{\pi}{2}} - \int_0^{\frac{\pi}{2}}(\cos^2\theta - \sin^2\theta)\sqrt{1-k^2\sin^2\theta}\,d\theta\right] && (2)\\
&= \frac{1}{k^2}\int_0^{\frac{\pi}{2}}(\cos^2\theta - \sin^2\theta)\sqrt{1-k^2\sin^2\theta}\,d\theta\\
&= \frac{1}{k^2}\int_0^{\frac{\pi}{2}}(2\cos^2\theta - 1)\sqrt{1-k^2\sin^2\theta}\,d\theta\\
&= \frac{2}{k^2}\int_0^{\frac{\pi}{2}}\cos^2\theta\sqrt{1-k^2\sin^2\theta}\,d\theta - \frac{1}{k^2}\int_0^{\frac{\pi}{2}}\sqrt{1-k^2\sin^2\theta}\,d\theta && (3)\\
&= \frac{1}{k^2}\int_0^{\frac{\pi}{2}}\cos^2\theta\sqrt{1-k^2\sin^2\theta}\,d\theta - \frac{1}{k^2}E(k)\\
&= \frac{2}{k^2}\int_0^{\frac{\pi}{2}}\cos^2\theta\frac{1-k^2\sin^2\theta}{\sqrt{1-k^2\sin^2\theta}}\,d\theta - \frac{1}{k^2}E(k)\\
&= \frac{2}{k^2}\int_0^{\frac{\pi}{2}}\frac{\cos^2\theta\,d\theta}{\sqrt{1-k^2\sin^2\theta}} - 2\int_0^{\frac{\pi}{2}}\frac{\sin^2\theta\cos^2\theta}{\sqrt{1-k^2\sin^2\theta}}\,d\theta - \frac{1}{k^2}E(k) && (4)\\
&= \frac{2}{k^2}\left[\left(1-\frac{1}{k^2}\right)K(k) + \frac{1}{k^2}E(k)\right] - 2I - \frac{1}{k^2}E(k) && (5)\\
3I &= \frac{2}{k^2}\left[\left(1-\frac{1}{k^2}\right)K(k) + \frac{1}{k^2}E(k)\right] - \frac{1}{k^2}E(k) && (6)\\
&= \frac{2}{k^2}\left(1-\frac{1}{k^2}\right)K(k) + \frac{1}{k^2}\left(\frac{2}{k^2}-1\right)E(k)\\
&= \frac{2}{k^2}\left(1-\frac{1}{k^2}\right)K(k) + \frac{1}{k^2}\left(\frac{2}{k^2}-1\right)E(k). && (7)
\end{aligned}$$

Let $k = 1/2$ and from (7),

$$3I = 8(1-4)K\left(\frac{1}{2}\right) + 4(8-1)E(k)$$

$$I = \frac{1}{3}\left[28E\left(\frac{1}{2}\right) - 24K\left(\frac{1}{2}\right)\right] \tag{8}$$

$$= \frac{28}{3} \times 1.467462 - 8 \times 1.68575 = 0.210. \tag{9}$$

4.2.8. Express the integral

$$I = \int_0^{\frac{\pi}{2}} \cos^2\theta\sqrt{1-k^2\sin^2\theta}\,d\theta, \quad k \in (0,1), \tag{1}$$

in terms of elliptic integrals.

Solution. The integral can be written

$$I = \int_0^{\frac{\pi}{2}} \frac{\cos^2\theta(1-k^2\sin^2\theta)}{\sqrt{1-k^2\sin^2\theta}}\,d\theta$$

$$= \int_0^{\frac{\pi}{2}} \frac{\cos^2\theta\,d\theta}{\sqrt{1-k^2\sin^2\theta}} - k^2\int_0^{\frac{\pi}{2}} \frac{\sin^2\theta\cos^2\theta}{\sqrt{1-k^2\sin^2\theta}}\,d\theta. \tag{2}$$

Using Secs.4.2.2 and 4.2.7, we have

$$I = \left(1-\frac{1}{k^2}\right)K(k) + \frac{1}{k^2}E(k) - \frac{2}{3}\left(1-\frac{1}{k^2}\right)K(k) - \frac{1}{3}\left(\frac{2}{k^2}-1\right)E(k)$$

$$= \frac{1}{3}\left[\left(1+\frac{1}{k^2}\right)E(k) + \left(1-\frac{1}{k^2}\right)K(k)\right], \quad k \in (0,1). \tag{3}$$

Particular case. Let $k = 1/5$ and from (3), we have

$$\int_0^{\frac{\pi}{2}} \cos^2\theta\sqrt{1-\frac{1}{25}\sin^2\theta}\,d\theta = \frac{1}{3}\left[(1+25)E\left(\frac{1}{5}\right) + (1-25)K\left(\frac{1}{5}\right)\right] \tag{4}$$

$$= \frac{1}{3}(26 \times 1.554969 - 24 \times 1.586868) = 0.781.$$

4.2.9. Evaluate the integral

$$I = \int_0^{\frac{\pi}{2}} \sin^2\theta\sqrt{1-k^2\sin^2\theta}\,d\theta, \quad k \in (0,1), \tag{1}$$

in terms of elliptic integrals.

Solution. The integral can be written

$$\begin{aligned} I &= \int_0^{\frac{\pi}{2}} \frac{\sin^2\theta(1-k^2\sin^2\theta)}{\sqrt{1-k^2\sin^2\theta}}\,d\theta \\ &= \int_0^{\frac{\pi}{2}} \frac{\sin^2\theta\,d\theta}{\sqrt{1-k^2\sin^2\theta}} - k^2\int_0^{\frac{\pi}{2}} \frac{(1-\cos^2\theta)\sin^2\theta}{\sqrt{1-k^2\sin^2\theta}}\,d\theta \\ &= (1-k^2)\int_0^{\frac{\pi}{2}} \frac{\sin^2\theta\,d\theta}{\sqrt{1-k^2\sin^2\theta}} + k^2\int_0^{\frac{\pi}{2}} \frac{\sin^2\theta\cos^2\theta}{\sqrt{1-k^2\sin^2\theta}}\,d\theta. \end{aligned} \tag{2}$$

Using Secs.4.2.2 and 4.2.7, we have

$$\begin{aligned} I &= \frac{1-k^2}{k^2}[K(k)-E(k)] + \frac{1}{3}\left[2\left(1-\frac{1}{k^2}\right)K(k) + \left(\frac{2}{k^2}-1\right)E(k)\right] \\ &= \frac{1}{3}\left(\frac{1}{k^2}-1\right)K(k) + \frac{1}{3}\left(2-\frac{1}{k^2}\right)E(k) \\ &= \frac{1}{3}\left[\left(\frac{1}{k^2}-1\right)K(k) + \left(2-\frac{1}{k^2}\right)E(k)\right], \quad k \in (0,1). \end{aligned} \tag{3}$$

Particular case. Let $k = 1/\sqrt{3}$,

$$\begin{aligned} I &= \frac{1}{3}\left[2K\left(\frac{1}{\sqrt{3}}\right) - E\left(\frac{1}{\sqrt{3}}\right)\right] \\ &= \frac{1}{3}(2\times 1.733917 - 1.430315) = 0.679. \end{aligned} \tag{4}$$

4.2.10. Evaluate the integral

$$I = \int_0^{\pi} \frac{d\theta}{\sqrt{1+k\cos\theta}}, \quad k \in (0,1), \tag{1}$$

in terms of elliptic integrals.

Solution. The integral can be written

$$I = \int_0^{\pi} \frac{d\theta}{\sqrt{1 + k - 2k \sin^2 \dfrac{\theta}{2}}} \tag{2}$$

and substituting $\theta = 2\phi$, the integral becomes

$$I = 2\int_0^{\frac{\pi}{2}} \frac{d\phi}{\sqrt{1 + k - 2k \sin^2 \phi}} = \frac{2}{\sqrt{1+k}} \int_0^{\frac{\pi}{2}} \frac{d\phi}{\sqrt{1 - \dfrac{2k}{1+k} \sin^2 \phi}}$$

or

$$\int_0^{\pi} \frac{d\theta}{\sqrt{1 + k \cos \theta}} = \frac{2}{\sqrt{1+k}} K\left(\sqrt{\frac{2k}{1+k}}\right), \quad 0 < \frac{2k}{1+k} < 1. \tag{3}$$

Particular cases.

a. Let $k = 1/4$ and from (3),

$$\int_0^{\pi} \frac{d\theta}{\sqrt{1 + \dfrac{1}{4} \cos \theta}} = \frac{4}{\sqrt{5}} K\left(\sqrt{\frac{2}{5}}\right) = \frac{4 \times 1.777519}{\sqrt{5}} = 3.18. \tag{4}$$

b. Let $k = 1/2$ and from (3),

$$\int_0^{\pi} \frac{d\theta}{\sqrt{1 + \dfrac{1}{2} \cos \theta}} = 2\sqrt{\frac{2}{3}} K\left(\sqrt{\frac{2}{3}}\right) = 2\sqrt{\frac{2}{3}} \times 2.02895 = 3.313. \tag{5}$$

4.2.11. Study the integral

$$I = \int_0^{\phi} \frac{d\theta}{\sqrt{\cos \theta - \cos \phi}}. \tag{1}$$

Solution. The integral can be written

$$I = \frac{1}{\sqrt{2}} \int_0^{\phi} \frac{d\theta}{\sqrt{\sin^2\frac{\phi}{2} - \sin^2\frac{\theta}{2}}}. \tag{2}$$

By substituting

$$\sin\frac{\theta}{2} = \sin\frac{\phi}{2}\sin\propto, \quad d\theta = \frac{2\sin\frac{\phi}{2}\cos\propto\, d\propto}{\sqrt{1-\sin^2\frac{\phi}{2}\sin^2\propto}} \tag{3}$$

from (2) we have

$$\begin{aligned} I &= \frac{1}{\sqrt{2}} \int_0^{\frac{\pi}{2}} \frac{2\sin\frac{\phi}{2}\cos\propto\, d\propto}{\sqrt{\sin^2\frac{\phi}{2} - \sin^2\frac{\phi}{2}\sin^2\propto}\cdot\sqrt{1-\sin^2\frac{\phi}{2}\sin^2\propto}} \\ &= \sqrt{2}\int_0^{\frac{\pi}{2}} \frac{d\propto}{\sqrt{1-\sin^2\frac{\phi}{2}\sin^2\propto}} \end{aligned}$$

or

$$\int_0^{\phi} \frac{d\theta}{\sqrt{\cos\theta - \cos\phi}} = \sqrt{2}K\left(\sin\frac{\phi}{2}\right), \quad 0 < \sin\frac{\phi}{2} < 1. \tag{4}$$

For application of this integral see Sec.4.2.29 (the pendulum).

Particular cases.

a. Let $\phi = \pi/2$ and from (4),

$$\begin{aligned} \int_0^{\frac{\pi}{2}} \frac{d\theta}{\sqrt{\cos\theta}} &= \sqrt{2}K\left(\sin\frac{\pi}{4}\right) = \sqrt{2}K(\sin 45^\circ) \\ &= \sqrt{2}\times 1.854075 = 2.622. \end{aligned}$$

See also Sec.4.3.2.

b. Let $\phi = \pi/3$ and from (4) we have

$$\int_0^{\frac{\pi}{3}} \frac{d\theta}{\sqrt{\cos\theta - \dfrac{1}{2}}} = \sqrt{2}K\left(\frac{1}{2}\right) = \sqrt{2} \times 1.68575 = 2.384.$$

Note that from (4) we also have

$$\int_0^{\frac{\pi}{3}} \frac{d\theta}{\sqrt{\cos\theta - \dfrac{1}{2}}} = \sqrt{2}K\left(\sin\frac{\pi}{6}\right) = \sqrt{2}K(\sin 30^\circ)$$
$$= \sqrt{2} \times 1.68575 = 2.384.$$

4.2.12. Evaluate the integral

$$I = \int_0^{\frac{\pi}{2}} \frac{d\theta}{\sqrt{1 + k^2 \sin^2\theta}}, \qquad k > 0, \tag{1}$$

in terms of elliptic integrals.

Solution. By substituting $\theta = \pi/2 - \phi$, we have

$$I = \int_0^{\frac{\pi}{2}} \frac{d\phi}{\sqrt{1 + k^2 \cos^2\phi}} = \int_0^{\frac{\pi}{2}} \frac{d\phi}{\sqrt{1 + k^2 - k^2 \sin^2\phi}} \tag{2}$$
$$= \frac{1}{\sqrt{1+k^2}} K\left(\frac{k}{\sqrt{1+k^2}}\right), \qquad k > 0. \tag{3}$$

Particular cases.

a. Let $k = 1$,

$$\int_0^{\frac{\pi}{2}} \frac{d\theta}{\sqrt{1 + \sin^2\theta}} = \frac{1}{\sqrt{2}} K\left(\frac{1}{\sqrt{2}}\right) = \frac{1}{\sqrt{2}} \times 1.854075 = 1.311. \tag{4}$$

b. Let $k = 1/2$,

$$\int_0^{\frac{\pi}{2}} \frac{d\theta}{\sqrt{1+\frac{1}{4}\sin^2\theta}} = \frac{2}{\sqrt{5}}K\left(\frac{1}{\sqrt{5}}\right) = \frac{2}{\sqrt{5}} \times 1.659623 = 1.484. \qquad (5)$$

4.2.13. Evaluate the integral

$$I = \int_0^{\frac{\pi}{2}} \sqrt{1+\sin^2 x}\, dx$$

in terms of elliptic integrals.

Solution. We have

$$I = \int_0^{\frac{\pi}{2}} \sqrt{1+\sin^2 x}\, dx = \int_0^{\frac{\pi}{2}} \sqrt{1+(1-\cos^2 x)}\, dx = \int_0^{\frac{\pi}{2}} \sqrt{2-\cos^2 x}\, dx$$

and with the substitution $x = \pi/2 - t$, $dx = -dt$,

$$\begin{aligned} I &= \int_0^{\frac{\pi}{2}} \sqrt{2-\cos^2\left(\frac{\pi}{2}-t\right)}\, dt = \int_0^{\frac{\pi}{2}} \sqrt{2-\sin^2 t}\, dt \\ &= \sqrt{2}\int_0^{\frac{\pi}{2}} \sqrt{1-\frac{1}{2}\sin^2 t}\, dt = \sqrt{2}E\left(\frac{1}{\sqrt{2}}\right) \\ &= \sqrt{2} \times 1.350644 = 1.910. \end{aligned}$$

4.2.14. Evaluate the integrals

$$I_1 = \int_0^{\frac{\pi}{2}} \frac{\sin^2 x\, dx}{\sqrt{1+\sin^2 x}}, \quad I_2 = \int_0^{\frac{\pi}{2}} \frac{\cos^2 x\, dx}{\sqrt{1+\sin^2 x}} \qquad (1)$$

in terms of elliptic integrals.

Solution. We have

$$I_1 = \int_0^{\frac{\pi}{2}} \frac{\sin^2 x\, dx}{\sqrt{1+\sin^2 x}} = \int_0^{\frac{\pi}{2}} \frac{\sin^2 x + 1}{\sqrt{1+\sin^2 x}}\, dx - \int_0^{\frac{\pi}{2}} \frac{dx}{\sqrt{1+\sin^2 x}}$$
$$= \int_0^{\frac{\pi}{2}} \sqrt{1+\sin^2 x}\, dx - \int_0^{\frac{\pi}{2}} \frac{dx}{\sqrt{1+\sin^2 x}}.$$

Using Secs.4.2.13 and 4.2.12 yields

$$\int_0^{\frac{\pi}{2}} \frac{\sin^2 x\, dx}{\sqrt{1+\sin^2 x}} = \sqrt{2}E\left(\frac{1}{\sqrt{2}}\right) - \frac{1}{\sqrt{2}}K\left(\frac{1}{\sqrt{2}}\right) = 0.599. \tag{2}$$

For the second integral,

$$I_2 = \int_0^{\frac{\pi}{2}} \frac{1-\sin^2 x}{\sqrt{1+\sin^2 x}}\, dx = \int_0^{\frac{\pi}{2}} \frac{dx}{\sqrt{1+\sin^2 x}} - \int_0^{\frac{\pi}{2}} \frac{\sin^2 x}{\sqrt{1+\sin^2 x}}\, dx,$$

and using Sec.4.2.12 and (2), we have

$$I_2 = \frac{1}{\sqrt{2}}K\left(\frac{1}{\sqrt{2}}\right) - \left[\sqrt{2}E\left(\frac{1}{\sqrt{2}}\right) - \frac{1}{\sqrt{2}}K\left(\frac{1}{\sqrt{2}}\right)\right]$$
$$= \sqrt{2}\left[K\left(\frac{1}{\sqrt{2}}\right) - E\left(\frac{1}{\sqrt{2}}\right)\right] = 0.712. \tag{3}$$

4.2.15. Evaluate the integral

$$I = \int_0^{\theta_0} \frac{\sqrt{1-k^2\sin^2\theta}}{\cos^2\theta}\, d\theta, \quad k^2 < 1, \quad \theta_0 \left[0, \frac{\pi}{2}\right) \tag{1}$$

in terms of the elliptic integrals.

Solution. Integrating by parts, see p. xi, with the notations

$$\begin{cases} u = \sqrt{1-k^2\sin^2\theta} \\ v' = \dfrac{1}{\cos^2\theta} \end{cases} \Rightarrow \begin{cases} u' = -\dfrac{k^2\sin\theta\cos\theta}{\sqrt{1-k^2\sin^2\theta}} \\ v = \tan\theta \end{cases}, \tag{2}$$

we have

$$I = (\tan\theta)\cdot\sqrt{1-k^2\sin^2\theta}\Big|_0^{\theta_0} + \int_0^{\theta_0}\frac{\sin\theta}{\cos\theta}\cdot\frac{k^2\sin\theta\cos\theta}{\sqrt{1-k^2\sin^2\theta}}\,d\theta$$

$$= (\tan\theta_0)\sqrt{1-k^2\sin^2\theta_0} + k^2\int_0^{\theta_0}\frac{\sin^2\theta\,d\theta}{\sqrt{1-k^2\sin^2\theta}} \tag{3}$$

and with Sec.4.2.2,

$$I = (\tan\theta_0)\sqrt{1-k^2\sin^2\theta_0} + F(k,\theta_0) - E(k,\theta_0). \tag{4}$$

Particular case. Let $k = 1/2$, $\theta_0 = \pi/6$, $\sin\psi = 1/2$, $\psi = 30^\circ$,

$$I = \int_0^{\frac{\pi}{6}}\frac{\sqrt{1-\frac{1}{4}\sin^2\theta}}{\cos^2\theta}\,d\theta = \left(\tan\frac{\pi}{6}\right)\sqrt{1-\frac{1}{4}\cdot\frac{1}{4}} + F\left(\frac{1}{2},\frac{\pi}{6}\right) - E\left(\frac{1}{2},\frac{\pi}{6}\right)$$

$$= \frac{\sqrt{5}}{4} + 0.5294 - 0.5179 = 0.571.$$

4.2.16. Express the integral

$$I = \int_{\phi_0}^{\phi}\frac{\sqrt{1-k^2\sin^2\theta}}{\sin^2\theta}\,d\theta, \quad k^2 < 1, \quad 0 < \phi_0 < \phi < \frac{\pi}{2} \tag{1}$$

in terms of elliptic integrals.

Solution. Integrating by parts, with the notations

$$\begin{cases} u = \sqrt{1-k^2\sin^2\theta} \\ v' = \dfrac{1}{\sin^2\theta} \end{cases} \qquad \begin{cases} u' = \dfrac{-k^2\sin\theta\cos\theta}{\sqrt{1-k^2\sin^2\theta}} \\ v = -\cot\theta \end{cases},$$

we have

$$
\begin{aligned}
I &= -\cot\theta \cdot \sqrt{1-k^2\sin^2\theta}\Big|_{\phi_0}^{\phi} + \int_{\phi_0}^{\phi} \cot\theta \cdot \frac{-k^2\sin\theta\cos\theta}{\sqrt{1-k^2\sin^2\theta}}\,d\theta \\
&= -\left(\cot\theta \cdot \sqrt{1-k^2\sin^2\theta}\right)\Big|_{\phi_0}^{\phi} - \int_{\phi_0}^{\phi} \frac{\cos\theta}{\sin\theta} \cdot \frac{k^2\sin\theta\cos\theta}{\sqrt{1-k^2\sin^2\theta}}\,d\theta \\
&= -\left(\cot\theta \cdot \sqrt{1-k^2\sin^2\theta}\right)\Big|_{\phi_0}^{\phi} - \int_{\phi_0}^{\phi} \frac{k^2\cos^2\theta}{\sqrt{1-k^2\sin^2\theta}}\,d\theta \\
&= -\left(\cot\theta \cdot \sqrt{1-k^2\sin^2\theta}\right)\Big|_{\phi_0}^{\phi} - \int_{\phi_0}^{\phi} \frac{k^2(1-\sin^2\theta)}{\sqrt{1-k^2\sin^2\theta}}\,d\theta, \qquad (2)
\end{aligned}
$$

$$
\begin{aligned}
I_1 &= \int_{\phi_0}^{\phi} \frac{k^2(1-\sin^2\theta)}{\sqrt{1-k^2\sin^2\theta}}\,d\theta = \int_{\phi_0}^{\phi} \frac{k^2-k^2\sin^2\theta}{\sqrt{1-k^2\sin^2\theta}}\,d\theta \\
&= \int_{\phi_0}^{\phi} \frac{k^2+(1-k^2\sin^2\theta)-1}{\sqrt{1-k^2\sin^2\theta}}\,d\theta \\
&= \int_{\phi_0}^{\phi} \frac{(k^2-1)+(1-k^2\sin^2\theta)}{\sqrt{1-k^2\sin^2\theta}}\,d\theta \\
&= (k^2-1)F_0(k,\phi) + \int_{\phi_0}^{\phi} \sqrt{1-k^2\sin^2\theta}\,d\theta \\
&= (k^2-1)F_0(k,\phi) + E_0(k,\phi) \qquad (3)
\end{aligned}
$$

and (2) becomes

$$
\begin{aligned}
I &= -\cot\theta\sqrt{1-k^2\sin^2\theta}\Big|_{\phi_0}^{\phi} - \left[(k^2-1)F_0(k,\phi) + E_0(k,\phi)\right] \\
&= -\cot\theta\sqrt{1-k^2\sin^2\theta}\Big|_{\phi_0}^{\phi} + (1-k^2)F_0(k,\phi) - E_0(k,\phi). \qquad (4)
\end{aligned}
$$

Particular case. Let $k = 1/2$, $\phi_0 = \pi/6 = 30^\circ$, $\phi = \pi/3 = 60^\circ$, $\sin\psi = 1/2$, $\psi = 30^\circ$,

$$(1-k^2)F(k,\theta)\Big|_{\frac{\pi}{6}}^{\frac{\pi}{3}} = \left(1-\frac{1}{4}\right)\left[F\left(\frac{1}{2},\frac{\pi}{3}\right) - F\left(\frac{1}{2},\frac{\pi}{6}\right)\right]$$
$$= \frac{3}{4}(1.0896 - 0.5294) = \underline{0.42015}.$$
$$E(k,\theta)\Big|_{\frac{\pi}{6}}^{\frac{\pi}{3}} = E\left(\frac{1}{2},\frac{\pi}{3}\right) - E\left(\frac{1}{2},\frac{\pi}{6}\right) = 1.0076 - 0.5179 = \underline{0.4897}.$$
$$(\cot\theta)\sqrt{1-k^2\sin^2\theta}\Big|_{\frac{\pi}{6}}^{\frac{\pi}{3}} = \cot 60^\circ\cdot\sqrt{1-\frac{1}{4}\sin^2 60^\circ} - \cot 30^\circ\cdot\sqrt{1-\frac{1}{4}\sin^2 30^\circ}$$
$$= \frac{1}{\sqrt{3}}\sqrt{1-\frac{1}{4}\cdot\frac{3}{4}} - \sqrt{3}\sqrt{1-\frac{1}{4}\cdot\frac{1}{4}}$$
$$= \frac{1}{4}\left(\sqrt{\frac{13}{3}} - 3\sqrt{5}\right) = -\underline{1.1566}.$$

Substituting in (4),

$$I = 0.42015 - 0.4897 + 1.1566 = 1.087. \tag{5}$$

4.2.17. Evaluate the integral

$$I = \int_0^{\theta_0} \frac{d\theta}{\cos^2\theta\sqrt{1-k^2\sin^2\theta}}, \quad \theta\in\left[0,\frac{\pi}{2}\right), \quad k\in(0,1) \tag{1}$$

in terms of elliptic integrals.

Solution. The integral can be written

$$I = \int_0^{\theta_0} \frac{(\sin^2\theta+\cos^2\theta)\,d\theta}{\cos^2\theta\sqrt{1-k^2\sin^2\theta}} = F(k,\theta_0) + \int_0^{\theta_0}\frac{\sin^2\theta\,d\theta}{\cos^2\theta\sqrt{1-k^2\sin^2\theta}}. \tag{2}$$

Letting

$$\int_0^{\theta_0}\frac{\sin^2\theta\,d\theta}{\cos^2\theta\sqrt{1-k^2\sin^2\theta}} = \frac{1}{k^2}\int_0^{\theta_0}\frac{1-(1-k^2\sin^2\theta)}{\cos^2\theta\sqrt{1-k^2\sin^2\theta}}\,d\theta,$$

the integral becomes

$$I = F(k,\theta_0) + \frac{1}{k^2}\left(I - \int_0^{\theta_0} \frac{\sqrt{1-k^2\sin^2\theta}}{\cos^2\theta}\,d\theta\right) \tag{3}$$

and from this,

$$(k^2-1)\cdot I = k^2 F(k,\theta_0) - \int_0^{\theta_0} \frac{\sqrt{1-k^2\sin^2\theta}}{\cos^2\theta}\,d\theta \tag{4}$$

and using Sec.4.2.15, we have

$$(k^2-1)\cdot I = (k^2-1)F(k,\theta_0) + E(k,\theta_0) - \tan\theta_0\sqrt{1-k^2\sin^2\theta_0}$$

$$I = F(k,\theta_0) + \frac{1}{1-k^2}\left[\tan\theta_0\sqrt{1-k^2\sin^2\theta_0} - E(k,\theta_0)\right], \quad k\in(0,1). \tag{5}$$

Particular case. Let $\theta_0 = \pi/6$, $k = \sqrt{3}/2$ and the integral becomes

$$I = \int_0^{\frac{\pi}{6}} \frac{d\theta}{\cos^2\theta\sqrt{1-\frac{3}{4}\sin^2\theta}} \tag{6}$$

$$= F\left(\frac{\sqrt{3}}{2},\frac{\pi}{6}\right) + 4\left[\frac{1}{\sqrt{3}}\sqrt{1-\frac{3}{4}\cdot\frac{1}{4}} - E\left(\frac{\sqrt{3}}{2},\frac{\pi}{6}\right)\right]$$

$$= \sqrt{\frac{13}{3}} + F\left(\frac{\sqrt{3}}{2},\frac{\pi}{6}\right) - 4E\left(\frac{\sqrt{3}}{2},\frac{\pi}{6}\right) \tag{7}$$

$$= \sqrt{\frac{13}{3}} + 0.5422 - 4\times 0.5061 = 0.599. \tag{8}$$

4.2.18. Evaluate the integral

$$I = \int_{\phi_0}^{\phi} \frac{d\theta}{\sin^2\theta\cdot\sqrt{1-k^2\sin^2\theta}}, \qquad 0<\phi_0<\phi<\frac{\pi}{2}, \qquad k\in(0,1), \tag{1}$$

in terms of elliptic integrals.

Solution. The integral can be written

$$I = \int_{\phi_0}^{\phi} \frac{\sin^2\theta + \frac{1}{k^2}(1 - k^2\sin^2\theta) + 1 - \frac{1}{k^2}}{\sin^2\theta \cdot \sqrt{1 - k^2\sin^2\theta}}\, d\theta \tag{2}$$

$$\frac{1}{k^2} I = F(k,\theta)\Big|_{\phi_0}^{\phi} + \frac{1}{k^2}\int_{\phi_0}^{\phi} \frac{\sqrt{1 - k^2\sin^2\theta}}{\sin^2\theta}\, d\theta$$

$$I = k^2 F(k,\theta)\Big|_{\phi_0}^{\phi} + \int_{\phi_0}^{\phi} \frac{\sqrt{1 - k^2\sin^2\theta}}{\sin^2\theta}\, d\theta. \tag{3}$$

Using Sec.4.2.16 for the last integral, we have

$$I = k^2 F(k,\theta)\Big|_{\phi_0}^{\phi} + (1 - k^2)\cdot F(k,\theta)\Big|_{\phi_0}^{\phi} - E(k,\theta)\Big|_{\phi_0}^{\phi} - \left[\cot\theta \cdot \sqrt{1 - k^2\sin^2\theta}\right]\Big|_{\phi_0}^{\phi}$$

$$= F_0(k,\phi) - E_0(k,\phi) - \left[\cot\theta\sqrt{1 - k^2\sin^2\theta}\right]_{\phi_0}^{\phi} \tag{4}$$

where $0 < k < 1$ and $0 < \phi_0 < \phi < \frac{\pi}{2}$.

Particular case. Let $\phi_0 = \pi/6$, $\phi = \pi/3$, $k = \sqrt{3}/2$, $\psi = 60^\circ$ and the integral (1) becomes

$$I = \int_{\frac{\pi}{6}}^{\frac{\pi}{3}} \frac{d\theta}{\sin^2\theta\sqrt{1 - \frac{3}{4}\sin^2\theta}}$$

$$= F\left(\frac{\sqrt{3}}{2}, \theta\right)\Big|_{\frac{\pi}{6}}^{\frac{\pi}{3}} - E\left(\frac{\sqrt{3}}{2}, \theta\right)\Big|_{\frac{\pi}{6}}^{\frac{\pi}{3}} - \left[\cot\theta \cdot \sqrt{1 - k^2\sin^2\theta}\right]\Big|_{\frac{\pi}{6}}^{\frac{\pi}{3}}. \tag{5}$$

$$F\left(\frac{\sqrt{3}}{2}, \theta\right)\Big|_{\frac{\pi}{6}}^{\frac{\pi}{3}} = F\left(\frac{\sqrt{3}}{2}, \frac{\pi}{3}\right) - F\left(\frac{\sqrt{3}}{2}, \frac{\pi}{6}\right) = 1.2125 - 0.5422 = 0.6703$$

$$E\left(\frac{\sqrt{3}}{2}, \theta\right)\Big|_{\frac{\pi}{6}}^{\frac{\pi}{3}} = E\left(\frac{\sqrt{3}}{2}, \frac{\pi}{3}\right) - E\left(\frac{\sqrt{3}}{2}, \frac{\pi}{6}\right) = 0.9184 - 0.5061 = 0.4123$$

$$\cot\theta \cdot \sqrt{1 - k^2\sin^2\theta}\Big|_{\frac{\pi}{6}}^{\frac{\pi}{3}} = \frac{1}{\sqrt{3}}\sqrt{1 - \frac{3}{4}\cdot\frac{3}{4}} - \sqrt{3}\cdot\sqrt{1 - \frac{3}{4}\cdot\frac{1}{4}} = -1.1794$$

and the evaluation of integral (1) is

$$\int_{\frac{\pi}{6}}^{\frac{\pi}{3}} \frac{d\theta}{\sin^2\theta\sqrt{1-\frac{3}{4}\sin^2\theta}} = 0.6703 - 0.4123 + 1.1794 = 1.4374. \tag{6}$$

4.2.19. Evaluate the integral

$$I = \int_0^a \frac{dx}{\sqrt{1+x^4}}, \quad a \in R^+ \tag{1}$$

in terms of the elliptic integrals.

Solution. By substituting

$$x = \tan\frac{\theta}{2}, \quad dx = \frac{1}{2}\frac{d\theta}{\cos^2\frac{\theta}{2}} \tag{2}$$

we can write

$$1 + x^4 = 1 + \tan^4\frac{\theta}{2} = \left(1 + \sqrt{2}\tan\frac{\theta}{2} + \tan^2\frac{\theta}{2}\right)\left(1 - \sqrt{2}\tan\frac{\theta}{2} + \tan^2\frac{\theta}{2}\right)$$

$$= \frac{1}{\cos^4\frac{\theta}{2}}\left(1 - 2\sin^2\frac{\theta}{2}\cos^2\frac{\theta}{2}\right) = \frac{1}{\cos^4\frac{\theta}{2}}\left(1 - \frac{1}{2}\sin^2\theta\right). \tag{3}$$

From (2), the inferior integral limit, for $x = 0$, $\theta = 0$, and for the superior limit, $\tan\frac{\theta}{2} = a$ or $\theta = 2\tan^{-1}a$, and the integral becomes

$$\int_0^a \frac{dx}{\sqrt{1+x^4}} = \frac{1}{2}\int_0^{\theta} \frac{d\theta}{\sqrt{1-\frac{1}{2}\sin^2\theta}} = \frac{1}{2}F\left(\frac{1}{\sqrt{2}}, 2\tan^{-1}a\right), \quad a > 0. \tag{4}$$

The integral can also be written in another form. Dividing, member by member, the formulas

$$\sin^2\frac{\theta}{2} = \frac{1-\cos\theta}{2} \quad \text{and} \quad \cos^2\frac{\theta}{2} = \frac{1+\cos\theta}{2}$$

we have

$$\tan^2\frac{\theta}{2} = \frac{1-\cos\theta}{1+\cos\theta} \quad \text{or} \quad a^2 = \frac{1-\cos\theta}{1+\cos\theta}$$

or

$$\cos\theta = \frac{1-a^2}{1+a^2} \quad \text{and} \quad \theta = \cos^{-1}\frac{1-a^2}{1+a^2}.$$

Therefore,

$$\int_0^a \frac{dx}{\sqrt{1+x^4}} = \frac{1}{2}F\left(\frac{1}{\sqrt{2}}, \theta\right) = \frac{1}{2}F\left(\frac{1}{\sqrt{2}}, \cos^{-1}\frac{1-a^2}{1+a^2}\right), \quad a > 0. \tag{5}$$

Particular cases.

a. Let $a = 1$ and from (4) we have

$$\int_0^1 \frac{dx}{\sqrt{1+x^4}} = \frac{1}{2}F\left(\frac{1}{\sqrt{2}}, 2\tan^{-1}1\right) = \frac{1}{2}F\left(\frac{1}{\sqrt{2}}, \frac{\pi}{2}\right)$$
$$= \frac{1}{2}K\left(\frac{1}{\sqrt{2}}\right) = \frac{1}{2} \times 1.854075 = 0.927. \tag{6}$$

See also Sec.2.2.4a, with the integral evaluation in terms of the gamma and beta functions.

b. Let $a = +\infty$, and from (5) we have

$$\int_0^\infty \frac{dx}{\sqrt{1+x^4}} = \frac{1}{2}\lim_{a\to\infty} F\left(\frac{1}{\sqrt{2}}, \cos^{-1}\frac{1-a^2}{1+a^2}\right) = \frac{1}{2}F\left(\frac{1}{\sqrt{2}}, \pi\right)$$

and using Eq.(3.12), we have

$$\int_0^\infty \frac{dx}{\sqrt{1+x^4}} = K\left(\frac{1}{\sqrt{2}}\right) = 1.854. \tag{7}$$

Eq.(7) with Sec.2.3.2 (or 2.3.22 or 2.3.28) yields

$$K\left(\frac{1}{\sqrt{2}}\right) = \frac{1}{4\sqrt{\pi}}\left[\Gamma\left(\frac{1}{4}\right)\right]^2 = 1.854. \tag{8}$$

Also, from (6) and (7), we have

$$\int_1^\infty \frac{dx}{\sqrt{1+x^4}} = \int_0^\infty \frac{dx}{\sqrt{1+x^4}} - \int_0^1 \frac{dx}{\sqrt{1+x^4}}$$
$$= K\left(\frac{1}{\sqrt{2}}\right) - \frac{1}{2}K\left(\frac{1}{\sqrt{2}}\right)$$
$$= \frac{1}{2}K\left(\frac{1}{\sqrt{2}}\right) = 0.927. \tag{9}$$

See also Sec.2.3.2.

4.2.20. Evaluate the integral

$$I = \int_0^{\sqrt{2}} \frac{dx}{\sqrt{(2-x^2)(6-x^2)}} \tag{1}$$

in terms of elliptic integrals.

Test for convergence. According to Appendix A1.2, the integral converges at $x = \sqrt{2} - 0$, since we can write

$$\frac{1}{\sqrt{(2-x^2)(6-x^2)}} = \frac{1}{\sqrt{(\sqrt{2}-x)(\sqrt{2}+x)(6-x^2)}} = \underset{x=\sqrt{2}}{O}\left[\frac{1}{\sqrt{\sqrt{2}-x}}\right] \tag{2}$$

and hence, $\propto = 1/2 < 1$.

Integral evaluation. Letting

$$x = \sqrt{2}\sin\theta, \quad dx = \sqrt{2}\cos\theta \cdot d\theta, \tag{3}$$

$$I = \int_0^{\frac{\pi}{2}} \frac{\sqrt{2}\cos\theta\, d\theta}{\sqrt{(2-2\sin^2\theta)(6-2\sin^2\theta)}} = \int_0^{\frac{\pi}{2}} \frac{d\theta}{\sqrt{6-2\sin^2\theta}}$$
$$= \frac{1}{\sqrt{6}}\int_0^{\frac{\pi}{2}} \frac{d\theta}{\sqrt{1-\dfrac{1}{3}\sin^2\theta}} = \frac{1}{\sqrt{6}}K\left(\frac{1}{\sqrt{3}}\right) \tag{4}$$
$$= \frac{1}{\sqrt{6}} \times 1.733917 = 0.708. \tag{5}$$

4.2.21. Study the integral

$$\int_1^a \frac{dx}{\sqrt{x^4-1}}, \qquad a > 1. \tag{1}$$

Test for convergence. According to Appendix A1.2, the integral converges for $x = 1+$, since $1/\sqrt{x^4-1} = \underset{x=1+}{O}[1/\sqrt{x-1}]$ and hence $\alpha = 1/2 < 1$.

Integral evaluation. Let $x = 1/\cos\theta$, $dx = \dfrac{\sin\theta}{\cos^2\theta}\,d\theta$, $\cos^{-1}\dfrac{1}{a} = \phi$ and the integral becomes

$$\begin{aligned}\int_1^a \frac{dx}{\sqrt{x^4-1}} &= \int_0^\phi \frac{d\theta}{\sqrt{1+\cos^2\theta}} = \int_0^\phi \frac{d\theta}{\sqrt{2-\sin^2\theta}}\\ &= \frac{1}{\sqrt{2}}\int_0^\phi \frac{d\theta}{\sqrt{1-\frac{1}{2}\sin^2\theta}}\\ &= \frac{1}{\sqrt{2}}F\left(\frac{1}{\sqrt{2}}, \cos^{-1}\frac{1}{a}\right), \qquad a > 1.\end{aligned} \tag{2}$$

Particular cases.

a. Let $a = \infty$, and the integral converges at the upper limit, since $1/\sqrt{x^4-1} = \underset{x\to\infty}{O}[1/x^2]$ and hence $\alpha = 2 > 1$. From (2) we have

$$\int_1^\infty \frac{dx}{\sqrt{x^4-1}} = \frac{1}{\sqrt{2}}F\left(\frac{1}{\sqrt{2}}, \frac{\pi}{2}\right) = \frac{1}{\sqrt{2}}K\left(\frac{1}{\sqrt{2}}\right) = 1.311. \tag{3}$$

b. Let $a = 2$ and from (2),

$$\int_1^2 \frac{dx}{\sqrt{x^4-1}} = \frac{1}{\sqrt{2}}F\left(\frac{1}{\sqrt{2}}, \cos^{-1}\frac{1}{2}\right) = \frac{1.1424}{\sqrt{2}} = 0.808. \tag{4}$$

4.2.22. Evaluate the integral

$$I = \int_0^\infty \frac{dx}{\sqrt{(x^2+2)(x^2+4)}} \tag{1}$$

in terms of elliptic integrals.

Test for convergence. According to Appendix A1.1, the integral converges at the upper limit, since we can write

$$\frac{1}{\sqrt{(x^2+2)(x^2+4)}} = \underset{x=\infty}{O}\left[\frac{1}{x^2}\right] \tag{2}$$

and hence $\alpha = 2 > 1$.

Integral evaluation. Letting

$$x = \sqrt{2}\tan\theta, \quad dx = \sqrt{2}\frac{d\theta}{\cos^2\theta}, \tag{3}$$

the integral becomes

$$I = \int_0^{\frac{\pi}{2}} \frac{\sqrt{2}\dfrac{d\theta}{\cos^2\theta}}{\sqrt{(2\tan^2\theta+2)(2\tan^2\theta+4)}} = \frac{1}{\sqrt{2}}\int_0^{\frac{\pi}{2}} \frac{d\theta}{\sqrt{\sin^2\theta + 2\cos^2\theta}}$$

$$= \frac{1}{\sqrt{2}}\int_0^{\frac{\pi}{2}} \frac{d\theta}{\sqrt{\sin^2\theta + 2(1-\sin^2\theta)}} = \frac{1}{\sqrt{2}}\int_0^{\frac{\pi}{2}} \frac{d\theta}{\sqrt{2-\sin^2\theta}}$$

$$= \frac{1}{2}\int_0^{\frac{\pi}{2}} \frac{d\theta}{\sqrt{1-\dfrac{1}{2}\sin^2\theta}}$$

$$= \frac{1}{2}K\left(\frac{1}{\sqrt{2}}\right) = \frac{1}{2} \times 1.854075 = 0.927. \tag{4}$$

4.2.23. Evaluate the integral

$$I = \int_0^1 \frac{dx}{\sqrt{x^4+x^2+1}}. \tag{1}$$

Solution. The integral can be written

$$\int_0^1 \frac{dx}{\sqrt{(x^2+x+1)(x^2-x+1)}}$$

and we let

$$x = \frac{1-t}{1+t}, \quad dx = -\frac{2dt}{(1+t)^2},$$

$$I = 2\int_0^1 \frac{dt}{\sqrt{[(1-t)^2+(1-t^2)+(1+t)^2]\,[(1-t)^2-(1-t^2)+(1+t)^2]}}$$

$$= 2\int_0^1 \frac{dt}{\sqrt{(3+t^2)(1+3t^2)}}. \tag{2}$$

With the substitution $\sqrt{3}t = \tan\theta$, $dt = d\theta/\sqrt{3}\cos^2\theta$,

$$3+t^2 = 3+\frac{\tan^2\theta}{3} = \frac{9\cos^2\theta+\sin^2\theta}{3\cos^2\theta}$$

$$1+3t^2 = 1+\tan^2\theta = \frac{1}{\cos^2\theta},$$

the integral becomes

$$I = 2\int_0^{\frac{\pi}{3}} \frac{d\theta}{\sqrt{9\cos^2\theta+\sin^2\theta}} = 2\int_0^{\frac{\pi}{3}} \frac{d\theta}{\sqrt{9(1-\sin^2\theta)+\sin^2\theta}}$$

$$= 2\int_0^{\frac{\pi}{3}} \frac{d\theta}{\sqrt{9-8\sin^2\theta}} = \frac{2}{3}\int_0^{\frac{\pi}{3}} \frac{d\theta}{\sqrt{1-\frac{8}{9}\sin^2\theta}} \tag{3}$$

$$= \frac{2}{3}F\left(\sqrt{\frac{8}{9}}, \frac{\pi}{3}\right) = \frac{2}{3}\times 1.2619 = 0.8413. \tag{4}$$

4.2.24. Evaluate the length of a sinusoidal arc, $y = a\sin x$, for a complete cycle $x \in [0, 2\pi]$, $a \in R^+$. See Fig.4.1

Solution. The length of an elementary arc is given by the formula

$$ds = \sqrt{1+y'^2}\,dx. \tag{1}$$

However, $y' = a\cos x$ and $ds = \sqrt{1+a^2\cos^2 x}\,dx$ and because of symmetry,

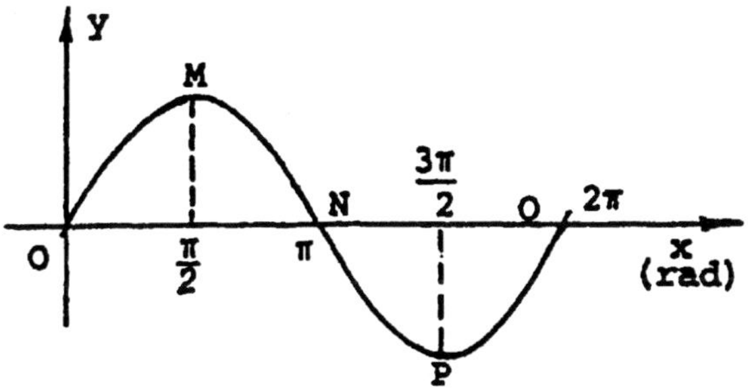

Fig.4.1

$$L = 4\int_{OM} ds = 4\int_0^{\frac{\pi}{2}} \sqrt{1 + a^2\cos^2 x}\, dx$$
$$= 4\int_0^{\frac{\pi}{2}} \sqrt{1 + a^2 - a^2\sin^2 x}\, dx$$
$$= 4\sqrt{1+a^2}\int_0^{\frac{\pi}{2}} \sqrt{1 - \frac{a^2}{1+a^2}\sin^2 x}\, dx$$

or

$$L = 4\sqrt{1+a^2}\,E\left(\frac{a}{\sqrt{1+a^2}}\right), \quad a \in R^+. \text{ [1]} \tag{4}$$

Particular case. Let $a = 1$ m; the curve is $y = \sin x$, and the length of a complete cycle is given by (2),

$$L = 4 \times \sqrt{2}E\left(\frac{1}{\sqrt{2}}\right) = 4\sqrt{2} \times 1.350644 = 7.64 \text{ m} \tag{3}$$

or, using Table 3.4,

$$L = 4\sqrt{2}E(\sin 45^\circ) = 4\sqrt{2} \times 1.350644 = 7.64 \text{ m}. \tag{4}$$

---

[1] Here k, from the definition of $E(k)$, is $a/\sqrt{1+a^2}$.

4.2.25. Evaluate the entire length of Bernoulli's lemniscate,

$$r^2 = a^2 \sin 2\theta, \quad a > 0. \tag{1}$$

Solution. Using the method given in Sec.2.2.12 the graph of the lemniscate is shown in Fig.4.2. The entire length of the lemniscate is given by

$$L = 4 \int_{OMB} ds \quad \text{where} \quad ds^2 = dr^2 + r^2\, d\theta^2. \tag{2}$$

From (1) we have

$$dr = \frac{a^2 \cos 2\theta}{r}\, d\theta = \frac{a \cos 2\theta}{\sqrt{\sin 2\theta}}\, d\theta \tag{3}$$

and

$$ds^2 = a^2 \left(\frac{\cos^2 2\theta}{\sin 2\theta} + \sin 2\theta \right) d\theta^2 = \frac{a^2}{\sin 2\theta}\, d\theta^2,$$

$$ds = \frac{a}{\sqrt{\sin 2\theta}}\, d\theta \tag{4}$$

and from (2)

$$L = 4a \int_0^{\frac{\pi}{4}} \frac{d\theta}{\sqrt{\sin 2\theta}}. \tag{5}$$

Change $2\theta = \varphi$, $2\, d\theta = d\varphi$,

$$L = 2a \int_0^{\frac{\pi}{2}} \frac{d\varphi}{\sqrt{\sin \varphi}} \tag{6}$$

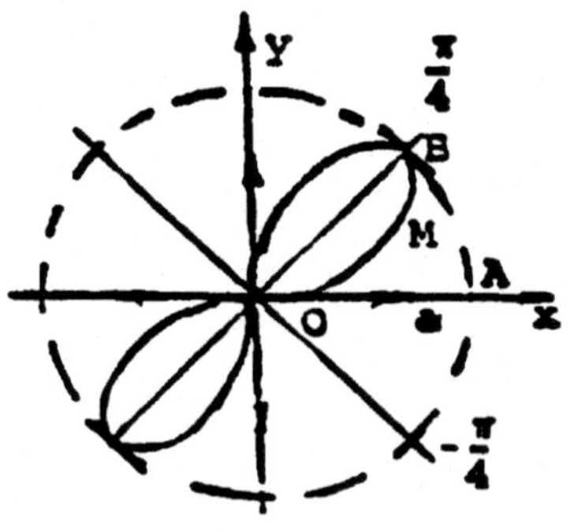

Fig.4.2

and with Sec.4.3.1, we have

$$L = 2\sqrt{2}K\left(\frac{1}{\sqrt{2}}\right)a = 2\sqrt{2}\,1.85407 \times a = 5.244 \times a \tag{7}$$

or

$$L = aB\left(\frac{1}{4}, \frac{1}{2}\right) = a\frac{1}{\sqrt{2\pi}}\left[\Gamma\left(\frac{1}{4}\right)\right]^2 = 5.244 \times a, \quad a > 0. \tag{8}$$

Let $a = 0.5$ m and $L = 5.244 \times 0.5 = 2.622$ m.

For the area of the lemniscate, see Secs. 2.2.12 and 2.2.13.

4.2.26. Evaluate the length of the arc of the ellipse

$$\frac{x^2}{a^2} + \frac{y^2}{b^2} = 1, \quad \{a, b \in R \mid a > b > 0\}. \tag{1}$$

Solution. A parametric representation of the ellipse is

$$x = a\sin\theta, \quad y = b\cos\theta \tag{2}$$

$$dx = a\cos\theta\, d\theta, \quad dy = -b\sin\theta\, d\theta$$

and the elementary arc length is given by $ds = \sqrt{(dx)^2 + (dy)^2}$.

Taking into account the symmetries, we have

$$\begin{aligned} L &= 4\int_0^{\frac{\pi}{2}} \sqrt{a^2\cos^2\theta + b^2\sin^2\theta}\, d\theta \\ &= 4\int_0^{\frac{\pi}{2}} \sqrt{a^2(1-\sin^2\theta) + b^2\sin^2\theta}\, d\theta \\ &= 4a\int_0^{\frac{\pi}{2}} \sqrt{1 - k^2\sin^2\theta}\, d\theta \\ &= 4a\cdot E(k), \quad k^2 = \frac{a^2 - b^2}{a^2}, \quad 0 < b < a. \end{aligned} \tag{3}$$

Note that the arc length of the ellipse cannot be expressed in terms of the elementary functions.

Particular cases.

a. Let $a = 4$ m, $b = 2$ m. The equation of the ellipse is

$$\frac{x^2}{4^2} + \frac{y^2}{2^2} = 1, \quad k^2 = \frac{3}{4}, \quad k = \frac{\sqrt{3}}{2} \tag{4}$$

and the total length of the arc of the ellipse is

$$L = 4 \times 4 \cdot E\left(\frac{\sqrt{3}}{2}\right) = 16 \times 1.211056 = 19.377 \text{ m} \tag{5}$$

or, using Table 3.4 for $\sin\psi = \sqrt{3}/2$ or $\psi = 60^\circ$,

$$L = 16E(\sin 60^\circ) = 16 \times 1.211056 = 19.377 \text{ m}. \tag{6}$$

b. The arc length of the ellipse from $A(a, 0)$ to a point $M(x, y)$ on the ellipse, in the first quadrant (see Fig.4.3), can be calculated with the formula

$$L_{AM} = a\int_0^{\phi} \sqrt{1 - k^2\sin^2\theta}\, d\theta = a \cdot E(k, \phi), \quad k^2 = \frac{a^2 - b^2}{a^2}, \quad 0 < b < a. \tag{7}$$

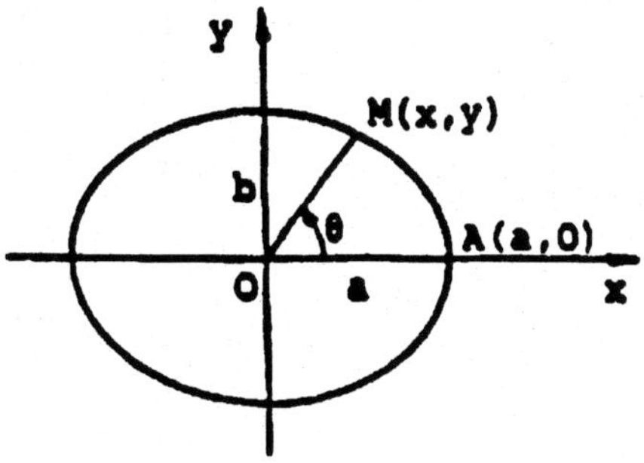

Fig.4.3

Let $a = 4$ m, $b = 2$ m, $\phi = 30^\circ$, $k^2 = 3/4$ and from (7),

$$L_{AM} = 4E\left(\frac{\sqrt{3}}{2}, 30^\circ\right) = 4 \times 0.5061 = 2.024 \text{ m}. \tag{8}$$

4.2.27. Evaluate the arc length of the hyperbola

$$\frac{x^2}{a^2} - \frac{y^2}{b^2} = 1, \quad a, b \in R^+. \tag{1}$$

Solution 1. The length of an arc from the point $A(a, 0)$ to the point $P(x_0, y_0)$ on the hyperbola, in the first quadrant (see Fig.4.4), is given by the line integral

$$L_{AP} = \int_{AP} ds, \quad \text{where} \quad ds = \sqrt{dx^2 + dy^2}. \tag{2}$$

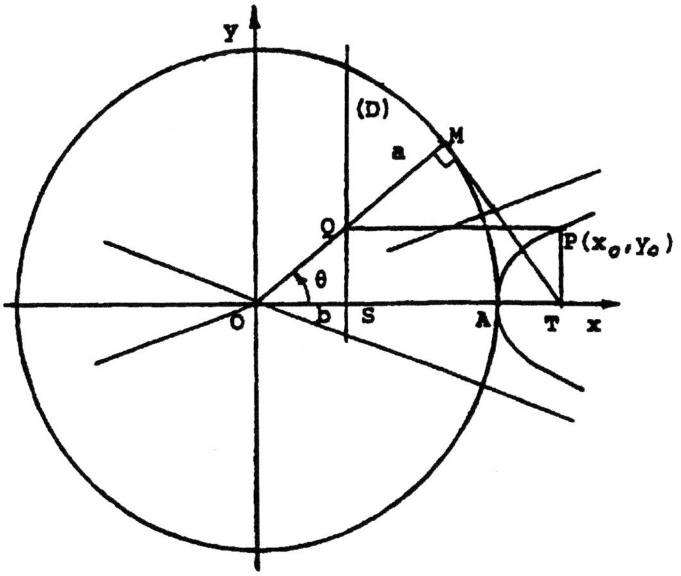

Fig.4.4

From Fig.4.4, we can obtain a parametric representation of hyperbola (1),

$$x = a\sec\theta, \quad y = b\tan\theta, \tag{3}$$

and we have

$$dx = \frac{a \sin\theta}{\cos^2\theta}\, d\theta, \quad dy = \frac{b}{\cos^2\theta}\, d\theta, \tag{4}$$

$$ds = \frac{\sqrt{a^2 \sin^2\theta + b^2}}{\cos^2\theta}\, d\theta \tag{5}$$

and from (2),

$$L_{AP} = \int_0^{\theta_0} \frac{\sqrt{a^2 \sin^2\theta + b^2}}{\cos^2\theta}\, d\theta. \tag{6}$$

With the substitution $\theta = \pi/2 - \phi$ we have

$$L_{AP} = \int_{\frac{\pi}{2}-\theta_0}^{\frac{\pi}{2}} \frac{\sqrt{a^2 \cos^2\phi + b^2}}{\sin^2\phi}\, d\phi = \int_{\frac{\pi}{2}-\theta_0}^{\frac{\pi}{2}} \frac{\sqrt{a^2 + b^2 - a^2 \sin^2\phi}}{\sin^2\phi}\, d\phi \tag{7}$$

$$= \sqrt{a^2 + b^2} \int_{\frac{\pi}{2}-\theta_0}^{\frac{\pi}{2}} \frac{\sqrt{1 - k^2 \sin^2\phi}}{\sin^2\phi}\, d\phi, \quad k^2 = \frac{a^2}{a^2 + b^2}. \tag{8}$$

Using the result from Sec.4.2.16, we have

$$L_{AP} = \sqrt{a^2 + b^2} \left[(1 - k^2) F(k, \phi) \Big|_{\frac{\pi}{2}-\theta_0}^{\frac{\pi}{2}} - E(k, \phi) \Big|_{\frac{\pi}{2}-\theta_0}^{\frac{\pi}{2}} - \left(\cot\phi \sqrt{1 - k^2 \sin^2\phi} \right) \Big|_{\frac{\pi}{2}-\theta_0}^{\frac{\pi}{2}} \right]. \tag{9}$$

Particular case. Let $a = 4\,\text{m}$, $b = 3\,\text{m}$, $x_0 = 8\,\text{m}$, and the equation of hyperbola, see Fig.4.5, is

$$\frac{x^2}{16} - \frac{y^2}{9} = 1, \tag{10}$$

from where

$$\frac{y_0^2}{9} = \frac{8^2}{16} - 1 \quad \text{and} \quad y_0 = 5.196\ \text{m}, \quad a^2 + b^2 = 25\ \text{m}^2,$$

$$k^2 = \frac{a^2}{a^2 + b^2} = \frac{16}{16 + 9} = 0.64, \quad k = 0.8, \quad 1 - k^2 = 1 - 0.64 = 0.36.$$

From (3), $\tan\theta_0 = y_0/b = 5.196/3 = 1.732$ and from this $\theta_0 = 60^\circ$ and

$$\frac{\pi}{2} - \theta_0 = 90^\circ - 60^\circ = 30^\circ, \quad \sin\psi = 0.8 \quad \text{and} \quad \psi = 53.13^\circ.$$

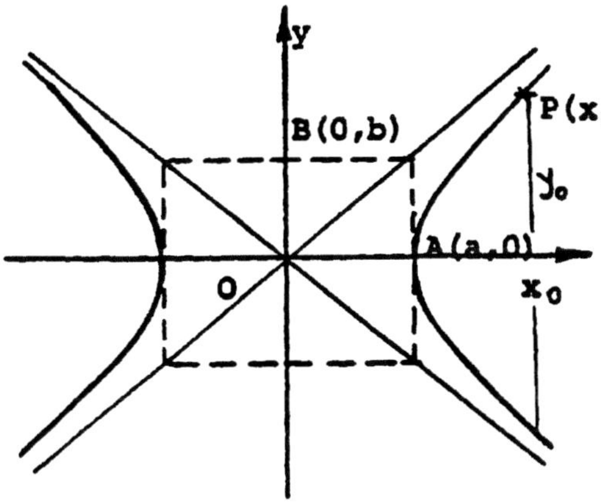

Fig.4.5

Using the tables given in [4],

$$F(k,\theta_0)\Big|_{30^\circ}^{90^\circ} = F(0.8, 90^\circ) - F(0.8, 30^\circ) = 1.995 - 0.537 = 1.458,$$

$$E(k,\theta_0)\Big|_{30^\circ}^{90^\circ} = F(0.8, 90^\circ) - F(0.8, 30^\circ) = 1.277 - 0.51 = 0.767,$$

$$\cot\phi\sqrt{1-k^2\sin^2\phi}\Big|_{30^\circ}^{90^\circ} = -\left[\cot(30^\circ)\right]\sqrt{1-0.64\times\sin^2(30^\circ)}$$

$$= -1.7321 \times 0.9165 = -1.5875$$

and from (9),

$$L_{AP} = 5(0.36 \times 1.458 - 0.767 + 1.5875) = 6.72 \text{ m}. \tag{11}$$

Solution 2. The arc length of the hyperbola is given by the line integral

$$L_{AP} = \int_{AP} ds = \int_0^y \sqrt{1 + \left(\frac{dx}{dy}\right)^2}\, dy. \tag{12}$$

From (1) we have

$$\frac{dx}{dy} = \frac{a^2}{b^2}\frac{y}{x} \tag{13}$$

and then

$$L_{AP} = \int_0^y \sqrt{\frac{(a^2+b^2)y^2+b^4}{b^2(y^2+b^2)}}\, dy. \tag{14}$$

Changing the variable,

$$(a^2+b^2)y^2 = b^4\tan^2\theta, \quad dy = \frac{b^2\,d\theta}{\sqrt{a^2+b^2}\cdot\cos^2\theta}, \qquad \theta\in\left[0,\frac{\pi}{2}\right), \tag{15}$$

we have

$$L_{AP} = \frac{b^2}{\sqrt{a^2+b^2}}\int_0^{\theta_0}\frac{d\theta}{\cos^2\theta\sqrt{1-k^2\sin^2\theta}}. \tag{16}$$

Using the result from Sec.4.2.17, we have

$$L_{AP} = \frac{b^2}{\sqrt{a^2+b^2}}\left\{F(k,\theta_0) - \frac{a^2+b^2}{b^2}\left[E(k,\theta_0) - \tan\theta_0\cdot\sqrt{1-k^2\sin^2\theta_0}\right]\right\} \tag{17}$$

where

$$k = \frac{a}{\sqrt{a^2+b^2}}, \qquad \theta_0\in\left[0,\frac{\pi}{2}\right).$$

Particular case. Evaluate the arc length of the hyperbola given in Solution 2:

$$a = 4\text{ m}, \quad b = 3\text{ m}, \quad x_0 = 8\text{ m}, \quad \text{see Fig.4.5},$$

$$\frac{y^2}{9} = \frac{64}{16} - 1 \quad\text{and}\quad y_0^2 = 27, \quad y_0 = 5.196\text{ m},$$

$$k = \frac{a}{\sqrt{a^2+b^2}} = \frac{4}{\sqrt{25}} = \frac{4}{5}, \qquad \sin\psi = \frac{4}{5}, \qquad \psi = 53.13^\circ$$

and from (15),

$$25y_0^2 = 3^4\tan^2\theta \to 25\times 27 = 81\times\tan^2\theta_0, \qquad \tan\theta_0 = 2.886,$$

$$\theta_0 = 70.89^\circ, \qquad \frac{a^2+b^2}{b^2} = \frac{25}{9},$$

and using the tables given in [4],

$$F(k,\theta_0) = F(0.8, 70.89^\circ) = 1.432,$$

$$E(k,\theta_0) = E(0.8, 70.89^\circ) = 1.05,$$

$$\sin\theta_0 = \sin 70.89^\circ = 0.945 \to \sin^2\theta_0 = 0.893,$$

$$\sqrt{1-k^2\sin^2\theta_0} = \sqrt{1-0.64\times 0.893} = 0.6546$$

and from (17),

$$L_{AP} = \frac{9}{5}\left[1.432 - \frac{25}{9}(1.05 - 2.886\times 0.6546)\right] = 6.773\text{ m}.$$

4.2.28. Assuming unit density, calculate the moment of inertia about the x axis, I_x, of the total arc of the ellipse,

$$\frac{x^2}{a^2} + \frac{y^2}{b^2} = 1, \quad \left\{a, b \in R^+ \mid a > b\right\}. \tag{1}$$

Solution. A parametric representation of the ellipse is

$$x = a\cos\theta, \quad y = b\sin\theta. \tag{2}$$

The arc element of the ellipse (see Fig.4.6) is

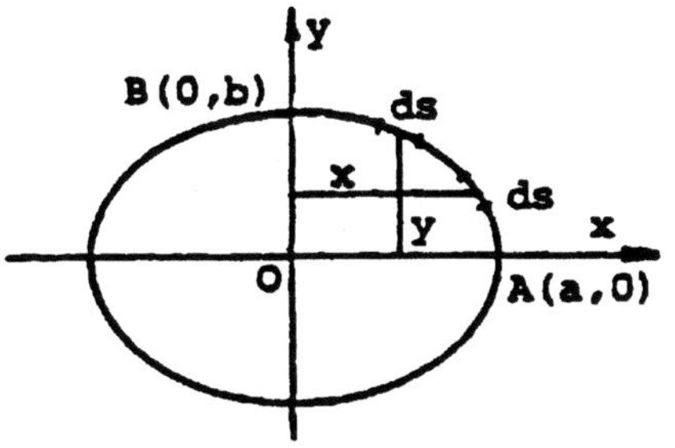

Fig.4.6

$$ds = \sqrt{a^2\sin^2\theta + b^2\cos^2\theta}\, d\theta, \tag{3}$$

and because of symmetries, the moment of inertia is

$$I_x = 4\int_{AB} y^2\, ds = 4b^2 \int_0^{\frac{\pi}{2}} \sin^2\theta \sqrt{a^2\sin^2\theta + b^2\cos^2\theta}\, d\theta. \tag{4}$$

Let $\theta = \pi/2 - \varphi$, $d\theta = -d\varphi$, $\sin\theta = \sin(\pi/2 - \varphi) = \cos\varphi$, $\cos\theta = \cos(\pi/2 - \varphi) = \sin\varphi$, and (4) becomes

$$\begin{aligned}
I_x &= 4b^2 \int_0^{\frac{\pi}{2}} \cos^2\varphi \sqrt{a^2\cos^2\varphi + b^2\sin^2\varphi}\, dy \\
&= 4b^2 \int_0^{\frac{\pi}{2}} \cos^2\varphi \sqrt{a^2(1-\sin^2\varphi) + b^2\sin^2\varphi}\, dy \\
&= 4ab^2 \int_0^{\frac{\pi}{2}} \cos^2\varphi \sqrt{1 - \sin^2\varphi + \frac{b^2}{a^2}\sin^2\varphi}\, dy \\
&= 4ab^2 \int_0^{\frac{\pi}{2}} \cos^2\varphi \sqrt{1 - \frac{a^2-b^2}{a^2}\sin^2\varphi}\, d\varphi \\
&= 4ab^2 \int_0^{\frac{\pi}{2}} \cos^2\varphi \sqrt{1 - k^2\sin^2\varphi}\, d\varphi, \quad k^2 = \frac{a^2-b^2}{a^2}, \quad 0 < b < a, \qquad (5)
\end{aligned}$$

and with Sec.4.2.8, we have

$$I_x = \frac{4ab^2}{3}\left[\left(1 + \frac{1}{k^2}\right)E(k) + \left(1 - \frac{1}{k^2}\right)K(k)\right]. \qquad (6)$$

Particular cases.

a. Let $a = 4$ cm, $b = 2$ cm,

$$\frac{x^2}{16} + \frac{y^2}{4} = 1, \quad k^2 = \frac{16-4}{16} = \frac{3}{4}, \qquad (7)$$

$$\begin{aligned}
I_x &= \frac{4^3}{3}\left[\left(1 + \frac{4}{3}\right)E\left(\frac{\sqrt{3}}{2}\right) + \left(1 - \frac{4}{3}\right)K\left(\frac{\sqrt{3}}{2}\right)\right] \\
&= \frac{64}{3}\left[\frac{7}{3}E\left(\frac{\sqrt{3}}{2}\right) - \frac{1}{3}K\left(\frac{\sqrt{3}}{2}\right)\right] \\
&= \frac{64}{9}(7 \times 1.211056 - 2.156516) = 44.948 \text{ cm}^3. \qquad (8)
\end{aligned}$$

b. Assuming unit density, calculate the moment of inertia about the x axis of the entire ellipse (see Fig.4.6). From the equation of ellipse we have

$x = f(y) = \dfrac{a}{b}\sqrt{b^2 - y^2}$ and the moment of inertia I_x for $1/4$ of an ellipse is

$$I_x = \int_0^b y^2 f(y)\,dy = \frac{a}{b}\int_0^b y^2\sqrt{b^2 - y^2}\,dy \tag{9}$$

and using Sec.2.3.9, we have

$$\begin{aligned} I_x &= \frac{a}{b}\frac{b^4}{2}B\left(\frac{3}{2}, 1 + \frac{1}{2}\right) \\ &= \frac{ab^3}{2}\frac{\left[\Gamma\left(\frac{3}{2}\right)\right]^2}{\Gamma(3)} = \frac{ab^3}{2}\frac{\left[\frac{1}{2}\Gamma\left(\frac{1}{2}\right)\right]^2}{2} \\ &= \frac{\pi}{16}ab^3 \end{aligned} \tag{10}$$

and because of symmetries, for the entire ellipse,

$$I_{xT} = 4\frac{\pi}{16}ab^3 = \frac{\pi}{4}ab^3. \tag{11}$$

c. Assuming unit density, calculate the moment of inertia about the y axis of the entire ellipse. The moment of inertia I_y for $1/4$ of an ellipse, see Sec.2.3.9, is

$$I_y = \frac{b}{a}\int_0^a x^2\sqrt{a^2 - x^2}\,dx = \frac{b}{a}\frac{a^4}{2}B\left(\frac{3}{2}, \frac{3}{2}\right) \tag{12}$$

$$\begin{aligned} &= \frac{a^3 b}{2}\frac{\left[\frac{1}{2}\Gamma\left(\frac{1}{2}\right)\right]^2}{2} \\ &= \frac{\pi}{16}a^3 b \end{aligned} \tag{13}$$

and for the entire ellipse,

$$I_{yT} = 4\frac{a^3 b}{16}\pi = \frac{\pi}{4}a^3 b. \tag{14}$$

d. The moment of inertia with respect to the pole, I_0, of the entire ellipse is

$$I_0 = I_{xT} + I_{yT} = \frac{\pi}{4}ab^3 + \frac{\pi}{4}a^3 b = \frac{\pi}{4}ab(a^2 + b^2). \tag{15}$$

4.2.29. The Pendulum. Let D be a fixed point from which a mass m is suspended by a rigid bar of length l, of negligible mass. Let θ be the angle between the vertical and the instantaneous position of mass m as shown in Fig.4.7. The distance s is measured from the equilibrium position DA. The forces acting on mass m are the gravity, mg, and the reaction, R, of the rigid bar. The resultant of these two forces is $mg\sin\theta$. Newton's second law of motion, $ma = F$, becomes

$$ma = -mg\sin\theta \tag{1}$$

where a is the acceleration of the motion, or

$$\frac{d^2s}{dt^2} = -g\sin\theta \tag{2}$$

and the minus sign indicates that the force F points in the direction of decreasing F.

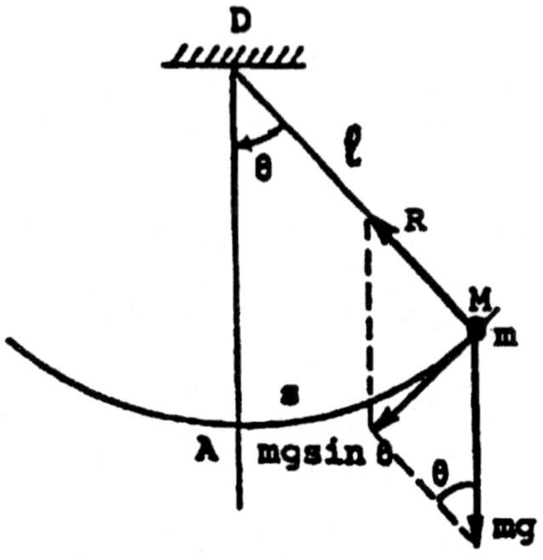

Fig.4.7

From Fig.4.7, $s = l\theta$ and (2) becomes

$$\ddot{\theta} = -\frac{g}{l}\sin\theta, \tag{3}$$

which is the differential equation of the motion of mass m. To integrate the differential equation (3), we make the notation

$$\frac{d\theta}{dt} = \dot{\theta}, \quad \ddot{\theta} = \frac{d\dot{\theta}}{dt} = \frac{d\dot{\theta}}{d\theta}\cdot\frac{d\theta}{dt} = \dot{\theta}\frac{d\dot{\theta}}{d\theta} \tag{4}$$

and (3) becomes

$$\dot{\theta}\frac{d\dot{\theta}}{d\theta} = -\frac{g}{l}\sin\theta, \tag{5}$$

$$\dot{\theta}\,d\dot{\theta} = -\frac{g}{l}\sin\theta\,d\theta. \tag{6}$$

A. Evaluate the period T of the pendulum in terms of a given initial angle θ_0. Integrating (6),

$$\frac{\dot{\theta}^2}{2} = \frac{g}{l}\cos\theta + C \tag{7}$$

where C is the constant of integration. To determine the constant C, let the initial condition be $t = 0$ when the angle is $\theta = \theta_0$ and the pendulum is released from rest, $\dot{\theta} = 0$. From (7),

$$C = -\frac{g}{l}\cos\theta_0$$

and (7) becomes

$$\dot{\theta}^2 = \frac{2g}{l}(\cos\theta - \cos\theta_0) \tag{8}$$

$$= \frac{2g}{l}(1 - 2\sin^2\frac{\theta}{2} - 1 + 2\frac{\sin^2}{2}\theta_0)$$

$$= \frac{4g}{l}\left(\sin^2\frac{\theta_0}{2} - \sin^2\frac{\theta}{2}\right)$$

$$= \frac{4g}{l}\sin^2\frac{\theta_0}{2}\left[1 - \left(\frac{\sin\frac{\theta}{2}}{\sin\frac{\theta_0}{2}}\right)^2\right]. \tag{9}$$

Letting $k = \sin\frac{\theta_0}{2}$ and defining a new variable ϕ by

$$\sin\phi = \frac{1}{k}\sin\frac{\theta}{2} \tag{10}$$

and differentiating (10),

$$\dot{\theta} = \frac{2k}{\cos\frac{\theta}{2}}\cos\phi \cdot \dot{\phi}$$
$$= \frac{2k\cos\phi}{\sqrt{1-k^2\sin^2\phi}} \cdot \dot{\phi}, \tag{11}$$

and from (11) and (9) we have

$$\frac{4k^2\cos^2\phi}{1-k^2\sin^2\phi}\dot{\phi}^2 = \frac{4g}{l}k^2\left(1-\frac{k^2\sin^2\phi}{k^2}\right)$$

or

$$\frac{\cos^2\phi}{1-k^2\sin^2\phi}\dot{\phi}^2 = \frac{g}{l}\cos^2\phi$$
$$\dot{\phi} = \sqrt{\frac{g}{l}(1-k^2\sin^2\phi)}$$
$$dt = \sqrt{\frac{l}{g}}\frac{d\phi}{\sqrt{1-k^2\sin^2\phi}}$$

and

$$t = \sqrt{\frac{l}{g}}\int_0^{\phi}\frac{d\phi}{\sqrt{1-k^2\sin^2\phi}}, \qquad k = \sin\frac{\phi_0}{2}. \tag{12}$$

The last integral is an incomplete elliptic integral of the first kind. For a period T of the pendulum, the lower and upper limits of the integral are calculated from (10), where $k = \sin(\theta_0/2)$ and in $1/4$ of a cycle, θ increases from 0 to θ_0, and from (10), when $\theta = 0$, $\phi = 0$, and when $\theta = \theta_0$, $\sin\phi = 1$ or $\phi = \pi/2$, and (12) becomes

$$T = 4\sqrt{\frac{l}{g}}\int_0^{\frac{\pi}{2}}\frac{d\phi}{\sqrt{1-k^2\sin^2\phi}}, \qquad k = \sin\frac{\theta_0}{2}, \tag{13}$$

where the last integral is a complete elliptic integral of the first kind, and hence,

$$T = 4\sqrt{\frac{l}{g}}K(k), \quad k \in (0,1), \tag{14}$$

or

$$T = 4\sqrt{\frac{l}{g}}K\left(\sin\frac{\theta_0}{2}\right), \quad \theta_0 \in [0, \pi). \tag{15}$$

Notes.

1. The time t does not depend on m (the mass of the bob).

2. The same result can be obtained from (8),

$$\dot{\theta} = \sqrt{\frac{2g}{l}(\cos\theta - \cos\theta_0)},$$

$$t = \sqrt{\frac{l}{2g}}\int_0^{\theta_0}\frac{d\theta}{\sqrt{\cos\theta - \cos\theta_0}}$$

and applying Sec.4.2.11,

$$t = \sqrt{\frac{l}{2g}}\sqrt{2}K\left(\sin\frac{\theta_0}{2}\right) = \sqrt{\frac{l}{g}}K\left(\sin\frac{\theta_0}{2}\right)$$

and

$$T = 4\sqrt{\frac{l}{g}}K\left(\sin\frac{\theta_0}{2}\right), \quad 0 < \sin\frac{\theta_0}{2} < 1. \tag{16}$$

3. One approximated solution can be obtained by expanding the integrand in (13) using the binomial formula and then integrating

$$\begin{aligned} T &= 4\sqrt{\frac{l}{g}}\int_0^{\frac{\pi}{2}}\left(1 + \frac{k^2}{2}\sin^2\phi + \cdots\right)d\phi \\ &= 4\sqrt{\frac{l}{g}}\left(\frac{\pi}{2} + \frac{k^2}{2}\cdot\frac{\pi}{4} + \cdots\right) \\ &= 2\pi\sqrt{\frac{l}{g}}\left(1 + \frac{k^2}{4} + \cdots\right). \end{aligned} \tag{17}$$

4. For small values of θ_0, we have

$$k^2 = \sin^2\frac{\theta_0}{2} \approx \frac{\theta_0^2}{4} \tag{18}$$

and (17) becomes

$$T \approx 2\pi\sqrt{\frac{l}{g}}\left(1+\frac{\theta_0^2}{16}\right), \quad \theta_0 \text{ in radians.} \tag{19}$$

5. If $\theta_0 \approx 0$, from (19) we get the *elementary pendulum formula*

$$T = 2\pi\sqrt{\frac{l}{g}}. \tag{20}$$

Particular cases. Let $l = 0.6$ m and $g = 9.8$ m/s^2.

a. Let $\theta_0 = \pi/2$. From (15), the period T is

$$T = 4\sqrt{\frac{l}{g}}K\left(\sin\frac{\pi}{4}\right) = 4\sqrt{\frac{l}{g}}K(\sin 45^\circ)$$
$$= 4\sqrt{\frac{0.6}{9.8}} \times 1.854075 = 1.835 \text{ s.} \tag{21}$$

Also, from (14) with Sec.4.3.1,

$$T = 4\sqrt{\frac{0.6}{9.8}}K\left(\frac{1}{\sqrt{2}}\right) = \sqrt{\frac{0.6}{9.8\pi}}\left[\Gamma\left(\frac{1}{4}\right)\right]^2 = 1.835 \text{ s.} \tag{22}$$

b. Let $\theta_0 = \pi/3$ and from (15),

$$T = 4\sqrt{\frac{l}{g}}K\left(\sin\frac{\pi}{6}\right) = 4\sqrt{\frac{l}{g}}K\left(\frac{1}{2}\right)$$
$$= 4\sqrt{\frac{0.6}{9.8}}1.68575 = 1.6684 \text{ s.} \tag{23}$$

c. Let the amplitude $\theta_0 = 14° = 0.244346$ radians, $k = \sin 7°$ and from (15)

$$T = 4\sqrt{\frac{l}{g}}K(\sin 7°)$$
$$= 4\sqrt{\frac{0.6}{9.8}} \times 1.576678 = 1.5605 \text{ s.} \tag{24}$$

From (19).

$$T = 2\pi\sqrt{\frac{0.6}{9.8}}\left(1 + \frac{0.244346^2}{16}\right) = 1.5605 \text{ s.} \tag{25}$$

Using the *elementary formula (20).*

$$T = 2\pi\sqrt{\frac{l}{g}} = 2\pi\sqrt{\frac{0.6}{9.8}} = 1.5546 \text{ s.} \tag{26}$$

B. Evaluate the angular velocity ω [rad/s] with the initial conditions θ_0 and ω_0. Let $\dot{\theta} = \omega$ and from (6),

$$\frac{d\omega}{d\theta} = -\frac{g}{l}\frac{\sin\theta}{\omega} \tag{27}$$

and integrating,

$$\omega^2 = \frac{2g}{l}\cos\theta + C \tag{28}$$

where C is the constant of integration. To determine the constant C, let the initial conditions $\theta = \theta_0$, $\omega = \omega_0$, and from (28),

$$C = \omega_0^2 - \frac{2g}{l}\cos\theta_0,$$

and substituting C in (28) we have the exact solution of (27),

$$\omega = \sqrt{\frac{2g}{l}(\cos\theta - \cos\theta_0) + \omega_0^2} \tag{29}$$

where

$g = 9.8$ m/s^2 is the acceleration of gravity,

l is the pendulum length in m (see Fig.4.7), in this example $l = 0.74$ m,

ω is the angular velocity in rad/s,

θ is the angle in degrees (see Fig.4.7),

$\theta = 35^\circ$ is the angle in degrees where we calculate ω,

$\theta_0 = 68^\circ, \omega_0 = 0.42$ rad/s are the initial conditions,

and from (29), the angular velocity for $\theta = 35^\circ$ is

$$\omega_{35} = \sqrt{\frac{2 \times 9.8}{0.74}(\cos 35^\circ - \cos 68^\circ) + 0.42^2} \\ = 3.457 \quad \text{rad/s.} \tag{30}$$

4.2.30. Evaluate the surface area of an ellipsoid of equation

$$\frac{x^2}{a^2} + \frac{y^2}{b^2} + \frac{z^2}{c^2} = 1, \quad 0 < c < b < a. \tag{1}$$

The surface area is given by formula, [3],

$$A = 2\pi ab \left[\sqrt{(1-m^2)(1-n^2)} + \left(\frac{1}{m} - m\right) \int_0^{\theta_0} \frac{d\theta}{\sqrt{1 - k^2 \sin^2 \theta}} \right. \\ \left. + m \int_0^{\theta_0} \sqrt{1 - k^2 \sin^2 \theta}\, d\theta \right] \tag{2}$$

or

$$A = 2\pi ab \left[\sqrt{(1-m^2)(1-n^2)} + \left(\frac{1}{m} - m\right) F(k, \theta_0) + mE(k, \theta_0) \right] \tag{3}$$

where

$$m^2 = \frac{a^2 - c^2}{a^2}, \quad n^2 = \frac{b^2 - c^2}{b^2}, \quad 0 < n < m < 1, \quad k = \frac{n}{m}, \quad \sin\theta_0 = m,$$

$$F(k, \theta_0) = \int_0^{\theta_0} \frac{d\theta}{\sqrt{1 - k^2 \sin^2 \theta}}, \quad E(k, \theta_0) = \int_0^{\theta_0} \sqrt{1 - k^2 \sin^2 \theta}\, d\theta.$$

To calculate the total surface area of an ellipsoid, we did a program with Microsoft QuickBASIC:

```
INPUT A, B, C, F, E
  LET P = 3.141592654#
  M = ((A ^ 2 - C ^ 2) / A ^ 2) ^ (1 / 2)
  N = ((B ^ 2 - C ^ 2) / B ^ 2) ^ (1 / 2)
  G = ((1 - M ^ 2) * (1 - N ^ 2)) ^ (1 / 2)                    (4)
  Q = (1 / M) - M
  R = 2 * A * B * P * (G + Q * F + M * E)
 LPRINT A, B, C, F, E, M, N, G, Q, R
```

- Let $a = 4$ m, $b = 3$ m, $c = 2$ m, $\sin\theta_0 = \sqrt{\dfrac{4^2 - 2^2}{4^2}} = \dfrac{\sqrt{3}}{2}$, $\theta_0 = 60°$,

$$k = \frac{n}{m} = \frac{2}{3}\sqrt{\frac{5}{3}} = 0.8606, \quad F(k, \theta_0) = 1.2125, \quad E(k, \theta_0) = 0.9184$$

and from (3) the ellipsoid area is

$$A = 111.4921 \text{ m}^2 \tag{5}$$

or using the program (4), the computer printout is

```
?      4,3,2, 1.2125, 0.9184
 4             3             2              1.2125        .9184
 .8660254      .745356       .3333334       .2886752      111.4921
```

4.3. Elliptic Integrals and the Gamma Function

4.3.1. Evaluate the integral

$$I = \int_0^{\frac{\pi}{2}} \frac{dx}{\sqrt{\sin x}} \tag{1}$$

and prove the relation

$$K\left(\frac{1}{\sqrt{2}}\right) = \frac{1}{4\sqrt{\pi}}\left[\Gamma\left(\frac{1}{4}\right)\right]^2. \tag{2}$$

Test for convergence. According to Appendix A1.2, the integral converges for $x = 0+$, since we can write

$$\frac{1}{\sqrt{\sin x}} = \underset{x=0}{O}\left[\frac{1}{x^{\frac{1}{2}}}\right] \tag{3}$$

and hence $\alpha = 1/2 < 1$.

Integral evaluation.

Solution 1. In terms of the elliptic integrals. Letting

$$\sin x = \cos^2\theta, \quad dx = -\frac{2\cos\theta}{\sqrt{1+\cos^2\theta}}\,d\theta, \tag{4}$$

the integral becomes

$$\begin{aligned} I &= 2\int_0^{\frac{\pi}{2}} \frac{1}{\sqrt{\cos^2\theta}} \cdot \frac{\cos\theta\,d\theta}{\sqrt{1+\cos^2\theta}} = 2\int_0^{\frac{\pi}{2}} \frac{d\theta}{\sqrt{1+\cos^2\theta}} \\ &= 2\int_0^{\frac{\pi}{2}} \frac{d\theta}{\sqrt{2-\sin^2\theta}} = \sqrt{2}\int_0^{\frac{\pi}{2}} \frac{d\theta}{\sqrt{1-\frac{1}{2}\sin^2\theta}} \\ &= \sqrt{2}K\left(\frac{1}{\sqrt{2}}\right) = \sqrt{2}\times 1.854075 = 2.622. \end{aligned} \tag{5}$$

The integral can also be evaluated with the substitution $\sin x = \sin^2\theta$.

Solution 2. In terms of the gamma and beta functions. The integral (1) can be written

$$I = \int_0^{\frac{\pi}{2}} \sin^{-\frac{1}{2}} dx \tag{6}$$

and using Eq.(1.120), we have

$$\int_0^{\frac{\pi}{2}} \frac{dx}{\sqrt{\sin x}} = \frac{1}{2}B\left(\frac{1}{4},\frac{1}{2}\right) = \frac{\sqrt{\pi}}{2}\frac{\Gamma\left(\frac{1}{2}\right)}{\Gamma\left(\frac{3}{4}\right)} = \frac{1}{2\sqrt{2\pi}}\left[\Gamma\left(\frac{1}{4}\right)\right]^2 = 2.622. \tag{7}$$

From (5) and (7) we have

$$K\left(\frac{1}{\sqrt{2}}\right)=\frac{1}{4\sqrt{\pi}}\left[\Gamma\left(\frac{1}{4}\right)\right]^2=1.854. \tag{8}$$

Note that the integral (1) cannot be evaluated in terms of the elementary functions. See also Sec.4.2.25.

4.3.2. Evaluate the integral

$$I=\int_0^{\frac{\pi}{2}}\frac{dx}{\sqrt{\cos x}}. \tag{1}$$

Test for convergence. According to Appendix A1.2, the integral converges for $x=\frac{\pi}{2}-0$, since we can write

$$\frac{1}{\sqrt{\cos x}}=\frac{1}{\sqrt{\sin\left(\frac{\pi}{2}-x\right)}}=\underset{x=\frac{\pi}{2}}{O}\left[\frac{1}{\sqrt{\frac{\pi}{2}-x}}\right] \tag{2}$$

and hence $\alpha=1/2<1$.

Integral evaluation.

Solution 1. Letting

$$\cos x=\sin^2\theta,\quad dx=-\frac{2\sin\theta}{\sqrt{1+\sin^2\theta}}d\theta, \tag{3}$$

the integral becomes

$$I=\int_{\frac{\pi}{2}}^{0}\frac{1}{\sqrt{\sin^2\theta}}\cdot\frac{-2\sin\theta}{\sqrt{1+\sin^2\theta}}d\theta=2\int_0^{\frac{\pi}{2}}\frac{d\theta}{\sqrt{1+\sin^2\theta}} \tag{4}$$

and using Sec.4.2.12,

$$I=\sqrt{2}K\left(\frac{1}{\sqrt{2}}\right) \tag{5}$$

and with Sec.4.3.1,

$$\int_0^{\frac{\pi}{2}} \frac{dx}{\sqrt{\cos x}} = \int_0^{\frac{\pi}{2}} \frac{dx}{\sqrt{\sin x}} = \sqrt{2} K\left(\frac{1}{\sqrt{2}}\right) = \sqrt{2} \times 1.854074 = 2.622. \tag{6}$$

The integral can also be evaluated with the substitution $\cos x = \cos^2 \theta$.

Solution 2. The integral can be written

$$I = \int_0^{\frac{\pi}{2}} \cos^{-\frac{1}{2}} x \, dx \tag{7}$$

and using Eq.(1.120), we have

$$\int_0^{\frac{\pi}{2}} \frac{dx}{\sqrt{\cos x}} = \frac{1}{2} B\left(\frac{1}{4}, \frac{1}{2}\right) = \frac{\sqrt{\pi}}{2} \frac{\Gamma\left(\frac{1}{4}\right)}{\Gamma\left(\frac{3}{4}\right)} = \frac{1}{2\sqrt{2\pi}} \left[\Gamma\left(\frac{1}{4}\right)\right]^2 = 2.622 \tag{8}$$

and comparing with Sec.4.3.1,

$$\int_0^{\frac{\pi}{2}} \frac{dx}{\sqrt{\sin x}} = \int_0^{\frac{\pi}{2}} \frac{dx}{\sqrt{\cos x}} = \frac{1}{2} B\left(\frac{1}{4}, \frac{1}{2}\right). \tag{9}$$

4.3.3. Evaluate the integral

$$I = \int_0^{\frac{\pi}{2}} \sqrt{\cos x} \, dx \tag{1}$$

and prove the relation

$$2E\left(\frac{1}{\sqrt{2}}\right) - K\left(\frac{1}{\sqrt{2}}\right) = \frac{2\pi\sqrt{\pi}}{\left[\Gamma\left(\frac{1}{4}\right)\right]^2}. \tag{2}$$

Solution 1. Letting

$$\cos x = \sin^2 \theta, \quad dx = -\frac{2 \sin \theta \, d\theta}{\sqrt{1 + \sin^2 \theta}} \tag{3}$$

the integral (1) becomes

$$I = 2\int_0^{\frac{\pi}{2}} \frac{\sin^2\theta\, d\theta}{\sqrt{1+\sin^2\theta}} = 2\int_0^{\frac{\pi}{2}} \frac{(1+\sin^2\theta)-1}{\sqrt{1+\sin^2\theta}}\, d\theta$$

$$= 2\left(\int_0^{\frac{\pi}{2}} \sqrt{1+\sin^2\theta}\, d\theta - \int_0^{\frac{\pi}{2}} \frac{d\theta}{\sqrt{1+\sin^2\theta}}\right)$$

and using Secs.4.2.13 and 4.2.12,

$$I = 2\left[\sqrt{2}E\left(\frac{1}{\sqrt{2}}\right) - \frac{1}{\sqrt{2}}K\left(\frac{1}{\sqrt{2}}\right)\right]$$

or

$$\int_0^{\frac{\pi}{2}} \sqrt{\cos x}\, dx = \sqrt{2}\left[2E\left(\frac{1}{\sqrt{2}}\right) - K\left(\frac{1}{\sqrt{2}}\right)\right]. \tag{4}$$

Using Tables 3.4 and 3.2, we have

$$\int_0^{\frac{\pi}{2}} \sqrt{\cos x}\, dx = 2\sqrt{2} \times 1.350644 - \sqrt{2} \times 1.854075 = 1.198. \tag{5}$$

The integral can also be evaluated with the substitution $\cos x = \cos^2\theta$.

Solution 2. With Eq.(1.120), the integral becomes

$$\int_0^{\frac{\pi}{2}} \cos^{\frac{1}{2}} x\, dx = \frac{1}{2}B\left(\frac{3}{4}, \frac{1}{2}\right) = \frac{\sqrt{\pi}}{2}\frac{\Gamma\left(\frac{3}{4}\right)}{\Gamma\left(\frac{5}{4}\right)}$$

$$= 2\sqrt{\pi}\frac{\Gamma\left(\frac{3}{4}\right)}{\Gamma\left(\frac{1}{4}\right)} = 2\sqrt{\pi}\frac{1.225416}{3.62561} = 1.198. \tag{6}$$

From (4) and (6), we have

$$2E\left(\frac{1}{\sqrt{2}}\right) - K\left(\frac{1}{\sqrt{2}}\right) = \frac{2\pi\sqrt{\pi}}{\left[\Gamma\left(\frac{1}{4}\right)\right]^2} = 0.847. \tag{7}$$

4.3.4. Evaluate the integral

$$I = \int_0^{\frac{\pi}{4}} \frac{d\theta}{\sqrt{\cos 2\theta}} \tag{1}$$

using the gamma function and prove the relation

$$K\left(\frac{1}{\sqrt{2}}\right) = \frac{1}{4\sqrt{\pi}}\left[\Gamma\left(\frac{1}{4}\right)\right]^2. \tag{2}$$

Solution. For convergence, see Sec.4.2.6. Letting $2\theta = t$, $2\,d\theta = dt$, the integral (1) becomes

$$I = \frac{1}{2}\int_0^{\frac{\pi}{2}} \frac{dt}{\sqrt{\cos t}} = \frac{1}{2}\int_0^{\frac{\pi}{2}} \cos^{-\frac{1}{2}} t\,dt \tag{3}$$

and using Eq.(1.120), or Sec.4.3.2, we have

$$\int_0^{\frac{\pi}{4}} \frac{d\theta}{\sqrt{\cos 2\theta}} = \frac{1}{4}B\left(\frac{1}{4}, \frac{1}{2}\right) = \frac{1}{4}\frac{\Gamma\left(\frac{1}{4}\right)\Gamma\left(\frac{1}{2}\right)}{\Gamma\left(\frac{3}{4}\right)}$$

$$= \frac{\sqrt{\pi}}{4}\frac{\Gamma\left(\frac{1}{4}\right)}{\Gamma\left(\frac{3}{4}\right)} = \frac{1}{4\sqrt{2\pi}}\left[\Gamma\left(\frac{1}{4}\right)\right]^2 = 1.311. \tag{4}$$

From (4) and Sec.4.2.6, we have

$$K\left(\frac{1}{\sqrt{2}}\right) = \frac{1}{4\sqrt{\pi}}\left[\Gamma\left(\frac{1}{4}\right)\right]^2 = 1.854. \tag{5}$$

See also Sec.4.3.1.

4.3.5. Evaluate the integral

$$I = \int_0^{\frac{\pi}{2}} \sqrt{\sin x}\,dx \tag{1}$$

and prove the relation

$$2E\left(\frac{1}{\sqrt{2}}\right) - K\left(\frac{1}{\sqrt{2}}\right) = \frac{2\pi\sqrt{\pi}}{\left[\Gamma\left(\frac{1}{4}\right)\right]^2}. \tag{2}$$

Solution 1. Using the elliptic integrals. Letting

$$\sin x = \cos^2\theta, \quad dx = -\frac{2\cos\theta\, d\theta}{\sqrt{1+\cos^2\theta}}, \tag{3}$$

the integral (1) becomes

$$I = 2\int_0^{\frac{\pi}{2}} \frac{\cos^2\theta\, d\theta}{\sqrt{1+\cos^2\theta}} = 2\int_0^{\frac{\pi}{2}} \frac{1-\sin^2\theta}{\sqrt{2-\sin^2\theta}}\, d\theta$$

$$= \sqrt{2}\int_0^{\frac{\pi}{2}} \frac{(1-\sin^2\theta)\, d\theta}{\sqrt{1-\frac{1}{2}\sin^2\theta}} = \sqrt{2}\left[K\left(\frac{1}{\sqrt{2}}\right) - \int_0^{\frac{\pi}{2}} \frac{\sin^2\theta\, d\theta}{\sqrt{1-\frac{1}{2}\sin^2\theta}}\right].$$

The second integral is

$$I_2 = \int_0^{\frac{\pi}{2}} \frac{\sin^2\theta\, d\theta}{\sqrt{1-\frac{1}{2}\sin^2\theta}} = \int_0^{\frac{\pi}{2}} \frac{\left(1-\frac{1}{2}\sin^2\theta\right) - 1 + \frac{3}{2}\sin^2\theta}{\sqrt{1-\frac{1}{2}\sin^2\theta}}\, d\theta$$

$$= E\left(\frac{1}{\sqrt{2}}\right) - K\left(\frac{1}{\sqrt{2}}\right) + \frac{3}{2}I_2, \tag{4}$$

$$I_2 = 2\left[K\left(\frac{1}{\sqrt{2}}\right) - E\left(\frac{1}{\sqrt{2}}\right)\right]. \tag{5}$$

Thus,

$$I = \sqrt{2}\left[K\left(\frac{1}{\sqrt{2}}\right) - 2K\left(\frac{1}{\sqrt{2}}\right) + 2E\left(\frac{1}{\sqrt{2}}\right)\right] \tag{6}$$

$$= \sqrt{2}\left[2E\left(\frac{1}{\sqrt{2}}\right) - K\left(\frac{1}{\sqrt{2}}\right)\right]. \tag{7}$$

Using Tables 3.2 and 3.4, we have

$$\int_0^{\frac{\pi}{2}} \sqrt{\sin x}\, dx = \sqrt{2}(2 \times 1.350644 - 1.854075) = 1.198. \tag{8}$$

Comparing with Sec.4.3.3 yields

$$\int_0^{\frac{\pi}{2}} \sqrt{\sin x}\, dx = \int_0^{\frac{\pi}{2}} \sqrt{\cos x}\, dx = 1.198. \tag{9}$$

Evaluate the integral with the substitution $\sin x = \sin^2 \theta$.

Solution 2. Using the gamma and beta functions method. With Eq.(1.120) we have

$$I = \int_0^{\frac{\pi}{2}} \sin^{\frac{1}{2}} x\, dx = \frac{1}{2} B\left(\frac{3}{4}, \frac{1}{2}\right) = \frac{\sqrt{\pi}}{2} \frac{\Gamma\left(\frac{3}{4}\right)}{\Gamma\left(\frac{5}{4}\right)} \tag{10}$$

or

$$\int_0^{\frac{\pi}{2}} \sqrt{\sin x}\, dx = \frac{2\pi\sqrt{2\pi}}{\left[\Gamma\left(\frac{1}{4}\right)\right]^2} = 1.198. \tag{11}$$

From (7) and (11), we have

$$2E\left(\frac{1}{\sqrt{2}}\right) - K\left(\frac{1}{\sqrt{2}}\right) = \frac{1}{\sqrt{2}} \frac{2\pi\sqrt{2\pi}}{\left[\Gamma\left(\frac{1}{4}\right)\right]^2} = \frac{2\pi\sqrt{\pi}}{\left[\Gamma\left(\frac{1}{4}\right)\right]^2} = 0.847. \tag{12}$$

4.3.6. Evaluate the elliptic integral

$$I = \int_a^1 \frac{dx}{\sqrt{1 - x^4}}, \quad a < 1, \tag{1}$$

and prove the relation

$$K\left(\frac{1}{\sqrt{2}}\right) = \frac{1}{4\sqrt{\pi}} \left[\Gamma\left(\frac{1}{4}\right)\right]^2. \tag{2}$$

Test for convergence. According to Appendix A1.2, the integral converges at $x = 1 - 0$, since we can write

$$\frac{1}{\sqrt{1 - x^4}} = \frac{1}{\sqrt{(1 - x)(1 + x)(1 + x^2)}} \underset{x=1-}{=} O\left[\frac{1}{(1 - x)^{\frac{1}{2}}}\right]$$

and hence $\alpha = 1/2 < 1$.

Integral evaluation. Letting $x = \cos\theta$, $dx = -\sin\theta\, d\theta$ and $\cos^{-1} a = \phi$, the integral becomes

$$\int_a^1 \frac{dx}{\sqrt{1-x^4}} = \int_{\cos^{-1} a}^0 \frac{-\sin\theta\, d\theta}{\sqrt{1-\cos^4\theta}} = \int_0^\phi \frac{\sin\theta\, d\theta}{\sqrt{1-\cos^4\theta}}$$

$$= \int_0^\phi \frac{d\theta}{\sqrt{2-\sin^2\theta}} = \frac{1}{\sqrt{2}} \int_0^\phi \frac{d\theta}{\sqrt{1-\dfrac{1}{2}\sin^2\theta}}$$

$$= \frac{1}{\sqrt{2}} F\left(\frac{1}{\sqrt{2}}, \phi\right) = \frac{1}{\sqrt{2}} F\left(\frac{1}{\sqrt{2}}, \cos^{-1} a\right), \quad a < 1. \tag{3}$$

Particular case. Let $a = 0$, and from (3),

$$\int_0^1 \frac{dx}{\sqrt{1-x^4}} = \frac{1}{\sqrt{2}} F\left(\frac{1}{\sqrt{2}}, \frac{\pi}{2}\right) = \frac{1}{\sqrt{2}} K\left(\frac{1}{\sqrt{2}}\right) = 1.311. \tag{4}$$

From (4) and Sec.2.3.2 or Sec.2.3.10, we have

$$K\left(\frac{1}{\sqrt{2}}\right) = \frac{1}{4\sqrt{\pi}}\left[\Gamma\left(\frac{1}{4}\right)\right]^2 = 1.854. \tag{5}$$

Appendix

APPENDIX A1

A1.1 The Absolute Convergent Test for Improper Integrals of the First Kind

Let

$$I_{+\infty} = \int_a^{+\infty} f(x)\, dx.$$

If there is a number $\alpha > 1$ (condition a) for which

$$f(x) = \underset{x=+\infty}{O}\left[\frac{1}{x^{\alpha}}\right], \quad \text{(condition b)}$$

or in another form

$$x^{\alpha}|f(x)| < k < +\infty,$$

then the integral $I_{+\infty}$ is absolutely convergent at the upper limit.

Similarly for the integral

$$I_{-\infty} = \int_{-\infty}^{a} f(x)\, dx.$$

A1.2 The Absolute Convergent Test for Improper Integrals of the Second Kind

Let

$$I_a = \int_a^b f(x)\, dx$$

and $f(x)$ becomes unbounded only at the end point $x = a$ of the interval $x \in [a, b]$. Then, we define

$$\int_a^b f(x)\, dx = \lim_{\varepsilon \to O+} \int_{a+\varepsilon}^b f(x)\, dx.$$

If there is a number $\alpha < 1$ (condition a) for which

$$f(x) = \underset{x=a}{O}\left[\frac{1}{(x-a)^{\propto}}\right] \quad \text{(condition b)}$$

or in another form,

$$(x-a)^{\propto}|f(x)| < k < +\infty,$$

then the integral

$$I_{a+o} = \int_{a+o}^b f(x)\, dx$$

is absolutely convergent at $x = a$. Similarly, if the function $f(x)$ is unbounded at the upper limit, $x = b$.

Example. Using the Appendixes A1.1 and A1.2, find the convergence conditions of the improper integral,

$$\int_0^\infty \frac{x^{m-1}}{1+x^n}\, dx.$$

For the lower limit

$$f(x) = \underset{x\to\infty}{O}\left[\frac{1}{x^{1-m}}\right]$$

and the convergence condition is $1 - m < 1$ or $m > 0$.

For the upper limit

$$f(x) = \underset{x\to\infty}{O}\left[\frac{1}{x^{n-m+1}}\right]$$

and the convergence condition is $n - m + 1 > 1$, or $n > m$. Thus, the convergence conditions of the integral are $0 < m < n$.

Note that these convergence conditions could also be found from the evaluation of the integral expressed in terms of the gamma functions as shown in Sec.2.3.20.

APPENDIX A2

Transformations of Multiple Integrals

A2.1 Jacobian in 3-Dimensional Space

The transformation of a multiple integral in the 3-dimensional space, R^3, from rectangular coordinates (x, y, z) to curvilinear coordinates (u, v, w), is given by the formula

$$\iiint_D F(x, y, z)\, dx\, dy\, dz = \iiint_{D'} G(u, v, w) \cdot |J_3|\, du\, dv\, dw$$

where $|J_3|$ is the Jacobian or functional determinant of transformation of x, y, and z with respect to u, v, and w and it is given by

$$J_3 = \frac{\partial(x, y, z)}{\partial(u, v, w)} = \begin{vmatrix} \dfrac{\partial x}{\partial u} & \dfrac{\partial x}{\partial v} & \dfrac{\partial x}{\partial w} \\ \dfrac{\partial y}{\partial u} & \dfrac{\partial y}{\partial v} & \dfrac{\partial y}{\partial w} \\ \dfrac{\partial z}{\partial u} & \dfrac{\partial z}{\partial v} & \dfrac{\partial z}{\partial w} \end{vmatrix}$$

where x, y, and z are continuous functions of the variables u, v, and w. The element of volume is

$$dV = dx\, dy\, dz = |J_3|\, du\, dv\, dw.$$

A2.2 Jacobian in n-Dimensional Space

In an n-dimensional space, R^n, the Jacobian of the transformation from rectangular coordinates $x_1, x_2, \ldots, x_n$ to the curvilinear coordinates $u_1, u_2, \ldots, u_n$ is

$$J_n = \frac{\partial(x_1, x_2, \ldots, x_n)}{\partial(u_1, u_2, \ldots, u_n)} = \begin{vmatrix} \frac{\partial x_1}{\partial u_1} & \frac{\partial x_1}{\partial u_2} & \ldots & \frac{\partial x_1}{\partial u_n} \\ \frac{\partial x_2}{\partial u_1} & \frac{\partial x_2}{\partial u_2} & \ldots & \frac{\partial x_2}{\partial u_n} \\ \ldots & \ldots & \ldots & \ldots \\ \frac{\partial x_n}{\partial u_1} & \frac{\partial x_n}{\partial u_2} & \ldots & \frac{\partial x_n}{\partial u_n} \end{vmatrix}$$

where $x_1, x_2, \ldots, x_n$ are continuous functions of the variables $u_1, u_2, \ldots, u_n$.

The element of volume is given by

$$dV = dx_1\, dx_2 \ldots dx_n = |J_n|\, du_1\, du_2 \ldots du_n.$$

A2.3 Particular Case

If the system of functions $x_1, x_2, \ldots, x_n$ with respect to the variables $u_1, u_2, \ldots, u_n$ is of the following form,

$$\begin{cases} x_1 = f(u_1) \\ x_2 = f(u_1, u_2) \\ x_3 = f(u_1, u_2, u_3) \\ \dots\dots\dots \\ x_n = f(u_1, u_2, \dots, u_n), \end{cases}$$

the Jacobian of transformation reduces to its leading term,

$$J_n = \frac{\partial x_1}{\partial u_1}\frac{\partial x_2}{\partial u_2}\cdots\frac{\partial x_n}{\partial u_n}.$$

A2.4 Case of Functions of Functions

If x_1, x_2, x_3 are functions of u_1, u_2, u_3, and u_1, u_2, u_3 are functions of t_1, t_2, t_3, we have

$$J_3 = \begin{vmatrix} \frac{\partial x_1}{\partial t_1} & \frac{\partial x_1}{\partial t_2} & \frac{\partial x_1}{\partial t_3} \\ \frac{\partial x_2}{\partial t_1} & \frac{\partial x_2}{\partial t_2} & \frac{\partial x_2}{\partial t_3} \\ \frac{\partial x_3}{\partial t_1} & \frac{\partial x_3}{\partial t_2} & \frac{\partial x_3}{\partial t_3} \end{vmatrix} = \begin{vmatrix} \frac{\partial x_1}{\partial u_1} & \frac{\partial x_1}{\partial u_2} & \frac{\partial x_1}{\partial u_3} \\ \frac{\partial x_2}{\partial u_1} & \frac{\partial x_2}{\partial u_2} & \frac{\partial x_2}{\partial u_3} \\ \frac{\partial x_3}{\partial u_1} & \frac{\partial x_3}{\partial u_2} & \frac{\partial x_3}{\partial u_3} \end{vmatrix} \cdot \begin{vmatrix} \frac{\partial u_1}{\partial t_1} & \frac{\partial u_1}{\partial t_2} & \frac{\partial u_1}{\partial t_3} \\ \frac{\partial u_2}{\partial t_1} & \frac{\partial u_2}{\partial t_2} & \frac{\partial u_2}{\partial t_3} \\ \frac{\partial u_3}{\partial t_1} & \frac{\partial u_3}{\partial t_2} & \frac{\partial u_3}{\partial t_3} \end{vmatrix}$$

or

$$J_3 = \frac{\partial(x_1, x_2, x_3)}{\partial(t_1, t_2, t_3)} = \frac{\partial(x_1, x_2, x_3)}{\partial(u_1, u_2, u_3)} \cdot \frac{\partial u_1, u_2, u_3)}{\partial(t_1, t_2, t_3)}.$$

Or, generalizing,

$$J_n = \frac{\partial(x_1, x_2, \dots, x_n)}{\partial(t_1, t_2, \dots, t_n)} = \frac{\partial(x_1, x_2, \dots, x_n)}{\partial(u_1, u_2, \dots, u_n)} \cdot \frac{\partial u_1, u_2, \dots, u_n)}{\partial(t_1, t_2, \dots, t_n)}.$$

A2.5 Jacobian of Implicit Functions

If x, y, z are regarded as implicit functions of u, v, w by the equations

$$F_1(x, y, z, u, v, w) = 0$$
$$F_2(x, y, z, u, v, w) = 0$$
$$F_3(x, y, z, u, v, w) = 0$$

the Jacobian of transformation is

$$J_3 = \frac{\partial(x, y, z)}{\partial(u, v, w)} = \frac{-\dfrac{\partial(F_1, F_2, F_3)}{\partial(u, v, w)}}{\dfrac{\partial(F_1, F_2, F_3)}{\partial(x, y, z)}}.$$

Or, generalizing, if $x_1, x_2, \ldots, x_n$ are regarded as implicit functions of $u_1, u_2, \ldots, u_n$ by the n equations of the form

$$\begin{cases} F_1(x_1, x_2, \ldots, x_n,\ u_1, u_2, \ldots, u_n) = 0 \\ F_2(x_1, x_2, \ldots, x_n,\ u_1, u_2, \ldots, u_n) = 0 \\ \cdots\cdots\cdots\cdots\cdots\cdots \\ F_n(x_1, x_2, \ldots, x_n,\ u_1, u_2, \ldots, u_n) = 0 \end{cases}$$

the Jacobian is

$$J_n = \frac{\partial(x_1, x_2, \ldots, x_n)}{\partial(u_1, u_2, \ldots, u_n)} = (-1)^n \frac{\dfrac{\partial(F_1, F_2, \ldots, F_n)}{\partial(u_1, u_2, \ldots, u_n)}}{\dfrac{\partial(F_1, F_2, \ldots, F_n)}{\partial(x_1, x_2, \ldots, x_n)}}.$$

A2.6 Particular Case of Implicit Functions

If the equations are of the form

$$\begin{cases} F_1(x_1, x_2, \ldots, x_n,\ u_1) = 0 \\ F_2(x_2, x_3, \ldots, x_n,\ u_1, u_2) = 0 \\ F_3(x_3, x_4, \ldots, x_n,\ u_1, u_2, u_3) = 0 \\ \cdots\cdots\cdots\cdots\cdots\cdots \\ F_n(x_n,\ u_1, u_2, \ldots, u_n) = 0 \end{cases}$$

the Jacobian is

$$J_n = \frac{\partial(x_1, x_2, \ldots, x_n)}{\partial(u_1, u_2, \ldots, u_n)} = (-1)^n \frac{\dfrac{\partial F_1}{\partial u_1}\dfrac{\partial F_2}{\partial u_2}\cdots\dfrac{\partial F_n}{\partial u_n}}{\dfrac{\partial F_1}{\partial x_1}\dfrac{\partial F_2}{\partial x_2}\cdots\dfrac{\partial F_n}{\partial x_n}}.$$

For the application of this transformation, see Sec.2.6.4.

APPENDIX A3

Jacobian for Polar and Spherical Coordinates

A3.1 Polar Coordinates in Plane

The transformation (see Fig.A3.1) is

$$x = r\cos\theta, \quad y = r\sin\theta$$

with $r \in R^+ \bigcup \{0\}$, $\theta \in [0, 2\pi)$. According to Appendix A2.2, the Jacobian of transformation is $J_2 = r$ and the element of area is

$$dA = dx\,dy = r\,dr\,d\theta.$$

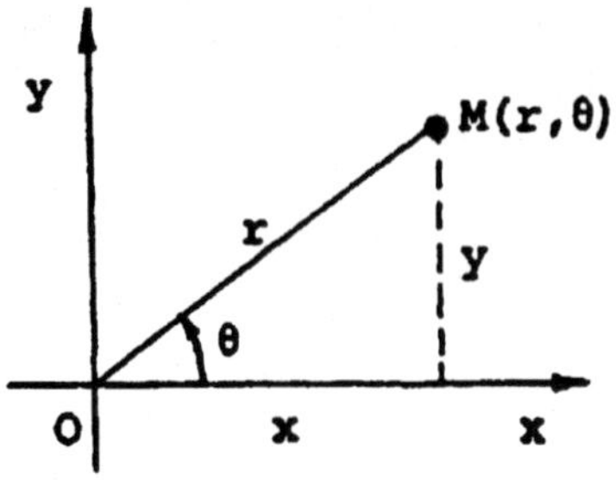

Fig.A3.1

A3.2 Spherical Coordinates in 3-Dimensional Space

The transformation (see Fig.A3.2) is

$$x = r\sin\phi\cos\theta, \quad y = r\sin\phi\sin\theta, \quad z = r\cos\phi$$

with

$$r \geq 0, \quad 0 \leq \phi \leq \pi, \quad 0 \leq \theta \leq 2\pi.$$

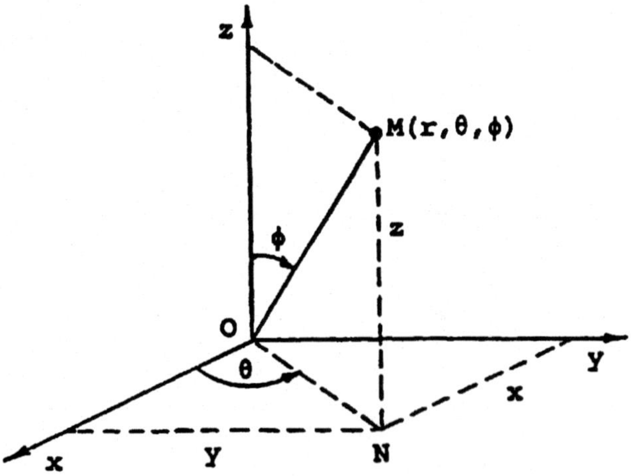

Fig.A3.2

According to Appendix A2.1, the Jacobian of transformation is

$$J_3 = r^2 \sin\phi$$

and the element of volume is given by

$$dV = dx\,dy\,dz = r^2 \sin\phi\,dr\,d\phi\,d\theta.$$

A3.3 Spherical Coordinates in the R^n-Space

The transformation in n-dimensional space of rectangular to hyper-spherical coordinates is given by the equations

$$\begin{cases} x_1 = r\cos\theta_1 \\ x_2 = r\sin\theta_1\cos\theta_2 \\ x_3 = r\sin\theta_1\sin\theta_2\cos\theta_3 \\ \dots\dots\dots\dots\dots \\ x_{n-1} = r\sin\theta_1\sin\theta_2\dots\sin\theta_{n-2}\cos\theta_{n-1} \\ x_n = r\sin\theta_1\sin\theta_2\dots\sin\theta_{n-2}\sin\theta_{n-1} \end{cases}$$

where

$$r \in R^+, \theta_1, \theta_2, \dots, \theta_{n-2} \in [0,\pi], \quad \theta_{n-1} \in [0, 2\pi).$$

The Jacobian of transformation is

$$J_n = r^{n-1}\sin^{n-2}\theta_1 \sin^{n-3}\theta_2 \dots \sin^2\theta_{n-3}\sin\theta_{n-2}$$

and the volume element is

$$\begin{aligned} dV_n &= dx_1\,dx_2\dots dx_n \\ &= r^{n-1}\sin^{n-2}\theta_1\sin^{n-3}\theta_2\dots\sin^2\theta_{n-3}\sin\theta_{n-2}\,dr\,d\theta_1\,d\theta_2\dots d\theta_{n-2}\,d\theta_{n-1}. \end{aligned}$$

For application of this transformation, see Sec.2.6.10.

A3.4 Element of a Sphere Surface in R^n-Space

The equation of a hypersphere in the R^n-space is

$$x_1^2 + x_2^2 + \cdots + x_n^2 = a^2, \quad a > 0,$$

and its surface is given by the integral

$$A_n = 2 \int \cdots \int_{S_{n-1}} \frac{a\, dx_1\, dx_2 \ldots dx_{n-1}}{x_n},$$

$S_{n-1}\{a, x_1, x_2, \ldots, x_n \in R \mid x_1, \ldots, x_{n-1} \geq 0,\ x_1^2 + x_2^2 + \cdots + x_{n-1}^2 = a^2\}$. The general transformation from rectangular to spherical coordinates in the R^n-space is

$$\begin{cases} x_1 = a \cos\theta_1 \\ x_2 = a \sin\theta_1 \cos\theta_2 \\ x_3 = a \sin\theta_1 \sin\theta_2 \cos\theta_3 \\ \ldots\ldots\ldots\ldots\ldots \\ x_{n-1} = a \sin\theta_1 \sin\theta_2 \ldots \sin\theta_{n-2} \cos\theta_{n-1} \\ x_n = a \sin\theta_1 \sin\theta_2 \ldots \sin\theta_{n-2} \sin\theta_{n-1} \end{cases}$$

where $\theta_1, \theta_2, \ldots, \theta_{n-2} \in [0, \pi]$, $\theta_{n-1} \in [0, 2\pi)$. The Jacobian of transformation is

$$J_{n-1} = (-a)^{n-1} \sin^{n-1}\theta_1 \sin^{n-2}\theta_2 \ldots \sin^2\theta_{n-2} \sin\theta_{n-1}$$

and the surface element is

$$\begin{aligned} dS_{n-1} &= dx_1\, dx_2 \ldots dx_{n-1} = |J_{n-1}|\, d\theta_1\, d\theta_2 \ldots d\theta_{n-1} \\ &= a^{n-1} \sin^{n-1}\theta_1 \sin^{n-2}\theta_2 \ldots \sin^2\theta_{n-2} \sin\theta_{n-1}\, d\theta_1\, d\theta_2 \ldots d\theta_{n-1}. \end{aligned}$$

For application of this formula, see Sec.2.5.3.

APPENDIX A4

Jacobian of a General Transformation

Let the transformation of the rectangular coordinates $x_1, x_2, \ldots, x_n$ to the curvilinear coordinates $u_1, u_2, \ldots, u_n$ be

$$\begin{cases} x_1 + x_2 + \cdots + x_n = u_1 \\ x_2 + \cdots + x_n = u_1 u_2 \\ \cdots\cdots\cdots\cdots\cdots\cdots \\ x_{n-1} + x_n = u_1 u_2 \ldots u_{n-1} \\ x_n = u_1 u_2 \ldots u_n . \end{cases}$$

Applying Appendix A2.6, with the notations

$$\begin{cases} F_1 = x_1 + x_2 + \cdots + x_n - u_1 \\ F_2 = x_2 + \cdots + x_n - u_1 u_2 \\ \cdots\cdots\cdots\cdots\cdots \\ F_{n-1} = x_{n-1} + x_n - u_1 u_2 \ldots u_{n-1} \\ F_n = x_n - u_1 u_2 \ldots u_{n-1} u_n \end{cases}$$

the Jacobian of transformation is

$$\begin{aligned} J_n &= (-1)^n \frac{\partial F_1}{\partial u_1} \frac{\partial F_2}{\partial u_2} \cdots \frac{\partial F_n}{\partial u_n} \\ &= (-1)^n u_1 (u_1 u_2) \ldots (u_1 u_2 \ldots u_{n-2})(u_1 u_2 \ldots u_{n-1}) \\ &= (-1)^n u_1^{n-1} u_2^{n-2} \ldots u_{n-2}^2 u_{n-1} \end{aligned}$$

and the element of volume is

$$dV = dx_1\, dx_2 \ldots dx_n = u_1^{n-1} u_2^{n-2} \ldots u_{n-2}^2 u_{n-1}\, du_1\, du_2 \ldots du_n .$$

For applications, see Secs.2.4.6, 2.6.4, 2.6.5, 2.6.8, and 2.6.9.

APPENDIX A5

Jacobians of Other General Transformations

Appendix A5.1 According to Appendix A2, the Jacobian of the general transformation

$$x_1 = a_1 u_1^{\frac{1}{m_1}}, \quad x_2 = a_2 u_2^{\frac{1}{m_2}}, \quad \ldots, \quad x_n = a_n u_n^{\frac{1}{m_n}}$$

is given by the functional determinant

$$J_n = \begin{vmatrix} \frac{a_1}{m_1} u_1^{\frac{1}{m_1}-1} & 0 & \ldots & 0 \\ 0 & \frac{a_2}{m_2} u_2^{\frac{1}{m_2}-1} & \ldots & 0 \\ \ldots & \ldots & \ldots & \ldots \\ 0 & 0 & \ldots & \frac{a_n}{m_n} u_n^{\frac{1}{m_n}-1} \end{vmatrix}$$

or

$$J_n = \frac{a_1 a_2 \ldots a_n}{m_1 m_2 \ldots m_n} u_1^{\frac{1}{m_1}-1} u_2^{\frac{1}{m_2}-1} \ldots u_n^{\frac{1}{m_n}-1}$$

and the element of volume is

$$dV_n = dx_1\, dx_2 \ldots dx_n = |J_n|\, du_1\, du_2 \ldots du_n.$$

For application of this transformation, see Secs.2.4.7, 2.6.4, 2.6.5, 2.6.10, 2.6.11, and 2.6.12.

Appendix A5.2 The Jacobian of the transformation

$$x_1 = a_1 u_1^{\frac{2}{m_1}}, x_2 = a_2 u_2^{\frac{2}{m_2}}, \ldots, x_n = a_n u_n^{\frac{2}{m_n}}$$

is

$$J_n = 2^n \frac{a_1 a_2 \ldots a_n}{m_1 m_2 \ldots m_n} u_1^{\frac{2}{m_1}-1} u_2^{\frac{2}{m_2}-1} \ldots u_n^{\frac{2}{m_n}-1}.$$

If $n = 3$, $a_1 = a$, $a_2 = b$, $a_3 = c$, $m_1 = m_2 = m_3 = m$,

$$J_3 = 8\frac{abc}{m^3} u_1^{\frac{2}{m}-1} u_2^{\frac{2}{m}-1} u_3^{\frac{2}{m}-1}.$$

For application of this transformation, see Sec.2.6.3.

Appendix A5.3 The Jacobian of the transformation

$$\frac{x_1}{a_1} = \frac{X_1}{R}, \frac{x_2}{a_2} = \frac{X_2}{R}, \ldots, \frac{x_n}{a_n} = \frac{X_n}{R}$$

or

$$x_1 = \frac{a_1}{R} X_1, x_2 = \frac{a_2}{R} X_2, \ldots, x_n = \frac{a_n}{R} X_n$$

is

$$J_n = \begin{vmatrix} \frac{a_1}{R} & 0 & \ldots & 0 \\ 0 & \frac{a_2}{R} & \ldots & 0 \\ \ldots & \ldots & \ldots & \ldots \\ 0 & 0 & \ldots & \frac{a_n}{R} \end{vmatrix} = \frac{a_1 a_2 \ldots a_n}{R^n}.$$

For application of this transformation, see Sec.2.6.7.

APPENDIX A6

Reference Formulas

Eq./Sec. No.

$$\Gamma(p) = \int_0^\infty x^{p-1} e^{-x}\, dx, \quad p > 0 \qquad (1.1)$$

(definition of the gamma function),

$$\Gamma(1) = 1, \qquad (1.3)$$

$$\Gamma(x) = \lim_{x\to\infty} \frac{n!n^x}{x(x+1)\cdots(x+n)}, \quad \text{(Gauss)}, \quad x \neq 0, -1, -2, \ldots, \qquad (1.4)$$

$$\Gamma(x) = e^{-\gamma x} \frac{1}{x} \prod_{n=1}^{\infty} \frac{e^{\frac{x}{n}}}{1+\frac{x}{n}}, \quad \begin{array}{l}\text{(Weierstrass)}\\ (\gamma = 0.57721566\ldots\\ \text{the Euler's constant}),\end{array} \qquad (1.6)$$

$$B(p,q) = \int_0^1 x^{p-1}(1-x)^{q-1}\, dx, \quad p, q > 0 \qquad (1.9)$$

(definition of beta function),

$$B(p,q) = 2\int_0^{\frac{\pi}{2}} \sin^{2p-1}\theta \cos^{2q-1}\theta\, d\theta, \quad p, q > 0 \qquad (1.12)$$

(the second definition of beta function),

$$\Gamma(p) = 2\int_0^\infty x^{2p-1} e^{-x^2}\, dx, \quad p > 0, \qquad (1.18)$$

$$\Gamma(p+1) = p\Gamma(p), \quad p > 0 \text{ (the recurrence formula)}, \qquad (1.20)$$

$$\Gamma\left(\frac{3}{2}\right) = \frac{\sqrt{\pi}}{2}, \quad \Gamma\left(\frac{7}{2}\right) = \frac{5}{2}\Gamma\left(\frac{5}{2}\right) = \frac{15}{8}\sqrt{\pi}, \qquad \text{Sec.1.5.2, 1.13}$$

$$\Gamma\left(\frac{5}{2}\right) = \frac{3\sqrt{\pi}}{4}, \quad \Gamma\left(\frac{9}{2}\right) = \frac{7}{2}\Gamma\left(\frac{7}{2}\right) = \frac{105}{16}\sqrt{\pi}, \qquad \text{Sec.1.5.2, 1.13}$$

$$\Gamma(n+1) = n!, \quad n = 0, 1, 2, \ldots, \tag{1.24}$$

$$\Gamma(n) = (n-1)!, \quad n \in N, \tag{1.25}$$

$$\Gamma(p) = (p-1)(p-2)\cdots(p-n)\Gamma(p-n), \quad p-n \in (0,1], \quad n \in N, \tag{1.26}$$

$$\Gamma(p+n) = p(p+1)\cdots(p+n-1)\Gamma(p), \quad p \in (0,1], \quad n \in N, \tag{1.27}$$

$$\Gamma(p) = \frac{\Gamma(p+n)}{p(p+1)\cdots(p+n-1)}, \quad p \in (-n, -n+1), \quad n \in N \tag{1.28}$$

$$\Gamma\left(n+\frac{1}{3}\right) = \frac{1\cdot 4\cdot 7\cdot \cdots \cdot (3n-2)}{3^n}\Gamma\left(\frac{1}{3}\right), \quad n \in N, \tag{1.29}$$

$$\Gamma\left(n+\frac{2}{3}\right) = \frac{2\cdot 5\cdot 8\cdot \cdots \cdot (3n-1)}{3^n}\Gamma\left(\frac{2}{3}\right), \quad n \in N, \tag{1.30}$$

$$\Gamma\left(n+\frac{1}{4}\right) = \frac{1\cdot 5\cdot 9\cdot \cdots \cdot (4n-3)}{4^n}\Gamma\left(\frac{1}{4}\right), \quad n \in N, \tag{1.31}$$

$$\Gamma(x)\Gamma(1-x) = \frac{\pi}{\sin x\pi}, \quad 0 < x < 1 \quad \text{(the complement formula)}, \tag{1.35}$$

$$\Gamma\left(\frac{1}{2}\right) = \sqrt{\pi}, \qquad (1.37),\ (1.105),\ (1.115)$$

$$\Gamma\left(\frac{3}{4}\right) = \frac{\pi\sqrt{2}}{\Gamma\left(\frac{1}{4}\right)}, \tag{1.38}$$

$$\Gamma\left(\frac{1}{2}+p\right)\Gamma\left(\frac{1}{2}-p\right) = \frac{\pi}{\cos p\pi}, \quad p \neq n+\frac{1}{2}, \quad n = 0, 1, 2, \ldots, \tag{1.39}$$

$$\Gamma(1+p)\Gamma(1-p) = \frac{\pi p}{\sin p\pi}, \quad 0 < p < 1, \tag{1.40}$$

$$\Gamma\left(\frac{1}{2}+n\right)\Gamma\left(\frac{1}{2}-n\right) = (-1)^n\pi, \quad n = 0, 1, 2, 3, \ldots, \tag{1.42}$$

$$\Gamma\left(\frac{1}{2}+n\right) = \frac{\sqrt{\pi}}{2^n}(2n-1)!!, \quad n \in N, \tag{1.43}$$

$$\Gamma\left(\frac{1}{2}+n\right) = \frac{(2n-1)!!}{(2n)!!}\sqrt{\pi}\Gamma(n+1), \quad n \in N, \tag{1.44}$$

$$\Gamma\left(\frac{1}{2}-n\right) = (-1)^n\frac{\sqrt{\pi}\cdot 2^n}{(2n-1)!!}, \quad n \in N, \tag{1.45}$$

$$B(p,q) = B(q,p), \quad p, q \in R^+, \tag{1.55}$$

$$B(p,q) = \frac{q-1}{p+q-1}B(p,q-1), \quad p > 0, \quad q > 1, \tag{1.56}$$

$$B(p,q)=\frac{p-1}{p+q-1}B(p-1,q),\quad p>1,\quad q>0, \tag{1.57}$$

$$B(p,q+1)=\frac{q}{p+q}B(p,q),\quad p,q>0, \tag{1.58}$$

$$B(p+1,q)=\frac{p}{p+q}B(p,q),\quad p,q>0, \tag{1.59}$$

$$B(p,q)=B(p+1,q)+B(p,q+1),\quad p,q>0, \tag{1.60}$$

$$B(p,q+1)=\frac{q}{p}B(p+1,q),\quad p,q>0, \tag{1.61}$$

$$\int_0^\infty \frac{x^{q-1}}{(1+x)^{p+q}}\,dx=B(p,q),\quad p,q>0, \tag{1.62}$$

$$\int_0^\infty \frac{x^{p-1}}{(1+x)^{p+q}}\,dx=B(q,p),\quad p,q>0, \tag{1.63}$$

$$\int_0^1 \frac{x^{p-1}+x^{q-1}}{(1+x)^{p+q}}\,dx=B(p,q),\quad p,q>0, \tag{1.64}$$

$$\int_0^1 \frac{x^{p-1}\,dx}{(1+x)^{2p}}=\frac{1}{2}B(p,p),\quad p>0, \tag{1.64a}$$

$$\int_0^\infty \frac{x^{2q-1}}{(1+x^2)^{p+q}}\,dx=\frac{1}{2}B(p,q),\quad p,q>0, \tag{1.65}$$

$$\int_0^\infty \frac{x^{2p-1}}{(1+x^2)^{p+q}}\,dx=\frac{1}{2}B(q,p),\quad p,q>0, \tag{1.66}$$

$$\int_0^1 x^{2p-1}(1-x^2)^{q-1}\,dx=\frac{1}{2}B(p,q),\quad p,q>0, \tag{1.67}$$

$$\int_0^a x^{p-1}(a-x)^{q-1}\,dx=a^{p+q-1}B(p,q),\quad a>0,\quad p,q>0, \tag{1.68}$$

$$\int_0^\infty \frac{dx}{\sqrt{1+x^3}}=\frac{1}{3\sqrt{\pi}}\Gamma\left(\frac{1}{3}\right)\Gamma\left(\frac{1}{6}\right), \qquad \text{Sec.2.3.2}$$

$$\int_0^1 (1-x^2)^p\,dx=\frac{1}{2}B\left(\frac{1}{2},p+1\right),\quad p>-1, \qquad \text{Sec.2.3.8}$$

$$\int_0^a x^p(a^2-x^2)^q\,dx=\frac{a^{p+2q+1}}{2}B\left(\frac{p+1}{2},1+q\right),\quad p,q>-1, \qquad \text{Sec.2.3.9}$$

$$\int_0^1 \frac{x^{p-1}}{\sqrt{1-x^q}}\,dx=\frac{1}{q}B\left(\frac{p}{q},\frac{1}{2}\right),\quad p,q>0, \qquad \text{Sec.2.3.10}$$

$$\int_0^\infty \frac{x^{m-1}}{1+x}\,dx=B(1-m,m)=\frac{\pi}{\sin m\pi},\quad 0<m<1, \qquad \text{Sec.2.3.13}$$

$$\int_0^\infty \frac{dx}{1+x^n} = \frac{1}{n}B\left(1-\frac{1}{n},\frac{1}{n}\right) = \frac{\pi}{n\sin\dfrac{\pi}{n}}, \quad n>1,$$ Sec.2.3.16

$$\int_0^\infty \frac{x^{m-1}}{1+x^n}\,dx = \frac{1}{n}B\left(1-\frac{m}{n},\frac{m}{n}\right) = \frac{\pi}{n\sin\dfrac{m\pi}{n}}, \quad 0<m<n,$$ Sec.2.3.20

$$\int_0^\infty \frac{x^{p-1}}{(1+x^q)^2}\,dx = \frac{1}{q}B\left(2-\frac{p}{q},\frac{p}{q}\right) = \frac{q-p}{q^2}\cdot\frac{\pi}{\sin\left(\dfrac{p}{q}\pi\right)}, \quad 0<p<q,$$

Sec.2.3.25

$$\int_0^\infty \frac{x^{p-1}\,dx}{(1+x^q)^m} = \frac{1}{q}B\left(m-\frac{p}{q},\frac{p}{q}\right), \quad 0<\frac{p}{q}<m,$$ Sec.2.3.28

$$\int_0^\infty \frac{x^m\,dx}{(a+bx^n)^p} = \frac{1}{na^p}\left(\frac{a}{b}\right)^{\frac{m+1}{n}} B\left(p-\frac{m+1}{n},\frac{m+1}{n}\right),$$ Sec.2.3.28a

$pn > m+1 > 0,$

$$\int_0^\infty e^{-x^m}\,dx = \frac{1}{m}\Gamma\left(\frac{1}{m}\right), \quad m>0,$$ Sec.2.3.31

$$\int_0^\infty e^{-x^2}\,dx = \frac{\sqrt{\pi}}{2},$$ (1.76), Sec.2.7.1

$$\int_0^\infty e^{-a^2x^2}\,dx = \frac{\sqrt{\pi}}{2a}, \quad a>0,$$ (1.79)

$$\int_0^\infty e^{-x^2-\frac{a^2}{x^2}}\,dx = \frac{\sqrt{\pi}}{2a^{2a}}, \quad a>0,$$ (1.85)

$$\int_0^\infty e^{-x^2}x^{m-1}\,dx = \frac{1}{2}\Gamma\left(\frac{m}{2}\right), \quad m>0,$$ (1.87)

$$\Gamma(n) = 2a^n\int_0^\infty x^{2n-1}e^{-ax^2}\,dx, \quad n\in N, \quad a>0,$$ (1.88)

$$\int_0^\infty x^{2n-1}e^{-ax^2}\,dx = \frac{\Gamma(n)}{2a^n}, \quad n\in N, \quad a>0,$$ (1.89)

$$\int_0^\infty x^n e^{-ax^2}\,dx = \frac{1}{2a^{\frac{n+1}{2}}}\Gamma\left(\frac{n+1}{2}\right), \quad n>-1, \quad a>0,$$ (1.90)

$$\int_0^\infty x^n e^{-ax}\,dx = \frac{n}{a^{n+1}}\Gamma(n), \quad a,n>0,$$ Sec.2.3.11

$$\int_{-\infty}^{+\infty} x^{2n}e^{-ax^2}\,dx = \frac{1}{a^{n+\frac{1}{2}}}\Gamma\left(n+\frac{1}{2}\right), \quad n\in N\cup\{0\},$$ (1.91)

$$\int_{-\infty}^{+\infty} x^{2n-1} e^{-ax^2}\, dx = 0, \qquad (1.92)$$

$$\int_0^{\infty} t^{p-1} e^{-at}\, dt = \frac{\Gamma(p)}{a^p}, \quad a, p \in R^+, \qquad (1.94)$$

$$B(p,q) = \frac{\Gamma(p)\Gamma(q)}{\Gamma(p+q)}, \quad p, q \in R^+, \qquad (1.98), \text{Sec.2.7.2}$$

$$B(p,q+1) = \frac{q}{p+q} B(p,q), \quad p, q \in R^+ \qquad (1.106)$$

$$B(p+1,q) = \frac{p}{p+q} B(p,q), \quad p, q \in R^+, \qquad (1.107)$$

$$B(p,n+1) = \frac{n!}{(p+n)(p+n-1)\cdots(p+1)p}, \quad p \in (0,1), \quad n \in N, \qquad (1.108)$$

$$\Gamma\left(\frac{n}{2}\right) = \frac{\sqrt{\pi}(n-1)!}{2^{n-1}\left(\frac{n-1}{2}\right)!}, \quad n \text{ is an odd integer}, \qquad (1.111)$$

$$\int_0^{\frac{\pi}{2}} \sin^{2p-1}\theta \cos^{2q-1}\theta\, d\theta = \frac{1}{2} B(p,q), \quad p, q \in R^+, \qquad (1.112)$$

$$\int_0^{\frac{\pi}{2}} \sin^{m-1}\theta \cos^{n-1}\theta\, d\theta = \frac{1}{2} B\left(\frac{m}{2}, \frac{n}{2}\right), \quad m, n \in R^+, \qquad (1.113)$$

$$\int_0^{\frac{\pi}{2}} \sin^{p}\theta \cos^{q}\theta\, d\theta = \frac{1}{2} B\left(\frac{p+1}{2}, \frac{q+1}{2}\right), \quad p, q > -1, \qquad (1.116)$$

$$\int_0^{\frac{\pi}{2}} \sin^{2p-1}\theta\, d\theta = \int_0^{\frac{\pi}{2}} \cos^{2p-1}\theta\, d\theta = \frac{1}{2} B\left(p, \frac{1}{2}\right), \quad p \in R^+, \qquad (1.118)$$

$$\int_0^{\frac{\pi}{2}} \sin^{m-1}\theta\, d\theta = \int_0^{\frac{\pi}{2}} \cos^{m-1}\theta\, d\theta = \frac{1}{2} B\left(\frac{m}{2}, \frac{1}{2}\right), \quad m \in R^+, \qquad (1.119)$$

$$\int_0^{\frac{\pi}{2}} \sin^{p}\theta\, d\theta = \int_0^{\frac{\pi}{2}} \cos^{p}\theta\, d\theta = \frac{1}{2} B\left(\frac{p+1}{2}, \frac{1}{2}\right), \quad p > -1, \qquad (1.120)$$

$$\int_0^{\frac{\pi}{2}} \sin^{2p-1}\theta\, d\theta = 2^{2p-2} B(p,p), \quad p \in R^+, \qquad (1.121)$$

$$\int_0^{\frac{\pi}{2}} \cos^{2p-1}\theta\, d\theta = 2^{2p-2} B(p,p), \quad p \in R^+, \qquad (1.122)$$

$$\int_0^{\frac{\pi}{2}} \sin^{m-1}\theta\, d\theta = \int_0^{\frac{\pi}{2}} \cos^{m-1}\theta\, d\theta = 2^{m-2} B\left(\frac{m}{2}, \frac{m}{2}\right), \quad m \in R^+, \qquad (1.123)$$

$$\int_0^{\frac{\pi}{2}} \tan^n \theta \, d\theta = \frac{1}{2} B\left(\frac{1+n}{2}, \frac{1-n}{2}\right), \quad n \in (-1, +1), \qquad (1.124)$$

$$\int_0^{\frac{\pi}{2}} \tan^n \theta \, d\theta = \frac{\pi}{2} \sec \frac{n\pi}{2}, \quad |n| < 1, \qquad (1.125)$$

$$\int_0^{\frac{\pi}{2}} \sin^{2n} x \, dx = \int_0^{\frac{\pi}{2}} \cos^{2n} x \, dx = \frac{(2n-1)!!}{(2n)!!}\frac{\pi}{2}, \quad n \in N, \qquad (1.129)$$

$$\int_0^{\frac{\pi}{2}} \sin^{2n+1} \theta \, d\theta = \int_0^{\frac{\pi}{2}} \cos^{2n+1} \theta \, d\theta = \frac{(2n)!!}{(2n+1)!!}, \quad n \in N, \qquad (1.132)$$

$$\Gamma(2p) = \frac{2^{2p-1}}{\sqrt{\pi}} \Gamma(p) \Gamma\left(p + \frac{1}{2}\right), \quad p \in R^+ \qquad (1.135)$$

(Legendre duplication formula),

$$\Gamma\left(n + \frac{1}{2}\right) = \frac{\sqrt{\pi}}{2^{2n-1}} \frac{\Gamma(2n)}{\Gamma(n)} = \frac{\sqrt{\pi}}{2^{2n}} \frac{(2n)!}{n!}, \quad n \in N \cup \{0\}, \qquad (1.136)$$

$$\Gamma\left(n + \frac{1}{2}\right) = \frac{\sqrt{\pi}}{2^n}(2n-1)!!, \quad n \in N, \qquad (1.43), (1.137)$$

$$B(p_1, p_2, \ldots, p_n) = \frac{\Gamma(p_1)\Gamma(p_2)\cdots\Gamma(p_n)}{\Gamma(p_1 + p_2 + \cdots + p_n)}, \quad p_1, p_2, \ldots, p_n \in R^+, \qquad (1.142)$$

$$A_n = \frac{2\pi^{\frac{n}{2}}}{\Gamma\left(\frac{n}{2}\right)} R^{n-1}, \quad R > 0, \quad n \in N \qquad (1.149), \text{Secs.2.5.2, 2.6.6}$$

(the surface area of a hypersphere of radius R),

$$V_n = \frac{2\pi^{\frac{n}{2}}}{n\Gamma\left(\frac{n}{2}\right)} R^n, \quad R > 0, \quad n \in N \qquad \text{Sec.2.6.6}$$

(the volume of a hypersphere of radius R),

$$\mathcal{L}\{t^x\} = \frac{\Gamma(x+1)}{s^{x+1}}, \quad x > -1, \quad s > 0, \qquad (1.151)$$

$$\mathcal{L}\{t^n\} = \frac{n!}{s^{n+1}}, \quad s > 0, \quad n \in N, \qquad (1.152)$$

$$\mathcal{L}\{t^{n+\frac{1}{2}}\} = \frac{\sqrt{\pi}}{s^{n+\frac{3}{2}}} \cdot \frac{(2n+1)!!}{2^{n+1}}, \quad s > 0, \quad n \in N, \qquad (1.153)$$

$$\mathcal{L}\{\sin at\} = \frac{a}{s^2 + a^2}, \quad s > 0, \quad a \in R^+, \qquad (1.154)$$

$$\mathcal{L}\{\cos at\} = \frac{s}{s^2 + a^2}, \quad s > 0, \quad a \in R^+, \tag{1.155}$$

$$\mathcal{L}^{-1}\{\frac{1}{s^{x+1}}\} = \frac{t^x}{\Gamma(x+1)}, \quad x > -1, \tag{1.157}$$

$$\mathcal{L}^{-1}\{f(s) \cdot g(s)\} = \int_0^t F(u) \cdot G(t-u)\, du = F * G \tag{1.158}$$

(the convolution theorem),

$$\gamma(p, x) = \int_0^x t^{p-1} e^{-t}\, dt, \quad p \in R^+ \tag{1.159}$$

(the incomplete gamma function),

$$\Gamma(p, x) = \int_x^\infty t^{p-1} e^{-t}\, dt, \quad p \in R^+ \tag{1.160}$$

(the complementary incomplete gamma function),

$$\gamma(p, x) + \Gamma(p, x) = \Gamma(p), \quad p \in R^+, \tag{1.161}$$

$$B_x(p, q) = \int_0^x t^{p-1}(1-t)^{q-1}\, dt, \quad x \in [0, 1], \quad p, q \in R^+ \tag{1.162}$$

(the incomplete beta function),

$$\psi(x) = \frac{d}{dx} \ln \Gamma(x) = \frac{\Gamma'(x)}{\Gamma(x)}, \quad x \neq 0, -1, -2, -3, \ldots \tag{1.163}$$

(the definition of the function $\psi(x)$),

$$\psi(x+1) = \psi(x) + \frac{1}{x}, \quad x \neq 0, -1, -2, -3, \ldots \tag{1.165}$$

(the recurrence formula of the function $\psi(x)$),

$$\psi(x) = \psi(x+1) - \frac{1}{x}, \quad x \neq 0, -1, -2, -3, \ldots, \tag{1.166}$$

$$\beta(x) = \frac{1}{2}\left[\psi\left(\frac{x+1}{2}\right) - \psi\left(\frac{x}{2}\right)\right], \quad x > 0 \tag{1.171}$$

(the definition of the function $\beta(x)$),

$$\beta(p) = \int_0^1 \frac{x^{p-1}}{1+x}\, dx, \quad p > 0 \tag{1.172}$$

(an integral representation of the function $\beta(p)$),

$$\int_0^1 \frac{x^{p-1}}{1+x^q}\,dx = \frac{1}{q}\beta\left(\frac{p}{q}\right), \quad p, q > 0, \qquad (1.173)$$

$$\int_0^1 \frac{x^p\,dx}{(1+x^2)^2} = \frac{p-1}{4}\beta\left(\frac{p-1}{2}\right) - \frac{1}{4}, \quad p > 1, \qquad (1.175)$$

$$\int_0^\infty \sin x^2\,dx = \int_0^\infty \cos x^2\,dx = \frac{\sqrt{2\pi}}{4}, \quad \text{Fresnel's integrals,} \qquad \text{Sec.2.7.3}$$

$$\int_0^\infty \frac{\sin x}{x^p}\,dx = \frac{\pi}{2\Gamma(p)\sin\frac{p\pi}{2}}, \quad 0 < p < 1, \qquad \text{Sec.2.3.33, 2.7.6}$$

$$\int_0^\infty \frac{\cos x}{x^p}\,dx = \frac{\pi}{2\Gamma(p)\cos\frac{p\pi}{2}}, \quad 0 < p < 1, \qquad \text{Sec.2.3.33, 2.7.6}$$

$$\int_0^{\frac{\pi}{2}} \frac{d\theta}{\sqrt{1-k^2\sin^2\theta}} = K(k), \quad 0 < k < 1, \quad \sin\psi = k, \qquad (3.5),\ (4.1)$$

$$\int_0^{\frac{\pi}{2}} \sqrt{1-k^2\sin^2\theta}\,d\theta = E(k), \quad 0 < k < 1, \quad \sin\psi = k, \qquad (3.6),\ (4.2)$$

$$KE' + EK' - KK' = \frac{\pi}{2} \quad \text{(Legendre)}, \qquad (3.7),\ \text{Sec.4.2.1}$$

$$F(k,\pi) = \int_0^\pi \frac{d\theta}{\sqrt{1-k^2\sin^2\theta}} = 2K(k), \qquad (3.12)$$

$$E(k,\pi) = \int_0^\pi \sqrt{1-k^2\sin^2\theta}\,d\theta = 2E(k), \qquad (3.13)$$

$$\int_0^{\frac{\pi}{4}} \frac{d\theta}{\sqrt{\cos 2\theta}} = \frac{1}{\sqrt{2}}K\left(\frac{1}{\sqrt{2}}\right), \qquad \text{Sec.4.2.6}$$

$$\int_0^\pi \frac{d\theta}{\sqrt{1+k\cos\theta}} = \frac{2}{\sqrt{1+k}}K\left(\sqrt{\frac{2}{1+k}}\right), \quad k \in (0,1), \qquad \text{Sec.4.2.10}$$

$$\int_0^\phi \frac{d\theta}{\sqrt{\cos\theta - \cos\phi}} = \sqrt{2}K\left(\sin\frac{\phi}{2}\right), \quad 0 < \sin\frac{\phi}{2} < 1, \qquad \text{Sec.4.2.11}$$

$$\int_0^{\frac{\pi}{2}} \frac{d\theta}{\sqrt{1+k^2\sin^2\theta}} = \frac{1}{\sqrt{1+k^2}}K\left(\frac{k}{\sqrt{1+k^2}}\right), \quad k > 0, \qquad \text{Sec.4.2.12}$$

$$\int_0^{\frac{\pi}{2}} \sqrt{1+\sin^2 x}\,dx = \sqrt{2}E\left(\frac{1}{\sqrt{2}}\right), \qquad \text{Sec.4.2.13}$$

$$\int_0^1 \frac{dx}{\sqrt{1+x^4}} = \frac{1}{2}K\left(\frac{1}{\sqrt{2}}\right),$$ Sec.4.2.19

$$\int_0^\infty \frac{dx}{\sqrt{1+x^4}} = K\left(\frac{1}{\sqrt{2}}\right) = \frac{1}{4\sqrt{\pi}}\left[\Gamma\left(\frac{1}{4}\right)\right]^2,$$ Sec.2.3.22, 4.2.19

$$\int_1^\infty \frac{dx}{\sqrt{1+x^4}} = \frac{1}{2}K\left(\frac{1}{\sqrt{2}}\right),$$ Sec.4.2.19

$$\int_1^a \frac{dx}{\sqrt{x^4-1}} = \frac{1}{\sqrt{2}}F\left(\frac{1}{\sqrt{2}}, \cos^{-1}\frac{1}{a}\right), \quad a > 1,$$ Sec.4.2.21

$$\int_1^\infty \frac{dx}{\sqrt{x^4-1}} = \frac{1}{\sqrt{2}}K\left(\frac{1}{\sqrt{2}}\right),$$ Sec.4.2.21

$$K\left(\frac{1}{\sqrt{2}}\right) = \frac{1}{4\sqrt{\pi}}\left[\Gamma\left(\frac{1}{4}\right)\right]^2 = 1.854,$$ Sec.4.3.1, 4.3.4, 4.3.6

$$2E\left(\frac{1}{\sqrt{2}}\right) - K\left(\frac{1}{\sqrt{2}}\right) = \frac{2\pi\sqrt{\pi}}{\left[\Gamma\left(\frac{1}{4}\right)\right]^2} = 0.847.$$ Sec.4.3.3, 4.3.5

Bibliography

[1] M. Abramowitz and I. Stegun, *Handbook of Mathematical Functions with Formulas, Graphs and Mathematical Tables* (John Wiley & Sons, New York, 1972), pp. 257, 591–592.

[2] L.C. Andrews, *Special Functions of Mathematics for Engineers* (McGraw-Hill, New York, 1992), pp. 90–92.

[3] R. Badescu and C. Maican, *Integrals Used in Mechanics, Physics, Technique and their Calculation* (Editura Tehnica, Bucuresti, 1968), pp. 585–588.

[4] H.W. Beyer, *CRC Handbook of Mathematical Sciences*, 5th Edition (CRC Press, Inc., U.S.A., 1978).

[5] E. Goursat, *A Course in Mathematical Analysis* (Dover Publications, Inc., New York, 1959), pp. 308, 312–313.

[6] I.S. Gradshteyn and I.M. Ryzhik, *Table of Integrals, Series and Products* (Academic Press Inc., New York, 1965), pp. XIiii, 289, 292.

[7] R.K. Pathria, *Statistical Mechanics* (Pergamon Press, New York, 1977), pp. 501–502.

Made in United States
North Haven, CT
05 November 2026

21764943R10180